Lecture Notes in Mathematics 1718

Editors:
A. Dold, Heidelberg
F. Takens, Groningen
B. Teissier, Paris

W0106819

Springer
Berlin
Heidelberg
New York
Barcelona
Hong Kong
London
Milan
Paris
Singapore
Tokyo

Andreas Eberle

Uniqueness and Non-Uniqueness of Semigroups Generated by Singular Diffusion Operators

 Springer

Author

Andreas Eberle
Faculty of Mathematics
University of Bielefeld
P.O. Box 10 01 31
33501 Bielefeld, Germany
E-mail: eberle@mathematik.uni-bielefeld.de

Cataloging-in-Publication Data applied for

Die Deutsche Bibliothek - CIP-Einheitsaufnahme

Eberle, Andreas:
Uniqueness and non-uniqueness of semigroups generated by singular
diffusion operators / Andreas Eberle. - Berlin ; Heidelberg ; New
York ; Barcelona ; Hong Kong ; London ; Milan ; Paris ; Singapore ;
Tokyo : Springer, 1999
  (Lecture notes in mathematics ; 1718)
  Zugl: Bielefeld, Univ., Diss., 1998
  ISBN 3-540-66628-1

Mathematics Subject Classification (1991):
60J60, 35K10, 31C25, 47F05, 60J35, 47B25, 60H30, 60H15, 34B24

ISSN 0075-8434
ISBN 3-540-66628-1 Springer-Verlag Berlin Heidelberg New York

Typesetting: Camera-ready TeX output by the author
Printed on acid-free paper   SPIN: 10700296   41/3143-543210

Preface

This book is aimed at both probabilists working on diffusion processes and analysts interested in linear parabolic partial differential equations with singular coefficients. The central question discussed is whether a given diffusion operator, i.e., a second order linear elliptic differential operator without zeroth order term, which is a priori only defined on test functions over some (finite or infinite dimensional) state space, uniquely determines a strongly continuous semigroup on a corresponding weighted L^p space. This general uniqueness problem is related to several specific questions arising in different areas, e.g., uniqueness of martingale problems, existence of operator cores consisting of "nice" functions, essential self–adjointness and uniqueness problems in mathematical physics, as well as uniqueness problems for Dirichlet forms. We give particular emphasis to pointing out phenomena responsible for non–uniqueness, as well as to clarifying the relationships between the different notions of uniqueness appearing in analytic and probabilistic contexts.

A part of the new results presented in this work, in particular in Chapter 3, has been announced in the Comptes rendus note [Eb 95]. The results in Section f) of Chapter 2 are contained in the article [Eb 99].

I would like to thank Michael Röckner for constant encouragement, and Wilhelm Stannat, Johannes Brasche and Tu–Sheng Zhang for many stimulating discussions on the subject. Moreover, I would like to gratefully acknowledge financial support during the last years by Faculty of Mathematics and SFB 343, University of Bielefeld, the Mittag–Leffler Institute, Stockholm, the MSRI, Berkeley, and the European Union TMR project "Stochastic Analysis".

Contents

0 Introduction   **1**
  a) General introduction   1
  b) Results.   5

1 Uniqueness problems in various contexts   **9**
  a) Uniqueness of the martingale problem   10
  b) Cores   17
  c) Essential self–adjointness   21
  d) Uniqueness of Dirichlet forms   22
  e) Relations between the notions of uniqueness   24

  Appendix
  A Existence and uniqueness of C^0 semigroups on Banach spaces . .   30
  B Diffusion operators on L^p spaces   35

2 L^p uniqueness in finite dimensions   **41**
  a) The regular one–dimensional case   42
  b) Examples and counterexamples I : Regular operators   51
    1) Diffusion operators on $L^p(\mathbf{R}^n;\, dx)$ 51
    2) Operators on $L^p(\mathbf{R}^1;\, e^{-x^2/2}\, dx)$ 53
    3) Perturbations of generalized Schrödinger operators on $\mathbf{R}^1$ . .   54
  c) Regular diffusion operators on $\mathbf{R}^n$ 55
  d) Examples and counterexamples II : Singular operators   59
    1) Singular generalized Schrödinger operators on $\mathbf{R}^1$ 59
    2) Rotationally invariant generalized Schrödinger operators . . .   61
    3) Degeneracy of second order coefficients   65
  e) The singular one–dimensional case   66
  f) Singular diffusion operators on $\mathbf{R}^n$ 75

  Appendix
  C Regularity of distributional solutions of O.D.E.   85

3 Markov uniqueness   **89**

  a) Weak Sobolev spaces on $\mathbf{R}^n$ and on Banach spaces 92
  b) Weak and strong Sobolev spaces on general state spaces 104
  c) Maximal Dirichlet extensions 112
  d) Markov uniqueness in the one–dimensional case 118
  e) Density of smooth functions in weak Sobolev spaces over $\mathbf{R}^n$ . . 124
  f) Markov uniqueness in the finite dimensional case 129
  g) Ergodicity and extremality of symmetrizing measures 136

  Appendix

  D The geometry of diffusion operators 147
     1) Generalized differentials 148
     2) Measurable co–tangent bundles and differentials corresponding
           to diffusion operators 149
     3) Diffusion operators on manifolds and vector spaces 152
     4) Ornstein–Uhlenbeck geometries on path and loop spaces . . . 153
     5) Horizontal and vertical measure–valued diffusions 155
     6) Divergence operators and symmetrizing measures 159
     7) A representation theorem for diffusion operators on L^2 spaces 162

4 Probabilistic aspects of L^p and Markov uniqueness   **169**

  a) Feller classification and uniqueness 170
  b) Conservativity, ergodicity, and Markov uniqueness 178
  c) Probabilistic explanations for L^p uniqueness 179

5 First steps in infinite dimensions   **185**

  a) Infinite dimensional diffusion operators on linear spaces 187
  b) The generator of the Brownian string — A counterexample . . . 197
  c) Markov uniqueness of projective limits 215
  d) Approximative approaches to uniqueness and existence in L^p . . 225
  e) Applications to lattice systems in statistical mechanics 234
  f) Stability of L^p uniqueness under H–valued perturbations 238
     1) The general perturbation result 239
     2) Perturbations of operators with linear drift 240
     3) Infinite dimensional generalized Schrödinger operators 245
  g) Applications to perturbed operators 247
     1) Finite volume quantum fields 247
     2) Perturbations of the Ornstein–Uhlenbeck operator 250
     3) The Brownian string in a velocity field 251

Chapter 0

Introduction

a) General introduction

Diffusion operators over finite and infinite dimensional state spaces

Let E be an open subset in $\mathbf{R}^n$, and let $\mathcal{A}$ be a space of smooth functions on E, e.g., $\mathcal{A} = C_0^\infty(E)$. Suppose we are given a second order differential operator

$$\mathcal{L}f \;=\; \sum_{i,j=1}^n a_{ij}\,\frac{\partial^2 f}{\partial x_i \partial x_j} \;+\; \sum_{j=1}^n \beta_j\,\frac{\partial f}{\partial x_j}\,, \quad f \in \mathcal{A},$$

with measurable coefficients $a_{ij}, \beta_j : E \to \mathbf{R},\ 1 \le i, j \le n$, such that the matrix $(a_{ij}(x))$ is positive definite (in the non–strict sense) for all $x \in E$. We call such operators, as well as operators of a similar type on more general (in particular infinite dimensional) state spaces E, *diffusion operators*, cf. Appendix B for a general definition.

We are mainly interested in *singular*, *degenerate* and *infinite-dimensional* diffusion operators. In contrast to uniqueness problems for strictly elliptic diffusion operators with regular coefficients on $\mathbf{R}^n$, which are well understood by classical results, often surprisingly little is known about uniqueness of singular operators. Actually, below we will give several examples showing that in the singular case completely new phenomena occur, that are absent in the regular case. In particular, non-uniqueness can not only be caused by boundary effects (in the wide sense), but also by singularities or by infinite dimensional phenomena. In the finite dimensional case, one could try to consider the operator on the domain without the singularities, and to treat the singularities as boundaries. In fact, this way of thinking will sometimes enter into our considerations. However, often it fails to produce sufficiently strong uniqueness results, and L^p methods turn out to be more efficient.

Besides their theoretical importance in analysis and probability, singular finite dimensional diffusion operators occur in many applications, including in particular *stochastic mechanics*. Moreover, considering singular finite dimensional diffusion operators can be viewed as a pre-study for the more difficult infinite dimensional case. The interplay between finite and infinite dimensional analysis is very powerful: In this work we will both use techniques that have originally been developed in infinite dimensional analysis (e.g., the maximality result for Dirichlet forms on Banach spaces in [AlbKusRö 90]) to prove new finite dimensional results, and try to lift dimension independent finite dimensional results to infinite dimensions.

The uniqueness problems for infinite dimensional diffusion operators we are interested in play a crucial rôle in several areas of mathematical physics including *Euclidean quantum field theory*, and both *classical and quantum statistical mechanics*. Moreover, such uniqueness problems arise in many areas of stochastic analysis, e.g., *stochastic partial differential equations*, *particle systems* and *diffusion processes on path and loop spaces over Riemannian manifolds*.

Uniqueness problems on L^p spaces

Our main focus are uniqueness problems for diffusion operators on L^p spaces. Let $\mathcal{A}$ be a space of *test–functions* over the corresponding state space E, e.g., $\mathcal{A} = C_0^\infty(E)$, if E is an open subset in $\mathbf{R}^n$, or $\mathcal{A}$ is a space of smooth *cylinder functions* if E is infinite dimensional.

Let m be a σ–finite measure on E. Typically, the measure m we choose is related to the operator $(\mathcal{L}, \mathcal{A})$ e.g., it is a symmetrizing, an invariant, or, more generally, a sub-invariant measure for $(\mathcal{L}, \mathcal{A})$. Let $1 \leq p < \infty$. For the operators we are interested in, it is usually (though not always) known that there exists a C^0 semigroup $(T_t)_{t \geq 0}$ on $L^p(E \, ; m)$ such that its generator extends the operator $(\mathcal{L}, \mathcal{A})$. The semigroup is, for example, given as the transition semigroup of a diffusion process solving the corresponding martingale problem, or it has been constructed analytically. Typically, we also know that $(T_t)_{t \geq 0}$ is **sub–Markov**, i.e., $0 \leq T_t f \leq 1$ m-a.e. whenever $0 \leq f \leq 1$ m-a.e. This work deals with the following questions, which are sometimes much harder to answer than the existence problem:

- Is there at most one C^0 semigroup on $L^p(E \, ; m)$ such that its generator extends $(\mathcal{L}, \mathcal{A})$?

- Is there at most one sub–Markovian C^0 contraction semigroup on $L^1(E \, ; m)$ such that its generator extends $(\mathcal{L}, \mathcal{A})$?

If the answer is positive, we call the operator $(\mathcal{L}, \mathcal{A})$ $L^p(E \, ; m)$ **unique**, respectively **strongly Markov unique**. If the measure m is symmetrizing for $(\mathcal{L}, \mathcal{A})$, then it also makes sense to ask the following questions:

- Is there at most one symmetric C^0 semigroup on $L^2(E \, ; m)$ such that its generator extends $(\mathcal{L}, \mathcal{A})$?

- Is there at most one symmetric sub–Markovian C^0 semigroup on $L^2(E \, ; m)$ such that its generator extends $(\mathcal{L}, \mathcal{A})$?

The corresponding kinds of uniqueness of the operator $(\mathcal{L}, \mathcal{A})$ are called **essential self-adjointness** and **Markov uniqueness**. If m is symmetrizing, then essential self-adjointness and $L^2(E; m)$ uniqueness are known to be equivalent, cf. Corollary 1.2. Moreover, the most effective way to prove strong Markov uniqueness known so far is to show L^p uniqueness for some $p \in [1, \infty)$. Therefore, we will mainly discuss L^p uniqueness (which includes essential self-adjointness) and Markov uniqueness.

REMARK. From a probabilist's point of view, one may ask why we do not discuss the uniqueness problem in the class of Feller semigroups, which is usually considered in the classical probabilistic literature. However, for many of the singular and infinite-dimensional diffusion operators we are looking at, the framework of Feller semigroups does not seem to be appopriate. In fact, often not even the existence of a Feller semigroup generated by an extension of the operator is known. Similarly, in many examples, uniqueness of Feller semigroups can not be shown, or is even known not to hold. Very roughly speaking, uniqueness of Feller semigroups can be viewed as some limit of L^p uniqueness as $p \to \infty$. In many examples we will look at, however, L^p uniqueness holds or can be shown only for small p.

Reasons for studying uniqueness problems on L^p spaces

I will now briefly explain where my motivation for studying uniqueness problems for diffusion operators on L^p spaces comes from. There are several problems from different areas which can be rephrased as or reduced to uniqueness problems on L^p. I have listed some of them here, cf. in particular Chapter 1 below for more details :

- *Open uniqueness problems in mathematical physics* : E.g., it is not known whether essential self-adjointness holds in several non-trivial models from Euclidean quantum field theory, cf. e.g. [Alb 97, 6.4.2]. Similarly, for various lattice and particle systems in statistical mechanics, essential self-adjointness has only recently been shown or is still open, cf. below.

- *Operator cores*: Sometimes an inequality for a diffusion semigroup can be easily proven on test-functions, but it is needed for all functions in the domain of the generator. To extend it, one often has to know that the test-functions form a core for the generator, for example on some L^p space. The test-functions, however, are an operator core on L^p if and only if the generator defined on test-functions is L^p unique, cf. Chapter 1, Section b).

- *Uniqueness of martingale problems*: Since a diffusion process is uniquely determined by its transition semigroup, the analytic uniqueness problems described above are closely related to uniqueness of martingale problems, cf. Chapter 1, Section a).

- *Uniqueness of Dirichlet forms*: Markov uniqueness of a symmetric diffusion operator $(\mathcal{L}, \mathcal{A})$ on an L^2 space can be viewed as a kind of uniqueness of the corresponding pre–Dirichlet form $(\mathcal{E}, \mathcal{A})$. In the finite dimensional case, it is related to other uniqueness problems for Dirichlet forms, as e.g. Silverstein uniqueness, cf. Chapter 1, Section d).

Different aspects

The uniqueness problems raised above have several aspects:

- *Qualitative aspects*: Which phenomena cause non–uniqueness ? How does non–uniqueness arise, and how do the different extensions look like ?

- *Technical aspects*: Development of techniques for proving the different types of uniqueness in those cases, where the phenomena causing non–uniqueness are absent.

- *Quantitative aspects*: Precise determination of the "frontier" where uniqueness breaks down (e.g., in terms of integrability conditions on the operator coefficients).

- *Applications*: Proof of uniqueness or non–uniqueness in concrete models, e.g. from mathematical physics.

I have deliberately put the *qualitative aspects* at the top. In fact, in the considerations below our aim will always be not only to derive some sufficient conditions for uniqueness, but also to understand at which point uniqueness breaks down, and why this happens. This includes deriving very precise (sometimes even necessary and sufficient) conditions for uniqueness. Actually, we will point out three different phenomena causing non-uniqueness:

- Boundaries

- Singularities

- Infinite–dimensional effects

For sufficiently regular, non–degenerate diffusion operators on $\mathbf{R}^n$, non–uniqueness is always caused by some generalized kind of boundary. For singular and infinite–dimensionsal operators, however, non–uniqueness can arise, although the boundary behaviour of the coefficients is good.

In Chapter 4 and at other places of this work, we will try to obtain a better understanding of *analytic* L^p and Markov uniqueness results proven before by thinking probabilistically. This includes relations between uniqueness, conservativity, ergodicity, etc. The probabilistic picture can even be used to obtain an idea why some operators are strongly Markov unique and L^p unique for small p, but not L^p unique for large p. For these operators, there exists only one transition semigroup of an ordinary diffusion process such that its generator extends the operator defined on test-functions. Nevertheless, some of the other C^0 semigroups generated by extensions of the operator on L^p for large p can also be interpreted probabilistically by viewing them as transition semigroups of appropriate *particle systems*. The particle systems considered turn out to induce a semigroup of bounded operators on L^p precisely for those p where L^p uniqueness does not hold, cf. Section c) in Chapter 4.

For treating singular and infinite–dimensional operators, often completely new *techniques* have to be developed. For example, estimates in terms of Dirichlet forms play a crucial rôle in the study of L^p uniqueness problems even for *non–symmetric* operators, cf. Section f) in Chapter 2. For proving infinite dimensional L^p uniqueness results we use a variant of the Bochner technique, a method originally developed in analysis on manifolds, cf. the proof of Theorem 5.3. The newly introduced concept of weak Sobolev spaces associated with symmetric diffusion operators enables us to derive Markov uniqueness results for a large class of diffusion operators on $\mathbf{R}^n$, including singular and degenerate ones, cf. Chapter 3.

Quantitative results are in particular given in the one–dimensional case, where we precisely determine under which conditions on the coefficients of the operator L^p and Markov uniqueness hold, cf. Section e) in Chapter 2, Section d) in Chapter 3, and Section a) in Chapter 4. In the singular case, these results seem to be new. We also derive a necessary and sufficient condition for Markov uniqueness that applies to a large class of symmetric diffusion operators on domains in $\mathbf{R}^n$, cf. Chapter 3), Section f). This condition is, however, not as explicit as in the one-dimensional case. For operators of type $\Delta + \beta \cdot \nabla$, $\beta : \mathbf{R}^n \to \mathbf{R}^n$, we prove a condition on β guaranteeing L^p uniqueness which turns to be rather sharp in some sense, cf. Theorem 2.5 and the remarks below.

To keep the size of this work reasonable, we will only look at a limited number of *applications* to concrete models, e.g., from mathematical physics. In particular, we study generalized Schrödinger operators both in finite and infinite dimensions, cf. Sections d) and f) in Chapter 2, Section e) in Chapter 3, and Section f), 3) in Chapter 5. Moreover, we prove L^p uniqueness for the $P(\phi)_2$ quantum field in finite volume for $1 \le p < 2$, cf. Section g) in Chapter 5, and we show L^p uniqueness for a large class of lattice systems from classical statistical mechanics.

b) Results

For a detailed overview over the results obtained in this work, the reader is also referred to the introductions of the corresponding chapters. Chapter 1 explains in detail, where uniqueness problems arise. In Chapters 2, 3 and 4 we consider L^p uniqueness, Markov uniqueness, and probabilistic aspects of both for mainly finite dimensional diffusion operators. Chapter 5 contains some first results and applications in infinite dimensions. At this point, I just would like to point out briefly the main *new* results presented in this work:

- One dimensional diffusion operators are under very weak conditions symmetrizable, i.e., they can be written in Sturm–Liouville form. We give *complete characterizations of L^p and Markov uniqueness for singular Sturm–Liouville type diffusion operators*, cf. Section e) in Chapter 2, and Section d) in Chapter 3. For operators with singularities resp. points of degeneracy in the interior of the state space, I could not find corresponding results in the existing literature, cf. e.g. [Wei 87] for the regular case. We also look at some operators that are not in Sturm–Liouville form. Moreover, we discuss the relation of our results to Feller's boundary classification and a similar classification for singularities, cf. Section a) in Chapter 4. Our study of the one–dimensional case produces some interesting examples and counterexamples for higher–dimensional results, cf. e.g. Section b) and d) in Chapter 2. In particular, we demonstrate that generalized Schrödinger operators in $\mathbf{R}^n$ are not necessarily L^p unique (cf. Section d) 2) in Chapter 2), and degenerate symmetric diffusion operators in $\mathbf{R}^n$ are not necessarily Markov unique, cf. Theorem 3.2.

- For diffusion operators of type $\Delta + \beta \cdot \nabla$ on $\mathbf{R}^n$ with singular drift β, we prove two new conditions for L^p uniqueness, cf. Theorem 2.5 and Theorem 2.6. Both partially improve previous results of V. Liskevič and Y. Semenov [LiSem 92, 96], [Li 94], who showed L^p uniqueness in the symmetrizable case, provided β satisfies a global integrability condition. Our results show how to replace the global by a *local integrability condition plus* some *growth control.* They are also applicable in the non–symmetric case. Whereas Theorem 2.6 works best in the low–dimensional case (in particular $n = 2$), the condition in Theorem 2.5 does not depend on the dimension. The proof of Theorem 2.5 is surprisingly elementary, and requires no advanced regularity results.

- In Chapter 4, Section c), we demonstrate how some analytic L^p uniqueness results can be explained probabilistically by using *particle systems*, cf. also the comments above.

- For Markov uniqueness in finite dimensions, we obtain much more complete results than for L^p uniqueness. We prove a *necessary and sufficient condition for Markov uniqueness* of general symmetric diffusion operators over domains in $\mathbf{R}^n$, allowing even some degeneracy of the second order coefficients, cf. Theorem 3.6. This extends considerably previous results for generalized Schrödinger operators by M. Röckner and T.S. Zhang [RöZha 92, 94] and M. Takeda [Ta 92]. We show that *completeness* of the state space w.r.t. the intrinsic metric corresponding to the diffusion operator, as well as *conservativity* of the diffusion process generated by the Friedrichs extension of the operator both imply Markov uniqueness, provided the degeneracy of the second order coefficients is under control, cf. Corollary 3.3 and 3.4. Under the same condition, we prove that Markov uniqueness of the operator and Silverstein uniqueness of the corresponding Dirichlet form are equivalent, cf. Theorem 3.6. If the degeneracy is "too strong", this equivalence does, however, not hold, cf. the example in Section e) 2) of Chapter 3.

- Our Markov uniqueness results are based on a crucial *maximality result*, which shows the existence, and gives an explicit description, of a maximal Dirichlet operator extending a given symmetric diffusion operator, cf. Corollary 3.1. This result holds for arbitrary diffusion operators on general state spaces. It extends a previous maximality result for generators of so–called "classical Dirichlet forms" on Banach spaces by S. Albeverio, S. Kusuocka and M. Röckner [AlbKusRö 90]. The domain of the Dirichlet form corresponding to the maximal Dirichlet extension turns out to be a *weak Sobolev space* associated with the diffusion operator. These weak Sobolev spaces have been introduced here (respectively in the Comptes rendus note [Eb 95]) for the first time. Their definition requires the geometric representation $\mathcal{L} = -d^* d$ of the symmetric diffusion operator $\mathcal{L}$. We show in Appendix D, Section 7), that every symmetric diffusion operator can be reprensented in this way with a *generalized differential d* taking values in a *measurable co–tangent bundle $T'E$* over the state space E, cf. also [BouHi 91] for related considerations. Consequences of and examples for similar geometric representations are discussed in detail in Appendix D.

- We discuss in detail two examples of symmetric infinite–dimensional diffusion operators that are not even Markov unique, cf. Section b) in Chapter 5. Examples of this type have not been known so far, and might give new insight to infinite dimensional uniqueness problems.

- The second main contribution of this work in the infinite dimensional case is to unify two of the most effective approaches to prove uniqueness known so far: The *projective approach to Markov uniqueness* (cf. [AlbRöZha 92], and Section c) in Chapter 5 below), and the *approximative approach to essential self–adjointness* [AlbKoRö 95]. We present a method for proving L^p uniqueness of not necessarily symmetric infinite dimensional diffusion operators, which yields both a partial extension of the Markov uniqueness results in [AlbRöZha 92] to the stronger L^p uniqueness for $1 \le p < 2$, and an extension of the results in [AlbKoRö 95] to L^p uniqueness for non–symmetric operators, $1 \le p < \infty$, cf. Theorems 5.1 and 5.2. The only difference between both results turns out to be the way in which the resolvents (or, equivalently, semigroups) of the finite dimensional approximations are estimated. Although our results are only stated for diffusion operators with "flat geometry" on linear spaces, they can probably be extended to non–flat situations. We apply the results to models in classical statistical mechanics that have been studied before in [AlbKoRö 95] and [AlbKoRö 97c], cf. Section e) in Chapter 5, and, in particular, to small perturbations of infinite dimensional diffusion operators with linear drift, cf. Chapter 5, Section f). The small perturbation case includes the $P(\phi)_2$

quantum field in finite volume, as well as other examples studied in Section g) of Chapter 5.

Chapter 1

Motivation and basic definitions : Uniqueness problems in various contexts

In this introductory chapter, we present some uniqueness problems that arise in probability theory, analysis and mathematical physics. Somehow, they are all variations of one fundamental problem. When trying to solve these problems, the basic analytic questions of L^p uniqueness, essential self–adjointness, Markov uniqueness, and strong Markov uniqueness of generators of semigroups, as well as Silverstein uniqueness of quadratic forms, arise naturally. These analytic uniqueness questions are the ones, that will be investigated and applied in detail in the subsequent chapters.

After briefly describing our general framework, we discuss different kinds of uniqueness of martingale problems in Section a). In Section b), we look at the analytic problems of operator cores and form cores, which play a crucial rôle in many proofs of results on diffusion operators. We then briefly consider the notion of essential self–adjointness of a symmetric operator, that arises in particular in mathematical physics, cf. Section c). In Section d), we look at uniqueness problems from the theory of Dirichlet forms. Finally, in Section e) we compare the different analytic notions of uniqueness introduced in the preceeding sections.

We now start by describing our framework. Let E be some topological state space. Suppose we are given a linear space $\mathcal{A}$ consisting of "nice" functions on E. When speaking of *functions* we always mean (at least) Borel-measurable functions. Although for a part of our considerations, E and $\mathcal{A}$ may be rather general, E is usually supposed to be one of the following:

- An open subset of $\mathbf{R}^n$.

- A Riemannian manifold.

- A Banach space.

- A countable product of Riemannian manifolds.

- The space of continuous paths or loops over a Riemannian manifold.

- A space of measures on $\mathbf{R}^n$ or a manifold.

In the first two situations, $\mathcal{A}$ typically consists of smooth functions on E, e.g. $\mathcal{A} = C_0^\infty(E)$ if E is an open subset of $\mathbf{R}^n$. In the infinite dimensional situations, $\mathcal{A}$ will usually be a space of smooth cylinder functions on E. We call the functions in $\mathcal{A}$ **test functions**.

Suppose $\mathcal{L}$ is a linear operator mapping the test functions to functions on E. Our main interest are cases where $\mathcal{L}$ is an elliptic (in the wide sense) second order differential operator without zeroth order term on one of the concrete spaces described above. Such operators will be called **diffusion operators**, since they appear as the generators of diffusion processes without killing in the inside. A precise definition of the notion "*diffusion operator*" in a general framework is given in Appendix B, but we will mainly consider concrete diffusion operators on one of the state spaces listed above. This work has been motivated in particular by lacking knowledge about singular and infinite dimensional diffusion operators arising e.g. in stochastic mechanics, statistical mechanics, and quantum field theory.

a)   Uniqueness of the martingale problem for a diffusion process

In this section, we consider martingale problems for generators of diffusion processes. To obtain suitable uniqueness results for the martingale problem in singular and infinite dimensional situations, one has to restrict the class of solutions considered. We look at several uniqueness problems arising in this way, and show how they are related to the analytic notions of strong Markov uniqueness and Markov uniqueness, which are defined below.

Let $E_\Delta := E \dot\cup \{\Delta\}$ be the topological space obtained by adding an isolated terminal point Δ to E. Functions f on E are extended to E_Δ by setting $f(\Delta) := 0$. Let Ω denote the space of continuous paths on E with possibly finite life–time, i.e., Ω consists of all functions $\omega : [0, \infty) \to E_\Delta$ such that, for some $\xi \in (0, \infty]$, ω is a continuous E–valued function on $[0, \xi)$, and $\omega(t) = \Delta$ for $t \geq \xi$. We equip Ω with the σ–algebra generated by the maps $X_t : \Omega \to E_\Delta$, $X_t(\omega) = \omega(t)$, $0 \leq t < \infty$. We fix a probability measure μ on E.

Definition   *A probability measure P on Ω (resp. the corresponding stochastic process $\left((X_t)_{t \geq 0}, P\right)$ is called a* **solution of the martingale problem for**

ons $f(X_t)$ and $(\mathcal{L}f)(X_t)$ are integrable w.r.t. P, the function $(\omega))$ is $P \otimes ds$ integrable for $0 \leq s \leq t$, and

$$M_t^{[f]} := f(X_t) - \int_0^t (\mathcal{L}f)(X_s)\, ds$$

nder P w.r.t. the filtration generated by X_t, $0 \leq t < \infty$.
tat P solves MP $(\mathcal{L}, \mathcal{A}, \mu)$.

ussion of martingale problems can be found in [EthKur 86]. For
erators, including very singular ones, the existence of solutions
n particular, the theory of Dirichlet forms provides powerful
istence of solutions to martingale problems for singular and
1al, symmetric (or, more generally, sectorial) diffusion opera-
e class of initial distributions, cf. e.g. [MaRö 92], [AlbRö 91],
[Eb 97]. The following question, however, is sometimes very

en an initial distribution μ, how can those diffusion operators
:erized, for which there is *exactly one* solution of MP $(\mathcal{L}, \mathcal{A}, \mu)$?

d S. Varadhan [StrVar 79] show uniqueness of the martingale
ad class of diffusion operators on $\mathbf{R}^n$, but they still assume that
e at least locally bounded. O. Okitaloshima and J. A. van Cast-
give a necessary and sufficient condition for the existence and
Markov process solving the martingale problem for any given
n, provided E is locally compact, and $\mathcal{A}$ and $\mathcal{L}(\mathcal{A})$ consist of
ons vanishing at infinity. Our main interest, however, are sin-
: dimensional diffusion operators defined on smooth functions,
:s are not applicable. In fact, it seems almost hopeless to answer
neral classes of singular diffusion operators as they appear e.g.
hanics, in infinite dimensional systems of statistical mechanics,
uantum field theory.
:re are good reasons not to ask for uniqueness of the martingale
ll continuous stochastic processes on E, but only among a cer-
: now look at some refinements of corresponding considerations
AlbRöZha 92]. Here, S. Albeverio, M. Röckner and T. S. Zhang
w analytic concepts as, for example, essential self–adjointness,
r Markov uniqueness, can be used to prove uniqueness of mar-
in restricted senses. Following them roughly, we now make dif-
s on the stochastic processes considered. Firstly, from now on
> account solutions of the martingale problem that are time–
liffusion processes, i.e., $\left((X_t)_{t\geq 0},\, P\right)$ is a strong Markov
ntinuous trajectories), and there exists a time–homogeneous

transition function[1] $p_t(x, dy)$ on E such that

(1.1) $E[f(X_{t+s}) \mid \sigma(X_u ; 0 \le u \le s)] \; = \; (p_t f)(X_s) \qquad P\text{--a.s.}$

for all $s, t \ge 0$, and all positive or bounded functions f on E. Here $E[\cdot|\cdot]$ denotes the conditional expectation w.r.t. P, and

$$(p_t f)(x) \; := \; \int f(y)\, p_t(x, dy) \qquad \text{for all } x \in E.$$

Note that usually, the function $p_t f$ is not uniquely determined by (1.1). However, it is unique up to coincidence $P \circ X_0^{-1}$ a.e. (– or, more generally, up to coincidence $P \circ X_s^{-1}$ a.e. for any $s \ge 0$).

In order to obtain satisfactory uniqueness results, we restrict the class of diffusion processes considered further. There are several more or less natural ways to do this. The first possibility is the following: Suppose we are given a σ-finite measure m on E. Typical choices for m might be Lebesgue measure if $E \subseteq \mathbf{R}^n$, the volume if E is a Riemannian manifold, or Wiener measure if $E = C([0,1]; M)$ with M a compact Riemannian manifold. Moreover, in singular cases it is often necessary to use other measures, e.g. absolutely continuous measures on $\mathbf{R}^n$, or Gibbs measures in infinite dimensions. In general, any measure m that is **invariant** w.r.t. $(\mathcal{L}, \mathcal{A})$, i.e., $\mathcal{L}f \in L^1(E; m)$ and $\int \mathcal{L}f\, dm = 0$ for all $f \in \mathcal{A}$, might be a good choice.

If m is a probability measure, then let $\mu := m$. Otherwise, let $\mu := \rho \cdot m$, where ρ is an m–integrable function such that $0 < \rho \le 1$ m–a.e., and $\int \rho\, dm = 1$. We consider all probability measures P on Ω such that $\left((X_t)_{t \ge 0}, P\right)$ is a time–homogeneous diffusion process with initial distribution μ, and

(1.2) $$\int p_t f\, dm \; \le \; \int f\, dm$$

for all positive functions f on E, and all $t \ge 0$.

Note that the validity of (1.2) does not depend on the chosen version of $p_t f$, because m is equivalent to the initial distribution μ. If $m = \mu$ then (1.2) means that

(1.3) $$E[f(X_t)] \; \le \; E[f(X_0)]$$

for all positive functions f on E, and all $t \ge 0$,

i.e., the process $\left((X_t)_{t \ge 0}, P\right)$ is **sub–stationary**.

Similarly, we can give a probabilistic interpretation of (1.2) if $m \ne \mu$. Let

$$P_m \; := \; \int P[\cdot \mid X_0 = z]\, m(dz).$$

---

[1] A function $p : [0, \infty) \times E \times \mathcal{B}(E) \to [0,1]$ is called a **time–homogeneous transition function** iff, for all $s, t \ge 0$, $x \in E$, and $\Gamma \in \mathcal{B}(E)$, $p_t(x, \bullet)$ is a probability measure, $p_0(x, \bullet) = \delta_x$, the map $(t, x) \mapsto p_t(x, \Gamma)$ is $\mathcal{B}([0, \infty) \times E)$–measurable, and $p_{t+s}(x, \Gamma) = \int p_s(x, \Gamma)\, p_t(x, dy)$.

Since m is equivalent to μ, P_m is a uniquely defined σ–finite measure. We may view $\left((X_t)_{t\geq 0}, P_m\right)$ as the diffusion $\left((X_t)_{t\geq 0}, P\right)$ conditioned to start with distribution m. Condition (1.2) is eqivalent to

$$(1.4) \qquad E_m\left[f(X_t)\right] \ \leq \ E_m\left[f(X_0)\right]$$
$$\text{for all positive functions } f \text{ on } E, \text{ and all } t \geq 0,$$

where E_m stands for integration w.r.t. P_m.

Problem 2: Let m and μ be as above. For which diffusion operators $(\mathcal{L}, \mathcal{A})$ exists exactly one solution P of MP $(\mathcal{L}, \mathcal{A}, \mu)$ such that $\left((X_t)_{t\geq 0}, P\right)$ is a time–homogeneous diffusion process satisfying (1.4) ?

REMARKS. (i) Fix m and μ as in Problem 2, and suppose ν is a probability measure equivalent to μ. By conditioning to start with initial distribution μ, one can construct from any time–homogeneous diffusion process $\left((X_t)_{t\geq 0}, P_\nu\right)$ solving MP $(\mathcal{L}, \mathcal{A}, \nu)$ a time–homogeneous diffusion process $\left((X_t)_{t\geq 0}, P_\mu\right)$ with the same transition function, which solves MP $(\mathcal{L}, \mathcal{A}, \mu)$, and vice versa. Hence uniqueness of MP $(\mathcal{L}, \mathcal{A}, \mu)$ in the sense of Problem 2 implies uniqueness of MP $(\mathcal{L}, \mathcal{A}, \nu)$ in the same sense for all probability measures ν equivalent to μ.

(ii) The methods we will use to solve Problem 2 do not imply uniqueness of the martingale problem w.r.t. each fixed initial distribution. This is consistent with the fact that for some singular and infinite–dimensional diffusion operators (e.g. the Ornstein–Uhlenbeck operator on the path space over a Riemannian manifold), the best known existence results do not yield the existence of a diffusion process $\left((X_t)_{t\geq 0}, (P_z)_{z\in E}\right)$ solving the corresponding martingale problem with initial distribution δ_z for *every* $z \in E$, but *at most for quasi–every* z.

By which methods can we treat Problem 2 ? Fix measures m and μ as above. Suppose p_t, $t \geq 0$, is a transition function satisfying (1.2). Then for m–integrable functions f, and $t \geq 0$, $(p_t f)(x) := \int f(y)\, p_t(x, dy)$ is defined for m–a.e. x. The map $f \mapsto p_t f$ respects m–classes, and induces a linear contraction both on $L^1(E; m)$ and on $L^\infty(E; m)$. We denote both contractions by T_t. Since p_t, $t \geq 0$, is a transition function, $(T_t)_{t\geq 0}$ is a measurable[2] semigroup of sub–Markovian[3] contractions on $L^1(E; m)$.

Lemma 1.1 *Let* $(\mathcal{L}, \mathcal{A})$ *be a diffusion operator on* E *such that* $\mathcal{A}$ *and* $\mathcal{L}(\mathcal{A})$ *consist of* m*–integrable functions, and* $\mathcal{A}$ *is dense in* $L^1(E; m)$. *Suppose* p_t, $t \geq$ 0, *is the transition function of a solution* $\left((X_t)_{t\geq 0}, P\right)$ *of* MP $(\mathcal{L}, \mathcal{A}, \mu)$, *and*

---

[2] A semigroup $(T_t)_{t\geq 0}$ of bounded linear operators on $L^p(E; m)$, $1 \leq p \leq \infty$, is called **measurable**, iff for each $f \in L^p(E; m)$ there exists a product measurable function u on $[0, \infty) \times E$ such that $u(t, \bullet)$ is an m–version of $(T_t f)$ for all $t \geq 0$.

[3] A linear operator S on $L^p(E; m)$, $1 \leq p \leq \infty$, is called **sub–Markov**, iff it is positivity preserving and L^∞–contractive, i.e., $0 \leq Sf \leq 1$ whenever $0 \leq f \leq 1$.

(1.4) resp. (1.2) holds. Then $(p_t)_{t \geq 0}$ *induces a sub–Markovian* C^0 *contraction semigroup[4]* $(T_t)_{t \geq 0}$ *on* $L^1(E; m)$, *such that the generator[5] of* $(T_t)_{t \geq 0}$ *extends* $(\mathcal{L}, \mathcal{A})$.

REMARK.   When saying that "the generator of $(T_t)_{t \geq 0}$ extends $(\mathcal{L}, \mathcal{A})$", we mean that every function f in $\mathcal{A}$ is a representative of an m–class $\bar{f}$ in the domain of the generator L, and $\mathcal{L}f$ is contained in the m–class $L\bar{f}$. In particular, the operator $(\mathcal{L}, \mathcal{A})$ respects m–classes, which is not clear a priori.

Since the lemma is crucial for our analytic treatment of the uniqueness of the martingale problem, we briefly sketch the proof, although it is essentially standard.

PROOF. We have already noted that $(T_t)_{t \geq 0}$ is a sub–Markovian measurable contraction semigroup on $L^1(E; m)$. Fix $f \in \mathcal{A}$. Since $\big((X_t)_{t \geq 0}, P\big)$ solves MP $(\mathcal{L}, \mathcal{A}, \mu)$, we have

$$E\left[f(X_t) \mid \sigma(X_0)\right] \;-\; E\left[\int_0^t (\mathcal{L}f)(X_s)\, ds \mid \sigma(X_0)\right] \;=\; f(X_0) \quad P\text{–a.s.}$$

for all $t \geq 0$. The first conditional expectation equals $(p_t f)(X_0)$ P–a.s. Moreover, one easily checks that the second conditional expectation is given P–a.s. by $\displaystyle \int_0^t p_s \mathcal{L}f (X_0)\, ds$. Note that this integral exists P–a.s., because by (1.2), and since $\mathcal{L}f$ is in $L^1(E; m)$,

$$E\left[\int_0^t |p_s \mathcal{L}f(X_0)|\, ds\right] \;\leq\; \int_0^t \int p_s |\mathcal{L}f|\, d\mu\, ds$$

$$\leq\; \int_0^t \int p_s |\mathcal{L}f|\, dm\, ds \;\leq\; t \cdot \int |\mathcal{L}f|\, dm \;<\; \infty.$$

We obtain

$$(p_t f)(X_0) \;-\; \int_0^t (p_s \mathcal{L}f)(X_0)\, ds \;=\; f(X_0) \qquad P\text{–a.s.,}$$

---

[4]A semigroup $(T_t)_{t \geq 0}$ of bounded linear operators on a Banach space $(B, \|\cdot\|)$ is called a C^0 semigroup, iff it is **strongly continuous**, i.e., the map $t \mapsto T_t f$ is continous w.r.t. $\|\cdot\|$ for each $f \in B$. Equivalently, the semigroup $(T_t)_{t \geq 0}$ is C^0 if and only if $T_t f$ converges weakly to f as $t \downarrow 0$ for all $f \in B$. It is called a **contraction semigroup**, iff T_t is a contraction for all $t \geq 0$. We refer to [Yo 80] or [Pa 85] for the basic notions of semigroup theory used here and below.

[5]The generator of a C^0 semigroup $(T_t)_{t \geq 0}$ on a Banach space B is the densely defined linear operator L on B given by

$$L f \;:=\; \lim_{t \downarrow 0} \frac{1}{t}\,(T_t f - f)$$

with domain $\mathrm{Dom}\,(L)$ consisting of all elements $f \in B$, for which the limit exists w.r.t. the norm in B.

and thus

$$(1.5) \qquad T_t f = f + \int_0^t T_s \mathcal{L} f \, ds \qquad m\text{–a.e.}$$

In particular, the map $t \mapsto T_t f$ is continuous w.r.t. the norm in $L^1(E; m)$ for every $f \in \mathcal{A}$. Since $\mathcal{A}$ is dense in $L^1(E; m)$, $(T_t)_{t \geq 0}$ is a C^0 semigroup. Hence $s \mapsto T_s \mathcal{L} f$ is also continuous for all $f \in \mathcal{A}$. Now (1.5) implies that the generator of $(T_t)_{t \geq 0}$ extends $(\mathcal{L}, \mathcal{A})$. ∎

The lemma motivates the following definition:

Definition 1.1 *A densely defined linear operator* $(\mathcal{L}, \mathcal{A})$ *on* $L^1(E; m)$ *is called* **strongly Markov–unique** *iff there exists at most one sub–Markovian* C^0 *contraction semigroup* $(T_t)_{t \geq 0}$ *on* $L^1(E; m)$ *such that its generator extends* $(\mathcal{L}, \mathcal{A})$.

A diffusion process on Ω is uniquely determined by its initial distribution and its transition function. A time–homogeneous diffusion that satisfies (1.4) and has initial distribution μ $(= \rho \cdot m, \; \rho \leq 1 \; m\text{–a.e.})$, is already uniquely determined by its operator semigroup $(T_t)_{t \geq 0}$ acting on $L^1(E; m)$. Hence we obtain:

Corollary 1.1 *If* $(\mathcal{L}, \mathcal{A})$ *is a strongly Markov unique linear operator on* $L^1(E; m)$, *then there is at most one probability measure* P *on* Ω, *such that* $\left((X_t)_{t \geq 0}, P\right)$ *is a time–homogeneous diffusion, which satisfies (1.4) and solves* MP $(\mathcal{L}, \mathcal{A}, \mu)$.

Another question, that is more difficult than Problem 2, arises naturally if we are interested in stationary[6] solutions of the martingale problem:

Problem 3: Is there a unique stationary solution $\left((X_t)_{t \geq 0}, P\right)$ of the martingale problem for $(\mathcal{L}, \mathcal{A})$?

Suppose $\left((X_t)_{t \geq 0}, P\right)$ is a stationary solution, and m is the invariant distribution, i.e., $m = P \circ X_t^{-1}$ for all t. Then the transition function p_t satisfies

$$(1.6) \qquad \int p_t f \, dm = \int f \, dm$$

for all $t \geq 0$, and all positive resp. bounded functions f on E. By the definition of a solution to the martingale problem, $f(X_t)$ and $(\mathcal{L}f)(X_t)$ are integrable w.r.t. P for all test functions f, and all $t \geq 0$. Hence f and $\mathcal{L}f$ are m–integrable, and we can apply Lemma 1.1 above. Thus $(p_t)_{t \geq 0}$ induces a C^0 contraction

---

[6] A stochastic process $\left((X_t)_{t \geq 0}, P\right)$ is called **stationary**, iff its marginals are invariant under time–shift, i.e., for any fixed reals $t_1, \ldots, t_n \geq 0$, $n \in \mathbb{N}$, the distributions of $(X_{t_1}, \ldots, X_{t_n})$ and $(X_{t+t_1}, \ldots, X_{t+t_n})$ under P coincide for all $t \geq 0$.

semigroup $(T_t)_{t \geq 0}$ on $L^1(E; m)$. By (1.6), (1.5), and the strong continuity of $(T_t)_{t \geq 0}$, we obtain

$$(1.7) \qquad \int \mathcal{L}f \, dm \;=\; 0 \qquad \text{for all } f \in \mathcal{A},$$

i.e., m is an invariant measure for $(\mathcal{L}, \mathcal{A})$. Thus Problem 3 can be divided into two parts:

  i) Is there a unique invariant probability measure m of $(\mathcal{L}, \mathcal{A})$?

  ii) Let m be a fixed invariant probability measure of $(\mathcal{L}, \mathcal{A})$. Is there a unique solution of MP $(\mathcal{L}, \mathcal{A}, m)$ such that (1.6) holds ?

The first question will not be considered here. A detailed treatment of this problem is given in [AlbBoRö 97].

REMARK. It is worth pointing out, however, that on a formal level, and on the level of some techniques that have been used, the problem of uniqueness of invariant measures is related to the L^p uniqueness problem considered below. In fact, uniqueness of invariant measures means that the equation $\mathcal{L}^* m = 0$ has a unique solution in a space of probability measures, whereas L^p uniqueness of a dissipative operator holds if and only if for some, or, equivalently, all $\lambda > 0$, the equation $\mathcal{L}^* u = \lambda u$ has only the trivial solution in L^q, $\frac{1}{p} + \frac{1}{q} = 1$, cf. Corollary 1.3 in Appendix A below.

The second question is a modification of Problem 2. By the corollary above, uniqueness in the sense of ii) holds, if $(\mathcal{L}, \mathcal{A})$ is strongly Markov unique on $L^1(E; m)$.

To state the fourth uniqueness problem, suppose that there exists a probability measure m on E such that $(\mathcal{L}, \mathcal{A})$ is a symmetric operator on $L^2(E; m)$, i.e., f and $\mathcal{L}f$ are square–integrable w.r.t. m, and

$$\int \mathcal{L}f \, g \, dm \;=\; \int f \, \mathcal{L}g \, dm \qquad \text{for all } f, g \in \mathcal{A}.$$

Then it is natural to look for reversible[7] solutions $\big((X_t)_{t \geq 0}, P\big)$ of the martingale problem.

Problem 4: Can we characterize those m–symmetric diffusion operators $(\mathcal{L}, \mathcal{A})$, for which there is precisely one reversible solution $\big((X_t)_{t \geq 0}, P\big)$ of MP $(\mathcal{L}, \mathcal{A}, m)$?

REMARK. By extending the definition of a solution $\big((X_t)_{t \geq 0}, P\big)$ of the martingale problem to the case where the initial distribution, and hence P, is a σ–finite measure, we may formulate the same problem for σ–finite measures m.

---

[7]A stochastic process $\big((X_t)_{t \geq 0}, P\big)$ is called **reversible**, iff $E\,[f(X_0)\,g(X_t)] = E\,[f(X_t)\,g(X_0)]$ for all $t \geq 0$, and all positive functions f, g on E.

If $(p_t)_{t\geq 0}$ is the transition semigroup of a reversible solution of MP $(\mathcal{L}, \mathcal{A}, m)$, then

$$\int p_t f \, g \, dm \;=\; \int f \, p_t g \, dm$$

for all positive functions f, g on E. In particular,

$$\int p_t f \, dm \;\leq\; \int f \, dm.$$

Hence in the same way as below Problem 2, we see that $(p_t)_{t\geq 0}$ induces a sub–Markovian C^0 semigroup of contractions on $L^1(E; m)$, and, more generally, on $L^p(E; m)$ for all $1 \leq p < \infty$. The semigroup $(T_t)_{t\geq 0}$ on $L^2(E; m)$ is symmetric, i.e., T_t is a symmetric operator for all t, and its generator extends $(\mathcal{L}, \mathcal{A})$.

Definition 1.2 *A symmetric linear operator $(\mathcal{L}, \mathcal{A})$ on $L^2(E; m)$ is called* **Markov unique,** *iff there exists at most one symmetric sub–Markovian C^0 contraction semigroup $(T_t)_{t\geq 0}$ on $L^2(E; m)$ such that its generator extends $(\mathcal{L}, \mathcal{A})$.*

REMARK. Note that any symmetric sub–Markovian C^0 contraction semigroup on $L^2(E; m)$ induces a C^0 contraction semigroup on each $L^p(E; m)$, $1 \leq p < \infty$, cf. Lemma 1.12 in Appendix B.

By the above, we have:

Lemma 1.2 *If $(\mathcal{L}, \mathcal{A})$ is a Markov unique symmetric linear operator on $L^2(E; m)$, then there exists at most one reversible diffusion $\left((X_t)_{t\geq 0}, P\right)$ solving MP $(\mathcal{L}, \mathcal{A}, m)$.*

The study of Markov uniqueness is one of the central topics of this work, cf. in particular Chapter 3 and 5 below.

b) Cores

The existence of cores consisting of "sufficiently nice" functions is a technical, but important problem in the analysis of diffusion operators. We first look at two of the many situations where this problem arises, and then describe the relations between existence of cores and uniqueness problems. We consider both operator cores and form cores.

EXAMPLE 1 (*Characteriztion of stationary distributions*). Let E be a domain in $\mathbf{R}^n$, $n \in \mathbf{N}$. Suppose we are given a transition function $(p_t)_{t\geq 0}$ on E, which induces a C^0 semigroup $(T_t)_{t\geq 0}$ of bounded lionear operators on $L^p(E; dx)$ for some $p \in [1, \infty)$. Moreover, assume that the generator L of $(T_t)_{t\geq 0}$ extends the diffusion operator $(\mathcal{L}, \mathcal{A})$, where $\mathcal{A} := C_0^\infty(E)$. For example, $(p_t)_{t\geq 0}$ may be the transition function of a suitable solution of the martingale problem for $(\mathcal{L}, \mathcal{A})$, cf. Section a) above. We want to know if there exists a **stationary distribution** m of $(p_t)_{t \geq 0}$, i.e., a probability measure m on E such that $m = \int p_t(x, \bullet) \, m \, (dx)$, or, equivalently,

$$(1.8) \qquad \int p_t f \, dm \; = \; \int f \, dm \qquad \text{for all } f \in \mathcal{A} \text{ and } t \geq 0.$$

By formally differentiating Equation (1.8), we might expect that m is a stationary distribution for $(p_t)_{t \geq 0}$ if and only if

$$(1.9) \qquad \int \mathcal{L} f \, dm \; = \; 0 \qquad \text{for all } f \in \mathcal{A},$$

i.e., if and only if m is an *invariant measure* for $(\mathcal{L}, \mathcal{A})$. This heuristic consideration is *false* in general. For example, Lebesgue measure on $(0,1)$ is an invariant measure for the operator $(\frac{d^2}{dx^2}, C_0^\infty(0,1))$, but it is not a stationary distribution for the transition function $(p_t)_{t \geq 0}$ of Brownian motion with absorption at 0 and 1, because $p_t 1 < 1$ for all $t > 0$.

On the other hand, suppose m is an absolutely continuous probability measure on E with density $\frac{dm}{dx} \in L^q(E; dx)$ (or, equivalently, $\frac{dm}{dx} \in L^{q-1}(E; m)$), $\frac{1}{p} + \frac{1}{q} = 1$. Then, by Hölder's inequality, $L^p(E; dx)$ is continuously embedded into $L^1(E; m)$. For $f \in \mathcal{A}$ and $t \geq 0$, we obtain

$$\int (p_t f - f) \, dm \; = \; \int (T_t f - f) \, dm$$

$$= \; \int_0^t \int T_s \mathcal{L} f \, dm \, ds \; = \; \int_0^t \int L T_s f \, dm \, ds.$$

The function $s \mapsto T_s \mathcal{L} f$ is continuous w.r.t. the $L^p(E; dx)$–norm, and hence w.r.t. the $L^1(E; m)$–norm. Therefore,

$$\int \mathcal{L} f \, dm \; = \; \lim_{t \downarrow 0} \frac{1}{t} \int_0^t \int T_s \mathcal{L} f \, dm \, ds \; = \; 0 \qquad \text{for all } f \in \mathcal{A},$$

provided m is a stationary distribution for $(p_t)_{t \geq 0}$, i.e., m is an invariant measure for $(\mathcal{L}, \mathcal{A})$ in this case. Conversely, m is a stationary distribution for $(p_t)_{t \geq 0}$, provided

$$(1.10) \qquad \int L f \, dm \; = \; 0 \qquad \text{for all } f \text{ in the domain of } L.$$

But this follows from (1.9), if $\mathcal{A}$ is an **operator core** for L, i.e., $\mathcal{A}$ is dense in the domain of L w.r.t. the graph norm

$$\|f\|_L \; := \; \|f\|_{L^p(E; dx)} + \|\mathcal{L} f\|_{L^p(E; dx)}, \qquad f \in \text{Dom}(L).$$

Summarizing, we have shown:

Lemma *Let m be an absolutely continuous probability measure on E such that $\frac{dm}{dx}$ is in $L^q(E; dx)$, $\frac{1}{p} + \frac{1}{q} = 1$. Suppose that $\mathcal{A}$ is a core for the generator L of the semigroup $(T_t)_{t \geq 0}$ on $L^p(E; dx)$. Then m is a stationary distribution for $(p_t)_{t \geq 0}$ if and only if it is an invariant measure for the operator $(\mathcal{L}, \mathcal{A})$.*

EXAMPLE 2 *(Sub–Markov property of diffusion semigroups)*. Suppose that $(\mathcal{L}, \mathcal{A})$ is a diffusion operator on an L^p space, and that $(T_t)_{t \geq 0}$ is a C^0 semigroup generated by an extension of $(\mathcal{L}, \mathcal{A})$. Since $(\mathcal{L}, \mathcal{A})$ is a diffusion operator, one might expect that $(T_t)_{t \geq 0}$ is sub–Markov. In general, this is false. If, however, $\mathcal{A}$ is a *core* for the generator of $(T_t)_{t \geq 0}$, and m is a *sub–invariant measure* of $(\mathcal{L} - \alpha, \mathcal{A})$ for some $\alpha \geq 0$, *then* we can prove that $(T_t)_{t \geq 0}$ is sub–Markov, cf. Lemma 1.9 in Appendix B.

In the examples above, and in many other situations, the following problem arises:

Problem 5: Let m be a σ–finite measure on E, and let $1 \leq p < \infty$. Suppose we are given a C^0 semigroup $(T_t)_{t \geq 0}$ on $L^p(E\,;\,m)$ such that its generator is an extension of $(\mathcal{L}, \mathcal{A})$. Under which conditions is $\mathcal{A}$ a *core* for the generator ?

From abstract semigroup theory it is known that under the assumptions from Problem 5, the following statements are equivalent, cf. Theorem 1.2 in Appendix A below:

(i) $\mathcal{A}$ is a core for the generator of $(T_t)_{t \geq 0}$.

(ii) The closure of the operator $(\mathcal{L}, \mathcal{A})$ on $L^p(E\,;\,m)$ is the generator of a C^0 semigroup.

(iii) $(T_t)_{t \geq 0}$ is the only C^0 semigroup on $L^p(E\,;\,m)$ which has a generator that extends $(\mathcal{L}, \mathcal{A})$.

This equivalence is one of the reasons for the importance of the following notion of uniqueness:

Definition 1.3 *A densely defined linear operator $(\mathcal{L}, \mathcal{A})$ on $L^p(E\,;\,m)$ is called $L^p(E\,;\,m)$ unique, if and only if there exists at most one C^0 semigroup $(T_t)_{t \geq 0}$ on $L^p(E\,;\,m)$ such that its generator extends $(\mathcal{L}, \mathcal{A})$.*

Sometimes, L^p uniqueness is also called **strong uniqueness**. "Strong" means that it is stronger than the notions of Markov uniqueness introduced above. Relations and differences between the various notions of uniqueness will be discussed in Section e) below. Detailed studies of L^p uniqueness for diffusion operators on finite and infinite dimensional state spaces are carried out in Chapter 2 and 5 below.

So far, we have pointed out the importance of knowing if a given test function space $\mathcal{A}$ is an operator core for the generator of a C^0 semigroup. In the symmetric case, however, it is sufficient for many applications to know that $\mathcal{A}$ is a *form core*.

Let $(\mathcal{L}, \mathcal{A})$ be a negative definite symmetric diffusion operator on $L^2(E\,;\,m)$ such that m is an invariant measure for $(\mathcal{L}, \mathcal{A})$. Suppose we are given a symmetric sub–Markovian C^0 contraction semigroup $(T_t)_{t \geq 0}$ on $L^2(E\,;\,m)$ such that its generator L extends $(\mathcal{L}, \mathcal{A})$. Since $(T_t)_{t \geq 0}$ is a symmetric contraction semigroup, L is a negative definite self-adjoint operator. Let $\mathcal{E}$ denote the positive definite quadratic form on $L^2(E; m)$ given by

$$\mathcal{E}(f,g) \;=\; -\int f\, Lg\, dm, \qquad f,\, g \in \mathrm{Dom}\,(L),$$

and let

$$\mathcal{E}_1(f,g) \;:=\; \mathcal{E}(f,g) + \int f\, g\, dm.$$

The space $\mathcal{A}$ is called a **form core** for L, iff it is dense in the domain of L w.r.t. the norm determined by the inner product $\mathcal{E}_1$.

Problem 6: Which conditions guarantee that $\mathcal{A}$ is a form core for the generator L of $(T_t)_{t \geq 0}$?

To answer Problem 6, we note that the bilinear form $(\mathcal{E}, \mathcal{A})$ is closable[8], and the closure[8] $(\bar{\mathcal{E}}, \bar{\mathcal{A}})$ is a closed[8] positive definite symmetric bilinear form. Since $(\mathcal{L}, \mathcal{A})$ is a diffusion operator with invariant measure m, and

$$\mathcal{E}(f,g) \;=\; -\int f\, \mathcal{L}g\, dm \qquad \text{for all } f,\, g \in \mathcal{A},$$

the form generator[9] $L^{(0)}$ of $(\bar{\mathcal{E}}, \bar{\mathcal{A}})$ generates a symmetric sub–Markovian C^0 contraction semigroup on $L^2(E; m)$, cf. Lemma 1.10 in Appendix B. Now suppose that $(\mathcal{L}, \mathcal{A})$ is Markov unique. Then $L^{(0)} = L$, whence $\mathcal{A}$ is a form core in this case. We have shown:

Lemma 1.3   *If $(\mathcal{L}, \mathcal{A})$ is Markov unique, then $\mathcal{A}$ is a form core for the generator of $(T_t)_{t \geq 0}$.*

Conversely, if $(\mathcal{L}, \mathcal{A})$ is not Markov unique, then there exists an extension of $(\mathcal{L}, \mathcal{A})$ which generates a sub–Markovian symmetric C^0 contraction semigroup on $L^2(E; m)$, but does not have $\mathcal{A}$ as a form core.

---

[8]A positive definite symmetric bilinear form $(\mathcal{E}, \mathcal{F})$ on $L^2(E; m)$ is called **closed**, iff $\mathcal{F}$ is complete w.r.t. the inner product $\mathcal{E}_1(u,v) := \mathcal{E}(u,v) + \int u\,v\,dm$. It is called **closable**, iff there exists a closed extension on $L^2(E; m)$, or, equivalently, iff $\mathcal{E}(u_n, u_n) \to 0$ as $n \to \infty$, whenever $(u_n)_{n \in \mathbb{N}}$ is a sequence in $L^2(E; m)$ such that $u_n \to 0$ in $L^2(E; m)$ and $\mathcal{E}(u_n - u_m, u_n - u_m) \to 0$ as $n,\, m \to \infty$. The closure $(\mathcal{E}, \bar{\mathcal{F}})$ of a closable form $(\mathcal{E}, \mathcal{F})$ is obtained by extending $\mathcal{E}$ continuously to the completion $\bar{\mathcal{F}}$ of $\mathcal{F}$ w.r.t. the $\mathcal{E}_1$ inner product. Closability guarantees that $\bar{\mathcal{F}}$ is embedded into $L^2(E; m)$.

[9]The **generator** L of a closed symmetric bilinear form $(\mathcal{E}, \mathcal{F})$ on $L^2(E; m)$ is the self-adjoint operator defined by

$$\mathrm{Dom}\,(L) \;=\; \{\, u \in L^2(E; m);\ \exists\, w \in L^2(E; m)\ \forall v \in \mathcal{F}\,:\ \mathcal{E}(u,v) = \int w\,v\,dm \,\},$$

$$\mathcal{E}(u,v) \;=\; -\int Lu\,v\,dm \qquad \text{for all } v \in \mathcal{F} \text{ and } u \in \mathrm{Dom}\,(L).$$

c) Essential self–adjointness and uniqueness of quantum dynamics

In non–relativistic quantum theory, Hamiltonians, i.e., the generators of quantum dynamics, are always self–adjoint[10] operators. Hence the question whether a given operator defined on test functions uniquely determines the dynamics, naturally leads to the following problem:

Problem 7: Suppose $(\mathcal{L}, \mathcal{A})$ is a densely defined symmetric linear operator on $L^2(E\,;\,m)$, where m is a σ–finite measure. Under which conditions exists exactly one self–adjoint extension of $(\mathcal{L}, \mathcal{A})$?

Every densely defined semibounded[11] symmetric linear operator L on a Hilbert space H has a self–adjoint extension, cf. e.g. [ReSi 75, Ch. X.3]. We recall the following well–known equivalence:

Lemma 1.4 *Suppose* $(\mathcal{L}, \mathcal{A})$ *is a semibounded symmetric linear operator on* $L^2(E\,;\,m)$. *Then the following assertions are equivalent:*

(i) There exists only one self–adjoint extension of $(\mathcal{L}, \mathcal{A})$.

(ii) There exists only one self–adjoint extension of $(\mathcal{L}, \mathcal{A})$ *which is semi–bounded.*

(iii) $(\mathcal{L}, \mathcal{A})$ *is* **essentially self–adjoint**, *i.e., the closure of* $(\mathcal{L}, \mathcal{A})$ *on* $L^2(E\,;\,m)$ *is a self–adjoint operator.*

The proof can be found e.g. in Chapter X of [ReSi 75], see in particular Theorem X.24 for the "hard" part (ii) $\Rightarrow$ (iii). Since the semi–bounded self–adjoint operators on a Hilbert space H are exactly the generators of the symmetric C^0 semigroups on H, Lemma 1.4 is a symmetric version of the equivalence mentioned below Problem 5. In particular, by combining the lemma and the equivalence, we obtain the following consequence, which is well–known but highly non–trivial:

Corollary 1.2 *A densely defined semi–bounded symmetric linear operator* $(\mathcal{L}, \mathcal{A})$ *on* $L^2(E\,;\,m)$ *is* $L^2(E\,;\,m)$ *unique if and only if it has only one self–adjoint extension.*

---

[10]A densely defined linear operator L on a Hilbert space H is called **self–adjoint** iff $L = L^*$, i.e., for a given element $v \in H$ there exists $w \in H$ such that

$$(Lu, v)_H \;=\; (u, w)_H \qquad \text{for all } u \in \text{Dom}\,(L)$$

if and only if v is in the domain of L, and $w = Lu$ in this case. Every self–adjoint operator is symmetric, but for unbounded operators, self–adjointness is a much stronger notion than symmetry.

[11]A symmetric linear operator L on a Hilbert space H is called **semibounded** (from above), iff there exists $\alpha \geq 0$ such that $(u, Lu)_H \leq \alpha\,(u, u)_H$ for all $u \in \text{Dom}\,(L)$.

Hence although at first sight, uniqueness of (semi–bounded) self–adjoint extensions seems to be a much weaker notion than uniqueness of extensions that generate a C^0 semigroup, both notions are actually equivalent for semi–bounded symmetric operators.

d)   Uniqueness of Dirichlet forms

In the theory of Dirichlet forms (cf. [FuOshTa 94], [MaRö 92]), uniqueness problems related to those described above naturally appear, too. In this section, we first show how the Markov uniqueness problem arises in Dirichlet form theory. Afterwards, we consider Silverstein uniqueness of Dirichlet forms, and discuss the relation to Markov uniqueness.

Fix a σ–finite measure m on E. A densely defined, positive definite symmetric bilinear form on $L^2(E\,;m)$ is called a **pre–Dirichlet form** iff it is closable, and the closure is a Dirichlet form, cf. Section b) for the definitions of closability and Dirichlet forms.

Suppose we are given a pre–Dirichlet form $(\mathcal{E},\,\mathcal{A})$ on $L^2(E\,;m)$ such that

$$(1.11) \qquad \mathcal{E}(f,g) \;=\; -\int \mathcal{L}f\,g\,dm \qquad \text{for all } f,\,g \in \mathcal{A},$$

where $(\mathcal{L},\,\mathcal{A})$ is a symmetric diffusion operator on $L^2(E\,;m)$ defined on a dense test function space $\mathcal{A}$. For given pre–Dirichlet forms, equations of type (1.11) can be derived by applying integration by parts identities, cf. the examples in [MaRö 92], and also Section a), 5) in Chapter 5 below.

The closure $(\bar{\mathcal{E}},\,\bar{\mathcal{A}})$ of the pre–Dirichlet form $(\mathcal{E},\,\mathcal{A})$ is a Dirichlet form, and its generator extends $(\mathcal{L},\,\mathcal{A})$, i.e., (1.11) holds for all $f \in \mathcal{A}$ and $g \in \bar{\mathcal{A}}$.

Problem 8: Is $(\bar{\mathcal{E}},\bar{\mathcal{A}})$ the only Dirichlet form extending $(\mathcal{E},\,\mathcal{A})$ such that $\mathcal{A}$ is contained in the domain of the corresponding generator ?

REMARK. Usually, Dirichlet forms have infinitely many extensions that are again Dirichlet forms. Only the additional requirement that $\mathcal{A}$ is contained in the domain of the generator of the extending form makes uniqueness in the sense of Problem 8 possible. This is illustrated by the following example:

EXAMPLE. Let

$$\mathcal{E}(u,v) \;=\; \int_{-\infty}^{\infty} u'\,v'\,dx, \qquad u,\,v \in H^{1,2}(\mathbf{R}^1;\,dx),$$

be the Dirichlet form of one–dimensional Brownian motion on $L^2(\mathbf{R}^1;\,dx)$. It is the closure of the corresponding pre–Dirichlet form with domain $\mathcal{A} := C_0^\infty(\mathbf{R}^1)$. For $y \in \mathbf{R}^1$, we consider the Dirichlet form $(\mathcal{E}^y,\,\mathcal{F}^y)$, where

$$\mathcal{E}^y(u,v) \;:=\; \int_{-\infty}^{y} u'\,v'\,dx \;+\; \int_{y}^{\infty} u'\,v'\,dx,$$

and $\mathcal{F}^y$ contains all square–integrable functions u on $\mathbf{R}^1$ that are absolutely continuous both on $(-\infty, y)$ and (y, ∞) with derivative u' in $L^2(-\infty, y; dx)$ resp. $L^2(y, \infty; dx)$. The forms $(\mathcal{E}^y, \mathcal{F}^y)$, $y \in \mathbf{R}^1$, are the Dirichlet forms of Brownian motion with reflection at y. All these forms extend $(\mathcal{E}, C_0^\infty(\mathbf{R}^1))$. Nevertheless, Problem 8 has a positive answer for this example. This follows e.g. from the essential self–adjointness of the corresponding diffusion operator $\mathcal{L}f = f''$ with domain $C_0^\infty(\mathbf{R}^1)$, cf. Lemma 1.5 below and the diagram in Section e), 2), below. In fact, the domain of the generator of $(\mathcal{E}^y, \mathcal{F}^y)$ contains only those functions f in $C_0^\infty(\mathbf{R}^1)$ that satisfy the Neumann condition $f'(y) = 0$.

Problem 8 is just a reformulation of the Markov uniqueness problem introduced in a). In fact, if $(\hat{\mathcal{E}}, \hat{\mathcal{A}})$ is a Dirichlet form extending $(\mathcal{E}, \mathcal{A})$, and $\mathcal{A}$ is in the domain of the generator $\hat{L}$, then by (1.11), $\hat{L}$ extends $(\mathcal{L}, \mathcal{A})$. Now recall that the map mapping a symmetric Dirichlet form to the semigroup $\left(e^{tL}\right)_{t \geq 0}$ generated by the form generator L is a one–to–one correspondence between the (symmetric) Dirichlet forms on $L^2(E; m)$, and the symmetric sub–Markovian C^0 contraction semigroups on $L^2(E; m)$, cf. e.g. [MaRö 92]. Hence the extensions of $(\mathcal{E}, \mathcal{A})$ in the sense of Problem 8 are in one–to–one correspondence to the symmetric sub–Markovian C^0 contraction semigroups on $L^2(E; m)$ that have a generator which extends $(\mathcal{L}, \mathcal{A})$. We obtain:

Lemma 1.5 *Uniqueness of $(\mathcal{E}, \mathcal{A})$ in the sense of Problem 8 holds if and only if the operator $(\mathcal{L}, \mathcal{A})$ is Markov–unique.*

We finally look at a second notion of uniqueness for Dirichlet forms over $\mathbf{R}^n$, or, more generally, locally compact spaces. Suppose E is a domain in $\mathbf{R}^n$, $n \in \mathbf{N}$, and $\mathcal{A} = C_0^\infty(E)$. Let m be a Radon measure on E. Motivated by his studies of the boundary theory of Markov processes, M. Silverstein considered the following class of extensions of a given pre–Dirichlet form $(\mathcal{E}, \mathcal{A})$ on $L^2(E; m)$, cf. [Sil 74], [Ta 96], and [KawTa 95].

Definition 1.4 *A symmetric Dirichlet form $(\hat{\mathcal{E}}, \hat{\mathcal{A}})$ extending $(\mathcal{E}, \mathcal{A})$ is called a Silverstein extension iff $u \cdot f$ is in the domain $\bar{\mathcal{A}}$ of the closure of $(\mathcal{E}, \mathcal{A})$ for all bounded functions $u \in \hat{\mathcal{A}}$ and $f \in \mathcal{A}$. The pre–Dirichlet form $(\mathcal{E}, \mathcal{A})$ is called* **Silverstein unique** *iff its closure $(\bar{\mathcal{E}}, \bar{\mathcal{A}})$ is the only Silverstein extension.*

Loosely speaking, a Dirichlet form $(\hat{\mathcal{E}}, \hat{\mathcal{A}})$ extending $(\mathcal{E}, \mathcal{A})$ is a Silverstein extension if it coincides locally with $(\bar{\mathcal{E}}, \bar{\mathcal{A}})$. Probabilistically, this means that $(\hat{\mathcal{E}}, \hat{\mathcal{A}})$ corresponds to a Markov process which coincides with the Markov process with absorption at the boundary generated by $(\bar{\mathcal{E}}, \bar{\mathcal{A}})$ up to the first hitting time of the complement of any fixed relatively compact open subset of E, cf. [Sil 74] for details.

A rather complete answer to the following problem is known already. Therefore, our interest concentrates not so much on the problem itself, but on the relation to the Markov uniqueness problem.

Problem 9: Under which conditions is $(\mathcal{E}, \mathcal{A})$ Silverstein unique ?

In Theorem 3.6 below, we will prove the highly non–trivial fact, that under very weak assumptions, Markov uniqueness for a symmetric diffusion operator $(\mathcal{L}, \mathcal{A})$ on $L^2(E; m)$, $\mathcal{A} = C_0^\infty(E)$, holds, if and only if the corresponding pre–Dirichlet form $(\mathcal{E}, \mathcal{A})$ is Silverstein unique. This allows us to apply the following results on Silverstein uniqueness to obtain far–reaching criteria for Markov–uniqueness, cf. Section f) in Chapter 3 below.

Results on Silverstein uniqueness: i) It is obvious from the definition of a Silverstein extension that $(\mathcal{E}, \mathcal{A})$ is Silverstein unique whenever the constant function 1 is in $\bar{\mathcal{A}}$. In particular, this is always the case, if $E = \mathbf{R}^n$, m is a finite measure, and $\mathcal{E}$ is a Dirichlet form of type

$$\mathcal{E}(u, v) = \int \sum_{i,j=1}^n a_{ij} \frac{\partial u}{\partial x_i} \frac{\partial v}{\partial x_j}\, dm$$

with bounded, locally strictly elliptic measurable coefficients $a_{ij} : \mathbf{R}^n \to \mathbf{R}$, $a_{ij} = a_{ji}$, $1 \leq i, j \leq n$. Note that the generators of such forms can have very singular first order coefficients, see [MaRö 92, Ch. 2] for examples.

ii) Results of M. Silverstein show that Silverstein uniqueness always holds if the Markov process associated with the Dirichlet form $(\bar{\mathcal{E}}, \bar{\mathcal{A}})$ is *conservative*, cf. also Section f), 3) in Chapter 3 below.

iii) T. Kawabata and M. Takeda [KawTa 95] have shown that *completeness* of E w.r.t. an "intrinsic metric" associated with the Dirichlet form $(\bar{\mathcal{E}}, \bar{\mathcal{A}})$ also implies Silverstein uniqueness.

REMARK. Silverstein extensions have been introduced to study the boundary behaviour of Markov processes on a **locally compact space**. They are related to localization techniques, and therefore the concept of Silverstein extensions only makes sense if the **test functions in $\mathcal{A}$ have compact support**. Because of the equivalence between Silverstein and Markov uniqueness for diffusion operators with controlled degeneracy on $\mathbf{R}^n$ that will be proven in Theorem 3.6, one might be tempted to try to use the approach via Silverstein uniqueness also to show Markov uniqueness for symmetric diffusion operators defined on cylinder functions over an infinite dimensional space. But here typically the constant function 1 is in $\mathcal{A}$, i.e., Silverstein uniqueness always holds, whereas Markov uniqueness does not necessarily hold, cf. Section b) in Chapter 5 below.

e) Relations between the different notions of uniqueness

In the preceeding sections we have shown how the analytic problems of L^p uniqueness, essential self–adjointness, strong Markov uniqueness, and Markov uniqueness of diffusion operators, as well as Silverstein uniqueness of Dirichlet forms, appear in different contexts. In this section, we discuss the relations between these analytic uniqueness concepts. We first comment on the relation between L^p uniqueness for different $p \in [1, \infty)$ and strong Markov uniqueness. Afterwards, we consider the relations to the other notions of uniqueness listed above in the symmetric case. In Chapter 4 below, we will give probabilistic explanations for some of the connections and differences between the various notions of uniqueness.

1) L^p uniqueness and strong Markov uniqueness

As before, we fix a σ-finite measure m on E. Suppose we are given a densely defined linear operator $(\mathcal{L}, \mathcal{A})$ on $L^1(E; m)$, and a sub–Markovian C^0 contraction semigroup $(T_t)_{t \geq 0}$ on $L^1(E; m)$, such that its generator extends $(\mathcal{L}, \mathcal{A})$. By an easy interpolation argument, it can be shown that $(T_t)_{t \geq 0}$ induces a C^0 contraction semigroup $\left(T_t^{(p)} \right)_{t \geq 0}$ on $L^p(E; m)$ for each $p \in [1, \infty)$, such that $T_t^{(p)} f = T_t f$ for all $t \geq 0$ and $f \in L^p(E; m) \cap L^1(E; m)$, cf. Lemma 1.11 in Appendix B. If $\mathcal{A}$ and $\mathcal{L}(\mathcal{A})$ are contained in $L^p(E; m)$, then the generator of $\left(T_t^{(p)} \right)_{t \geq 0}$ extends the operator $(\mathcal{L}, \mathcal{A})$.

Lemma 1.6 *Let $p \in [1, \infty)$. Suppose that both $\mathcal{A}$ and $\mathcal{L}(\mathcal{A})$ are contained in $L^p(E; m)$, and the operator $(\mathcal{L}, \mathcal{A})$ is $L^p(E; m)$ unique. Then:*

(i) $(T_t)_{t \geq 0}$ is the only sub–Markovian C^0 semigroup on $L^1(E; m)$, which has a generator that extends $(\mathcal{L}, \mathcal{A})$. In particular, $(\mathcal{L}, \mathcal{A})$ is strongly Markov unique.

(ii) If the measure m is finite, then $(\mathcal{L}, \mathcal{A})$ is $L^r(E; m)$ unique for all $r \in [1, p]$.

Hence if the measure m is *finite*, then the following relations between the different notions of uniqueness for the operator $(\mathcal{L}, \mathcal{A})$ hold for $1 \leq r \leq p$:

$$\begin{array}{ccccccc} L^p(E; m) & & L^r(E; m) & & L^1(E; m) & & \text{Strong Markov} \\ \text{uniqueness} & \Longrightarrow & \text{uniqueness} & \Longrightarrow & \text{uniqueness} & \Longrightarrow & \text{uniqueness} \end{array}$$

Proof of Lemma 1.6. (i) Let $(S_t)_{t \geq 0}$ be a sub–Markovian C^0 semigroup on $L^1(E; m)$, which is generated by an extension of the operator $(\mathcal{L}, \mathcal{A})$. By an interpolation argument, we see that there exists a C^0 semigroup $\left(S_t^{(p)} \right)_{t \geq 0}$ on $L^p(E; m)$ such that $S_t^{(p)} f = S_t f$ for all $f \in L^p(E; m) \cap L^1(E; m)$ and $t \geq 0$, cf. Lemma 1.11 in Appendix B. The generator of $\left(S_t^{(p)} \right)_{t \geq 0}$ extends $(\mathcal{L}, \mathcal{A})$ as well. Since the generator of $\left(T_t^{(p)} \right)_{t \geq 0}$ also extends the operator $(\mathcal{L}, \mathcal{A})$, which by assumption is $L^p(E; m)$ unique, the semigroups $\left(S_t^{(p)} \right)_{t \geq 0}$ and $\left(T_t^{(p)} \right)_{t \geq 0}$

coincide. It is now easy to conclude that $(S_t)_{t \geq 0}$ and $(T_t)_{t \geq 0}$ coincide as well.

(ii) Fix $r \in [1, p]$. By Theorem 1.2 in Appendix A, the $L^p(E; m)$ uniqueness of $(\mathcal{L}, \mathcal{A})$ implies that $\mathcal{A}$ is a core for the generator $L^{(p)}$ of $\left(T_t^{(p)}\right)_{t \geq 0}$. Since the measure m is finite, $L^p(E; m)$ is continuously embedded into $L^r(E; m)$. Hence for $t \geq 0$, the operator $T_t^{(r)}$ is an extension of $T_t^{(p)}$, and the generator $L^{(r)}$ of $\left(T_t^{(r)}\right)_{t \geq 0}$ extends $L^{(p)}$. The graph norm of $L^{(r)}$ on $\mathrm{Dom}\,(L^{(p)})$ can be estimated by the graph norm of $L^{(p)}$. In particular, $\mathcal{A}$ is dense in $\mathrm{Dom}\,(L^{(p)})$ w.r.t. the graph norm of $L^{(r)}$. On the other hand, the domain of $L^{(p)}$ is invariant under $T_t^{(r)}$ for all $t \geq 0$. Therefore, $\mathrm{Dom}\,(L^{(p)})$ is a core for the generator $L^{(r)}$, cf. Theorem 1.3 in Appendix A. Thus $\mathcal{A}$ is a core for $L^{(r)}$ as well, i.e., $(\mathcal{L}, \mathcal{A})$ is $L^r(E; m)$ unique, cf. Theorem 1.2 in Appendix A. ∎

REMARKS. (i) For *infinite measures* m, the second assertion in the lemma is false in general. There are examples, where L^p uniqueness holds for all $p > 1$ but not for $p = 1$, cf. Remark (ii) in Section c) of Chapter 2 below. Moreover, L^p uniqueness for some p does not imply L^r uniqueness for $r > p$, cf. the results and examples in Chapter 2.

(ii) There exist diffusion operators that are not L^p unique for any $p \in [1, \infty)$, but nevertheless strongly Markov unique, cf. Example 1 below.

Lemma 1.6 shows how strong Markov uniqueness can be derived from L^p uniqueness for some $p \in [1, \infty)$. In fact, in this work, we will not develop any special techniques for proving strong Markov uniqueness, but we will mainly focus on L^p uniqueness results, which, by the way, imply strong Markov uniqueness. If the measure m is finite, the easiest possibility to prove strong Markov uniqueness in this way is to show L^1 uniqueness, which is the weakest form of L^p uniqueness in this case, cf. the second assertion of Lemma 1.6. For infinite measures, it is often more appropriate to prove L^p uniqueness for $p > 1$, cf. Remark (i) above. The following example demonstrates, that one cannot always use L^p uniqueness to prove strong Markov uniqueness.

EXAMPLE 1 (*Strong Markov uniqueness does not imply L^p uniqueness*). We consider the diffusion operator $(\frac{d}{dx}, C_0^\infty(0, \infty))$ on $L^p(0, \infty; dx)$, $1 \leq p < \infty$. Let $(T_t)_{t \geq 0}$ be the transition semigroup of deterministic uniform motion to the right on $(0, \infty)$, i.e.,

$$(T_t f)\,(x) \;=\; f\,(x + t) \qquad \text{for all } f \in L^p(0, \infty; dx), \text{ and } t, x \geq 0.$$

One easily verifies that $(T_t)_{t \geq 0}$ is a sub–Markovian C^0 contraction semigroup on $L^p(0, \infty; dx)$ for each $p \in [1, \infty)$. The generator of $(T_t)_{t \geq 0}$ extends the operator $(\frac{d}{dx}, C_0^\infty(0, \infty))$.

Claim: The operator $(\frac{d}{dx}, C_0^\infty(0, \infty))$ is *strongly Markov unique* on $L^1(0, \infty; dx)$, but it is *not $L^p(0, \infty; dx)$ unique for any* $p \in [1, \infty)$.

PROOF OF THE CLAIM. Fix $p \in [1, \infty)$. It is not difficult to verify that the generator of the semigroup $(T_t)_{t \geq 0}$ on $L^p(0, \infty; dx)$ is $(\frac{d}{dx}, H^{1,p}(0, \infty; dx))$. On the other hand, the closure of $(\frac{d}{dx}, C_0^\infty(0, \infty))$ w.r.t. the graph norm on $L^p(0, \infty; dx)$ is the operator $(\frac{d}{dx}, H_0^{1,p}(0, \infty; dx))$. Since the Sobolev spaces $H_0^{1,p}(0, \infty; dx)$ and $H^{1,p}(0, \infty; dx)$ do not coincide, $C_0^\infty(0, \infty)$ is not a core for the generator of $(T_t)_{t \geq 0}$. Hence $(\frac{d}{dx}, C_0^\infty(0, \infty))$ is not $L^p(0, \infty; dx)$ unique, cf. Theorem 1.2 in Appendix A.

On the other hand, suppose L is the generator of a C^0 contraction semigroup $(S_t)_{t \geq 0}$ on $L^2(0, \infty; dx)$, and L extends $(\frac{d}{dx}, C_0^\infty(0, \infty))$. Then

$$\int_0^\infty u \, Lu \, dx \;=\; \lim_{t \downarrow 0} \frac{1}{t} \int_0^\infty u \, (T_t u - u) \, dx \;\leq\; 0$$

for all u in the domain of L. For $u \in \mathrm{Dom}\,(L)$, $f \in C_0^\infty(0, \infty)$, and $\lambda \in \mathbf{R}$, we obtain

$$
\begin{aligned}
0 \;\geq\; & \int_0^\infty (u + \lambda f) \, L(u + \lambda f) \, dx \\
=\; & \int_0^\infty u \, Lu \, dx \;+\; \lambda \left(\int_0^\infty f' u \, dx + \int_0^\infty f \, Lu \, dx \right) + \lambda^2 \int_0^\infty f' f \, dx \\
=\; & \int_0^\infty u \, Lu \, dx \;+\; \lambda \left(\int_0^\infty f' u \, dx + \int_0^\infty f \, Lu \, dx \right).
\end{aligned}
$$

This can only be true for all $\lambda \in \mathbf{R}$ if

$$\int_0^\infty f \, Lu \, dx \;=\; -\int_0^\infty f' u \, dx \qquad \text{for all } f \in C_0^\infty(0, \infty) \text{ and } u \in \mathrm{Dom}\,(L),$$

i.e., the operator $(\frac{d}{dx}, H^{1,2}(0, \infty; dx))$ extends L. As remarked above, the extending operator is the generator of the C^0 contraction semigroup $(T_t)_{t \geq 0}$ on $L^2(0, \infty; dx)$, whereas L is the generator of the C^0 contraction semigroup $(S_t)_{t \geq 0}$. Hence both $(1 - L, \mathrm{Dom}\,(L))$ and $(1 - \frac{d}{dx}, H^{1,2}(0, \infty; dx))$ are one-to-one maps onto $L^2(0, \infty; dx)$, cf. e.g. [Yo 80] or [Pa 85]. Therefore the two operators coincide, whence $(S_t)_{t \geq 0}$ and $(T_t)_{t \geq 0}$ coincide as well. We have shown $(T_t)_{t \geq 0}$ is the only C^0 contraction semigroup on $L^2(0, \infty; dx)$ which is generated by an extension of $(\frac{d}{dx}, C_0^\infty(0, \infty))$. By Lemma 1.11 in Appendix B, every sub–Markovian C^0 contraction semigroup on $L^1(0, \infty; dx)$ with a generator that extends $(\frac{d}{dx}, C_0^\infty(0, \infty))$ induces a sub–Markovain C^0 contraction semigroup on $L^2(0, \infty; dx)$, such that the generator again extends $(\frac{d}{dx}, C_0^\infty(0, \infty))$. Therefore, the operator $(\frac{d}{dx}, C_0^\infty(0, \infty))$ is strongly Markov unique. ∎

REMARK. Let $p \in [1, \infty)$, and let S_t, $t \geq 0$, be the bounded linear operators on $L^p(0, \infty; dx)$ defined by

$$(S_t f)\,(x) \;=\; \begin{cases} \sum_{n=0}^{[x+t]} f(x + t - n) & \text{if } x < 1 \\ f(x + t) & \text{if } x \geq 1 \end{cases},$$

where $[x + t]$ denotes the largest integer smaller or equal than $x + t$. Then it is not difficult to verify that $(S_t)_{t \geq 0}$ is a C^0 semigroup on $L^p(0, \infty; dx)$, and the generator of $(S_t)_{t\geq 0}$ extends the operator $(\frac{d}{dx}, C_0^\infty(0,\infty))$. Probabilistically, $(S_t)_{t\geq 0}$ can be interpreted as the transition semigroup of a particle system consisting of particles on $(0,\infty)$ which move uniformly to the right, and create a "child" at 0 when they pass through 1, see Chapter 4 for more about this point of view.

2) Uniqueness notions for symmetric operators

We fix a σ–finite measure m on E, and a densely defined, negative definite symmetric diffusion operator $(\mathcal{L}, \mathcal{A})$ on $L^2(E;m)$. We assume that $\mathcal{A}$ and $\mathcal{L}(\mathcal{A})$ consist of m–integrable functions, and $\mathcal{A}$ is dense in $L^1(E;m)$.

We have already shown in Section c), that $(\mathcal{L}, \mathcal{A})$ is $L^2(E;m)$ unique if and only if it is essentially self-adjoint. Moreover, we have noted above, that in this case, $(\mathcal{L}, \mathcal{A})$ is also strongly Markov unique. Strong Markov uniqueness implies Markov uniqueness, since every symmetric sub–Markovian C^0 semigroup on $L^2(E;m)$ with a generator that extends $(\mathcal{L}, \mathcal{A})$ induces a sub–Markovian C^0 contraction semigroup on $L^1(E;m)$ such that the generator again extends $(\mathcal{L}, \mathcal{A})$, cf. Lemma 1.12 in Appendix B. Finally, by a result of M. Takeda [Ta 96], Markov uniqueness implies Silverstein uniqueness of the corresponding pre–Dirichlet form $(\mathcal{E}, \mathcal{A})$. Summarizing, we have the following relations between the different notions of uniqueness for the operator $(\mathcal{L}, \mathcal{A})$:

$$
\begin{array}{ccccc}
\text{Essential} & \Longleftrightarrow & L^2(E;m) & \left(\Longrightarrow \quad L^1(E;m) \right) & \Longrightarrow \\
\text{self-adjointness} & & \text{uniqueness} & \text{uniqueness} &
\end{array}
$$

$$
\begin{array}{ccccc}
\Longrightarrow & \text{Strong Markov} & \Longrightarrow & \text{Markov} & \Longrightarrow & \text{Silverstein} \\
& \text{uniqueness} & & \text{uniqueness} & & \text{uniqueness for } (\mathcal{E}, \mathcal{A})
\end{array}
$$

Here $L^1(E;m)$ uniqueness only follows from $L^2(E;m)$ uniqueness if the measure m is finite, but strong Markov uniqueness always follows both from $L^2(E;m)$ uniqueness and from $L^1(E;m)$ uniqueness.

REMARKS. (i) We have already seen in Example 1 above that there are diffusion operators which are not $L^p(E;m)$ unique for any $p \geq 1$, but nevertheless strongly Markov unique. Example 2 below demonstrates that the same can happen if the diffusion operator is symmetric. However, it is not clear, whether there exist a *finite measure* m and a *symmetric diffusion operator* $(\mathcal{L}, \mathcal{A})$ on $L^2(E;m)$, such that $(\mathcal{L}, \mathcal{A})$ is Markov unique but not $L^1(E;m)$ unique. There is some evidence that this can not happen if $\mathcal{A} = C_0^\infty(E)$ for some open subset $E \subseteq \mathbf{R}^n$, cf. the remark above Table 1 in Chapter 4, Section a), below. Note, however, that if one considers second order differential operators with non–vanishing zeroth order term (i.e., generators of diffusion processes with killing) instead of diffusion operators, then one can easily give an example of a symmetric operator on a finite measure space, which is Markov unique but not $L^1(E\,;\,m)$ unique, cf. [Wu 97].

(ii) The question whether there exists a symmetric diffusion operator, which is Markov unique but not strongly Markov unique, is still open.

(iii) On non locally compact state spaces, Silverstein uniqueness does not imply Markov uniqueness, cf. the remark at the end of Section d). However, we will prove in Chapter 4, that for symmetric diffusion operators with domain $C_0^\infty(E)$, where E is an open subset in $\mathbf{R}^n$, Silverstein uniqueness and Markov uniqueness are equivalent if the diffusion matrix is non–degenerate.

EXAMPLE 2 (*Markov uniqueness does not imply L^p uniqueness*).
Let ρ be a strictly positive smooth function on $(0,1)$, such that $\rho(x) = x$ for all $x \le 1/4$, and $\rho(x) = (1-x)^{-1}$ for all $x \ge 3/4$. We consider the symmetric diffusion operator $(\mathcal{L}, C_0^\infty(0,1))$ on $L^2(0,1\,;\,\rho\,dx)$ given by

$$\mathcal{L}f = \frac{1}{\rho}\,(\rho f')' = f'' + \frac{\rho'}{\rho}\,f', \qquad \text{i.e.,}$$

$$(\mathcal{L}f)(x) = \begin{cases} f''(x) + x^{-1}f'(x) & \text{for } x \le 1/4. \\ f''(x) + (1-x)^{-1}f'(x) & \text{for } x \ge 3/4. \end{cases}$$

For each $p \in [1,\infty)$, $(\mathcal{L}, C_0^\infty(0,1))$ is a densely defined linear operator on $L^p(0,1\,;\,\rho\,dx)$, which has an extension that generates a sub–Markovian C^0 contraction semigroup. By Corollary 2.2 in Chapter 2, Section a), below, it is not $L^p(0,1\,;\,\rho\,dx)$ unique for any $p \in [1,\infty)$. In fact, we have

$$\int_0^{1/4} \left(\int_y^{1/4} (\rho(x))^{-1}\,dx \right)^q \rho(y)\,dy = \int_0^{1/4} (\log(1/4) - \log y)^q\, y\, dy \;<\; \infty$$

for all $q \in [1,\infty)$, which implies non $L^p(0,1\,;\,\rho\,dx)$ uniqueness for $p \in (1,\infty)$. Similarly,

$$\int_{3/4}^1 (\rho(x))^{-1} \int_{3/4}^1 \rho(z)\,dz\,dx = \int_{3/4}^1 (1-x)\,(\log(1/4) - \log(1-x))\,dx \;<\; \infty,$$

whence the operator is not $L^1(0,1\,;\,\rho\,dx)$ unique either.
However, $(\mathcal{L}, C_0^\infty(0,1))$ is Markov unique by Theorem 3.2 below. In fact, 0 is an entrance/ no exit boundary, and 1 is an exit/ no entrance boundary for $(\mathcal{L}, \mathcal{A})$. Hence there is no regular boundary, i.e., Markov uniqueness holds.

3) Open problems

The relations between the different notions of uniqueness for diffusion operators are still not completely clarified. We briefly mention some questions that remain open :

- Is there a symmetric diffusion operator which is Markov unique, but not strongly Markov unique ? This question could perhaps be answered by developping special techniques for proving, respectively disproving strong Markov uniqueness. In this work, we do not develop such techniques explicitly, but we use L^p uniqueness results to prove strong Markov uniqueness, respectively results on non–Markov uniqueness to disprove strong Markov uniqueness. That this is not too restrictive is demonstrated by the fact that the following question is also open :

- Is there a symmetric diffusion operator such that the semigroup generated by the Friedrichs extension is conservative, and the operator is Markov unique but not L^1 unique ? In Chapter 4, Section a), we demonstrate that such an operator does not exist in $\mathbf{R}^1$, cf. the remark above Table 1, and see also Remark (i) in Subsection 2) above.

- Is there a connection between strong Markov uniqueness and L^1 uniqueness for non–symmetric operators ? In contrast to the symmetric case, there is an example of a non–symmetric diffusion operator, which is strongly Markov unique but not L^1 unique, although the unique sub–Markovian C^0 contraction semigroup on L^1 generated by an extension of the operator is conservative, cf. Example 1 above. Is there any additional condition, under which strong Markov uniqueness implies L^1 uniqueness ?

- In Section a) above, we have shown that strong Markov uniqueness on $L^1(E; m)$, respectively Markov uniqueness on $L^2(E; m)$, implies uniqueness of the martingale problem among all sub–stationary, respectively reversible, time–homogeneous diffusion processes on E with initial distribution m. Is there a converse, i.e., does uniqueness of the martingale problem in an appropriate sense imply Markov uniqueness, strong Markov uniqueness, or even L^p uniqueness for some $p \in [1, \infty)$?

Appendix A   Some standard tools for proving existence and uniqueness of C^0 semigroups on Banach spaces

We briefly review some standard techniques for proving existence and uniqueness of C^0 semigroups generated by linear operators on Banach spaces. For basic definitions, and a more detailed treatment of semigroup theory, cf. Pazy's book [Pa 85].

Throughout this section, we fix a real Banach space $(B, \|\cdot\|)$, and a densely defined linear operator (L, A) on B. It is well–known that, if an extension of (L, A) is the generator of a C^0 semigroup $(T_t)_{t \geq 0}$ of contractions on B, then

(L, A) is **dissipative**, i.e., for any $u \in A$ there exists a linear functional $u^* \in B^*$ such that

$$(1.12) \qquad \langle u^*, u \rangle = \|u\|_B^2 = \|u^*\|_{B^*}^2, \qquad \text{and}$$
$$\langle u^*, Lu \rangle \leq 0.$$

Here B^* denotes the dual of B, and $\langle \cdot, \cdot \rangle$ the corresponding dualisation.

REMARK. Equivalently, (L, A) is dissipative if and only if $\|\lambda u - Lu\| \geq \lambda \|u\|$ for all $\lambda > 0$ and $u \in A$, cf. [Pa 85, Ch. 1, Thm. 4.2].

Now suppose in particular, that $B = L^p(E; m)$ where $(E, \mathcal{B}, m)$ is a σ-finite measure space, and $p \in [1, \infty)$. Then (L, A) is dissipative if and only if

$$\int \text{sgn}(u) |u|^{p-1} Lu \, dm \leq 0 \qquad \text{for all } u \in A,$$

where $\text{sgn}(x) := 1$ if $x > 0$, -1 if $x < 0$, and 0 if $x = 0$. Note that for $u \in L^p(E; m)$ such that $\|u\|_{L^p(E; m)} = 1$, the function $\text{sgn}(u) \cdot |u|^{p-1}$ is an element in the dual space $L^q(E; m)$, $\frac{1}{q} + \frac{1}{p} = 1$, which satisfies (1.12). An easy to check criterion for dissipativity on $L^p(E; m)$ is the following:

Lemma 1.7 *Let $\lambda \geq 0$. Suppose that for every smooth increasing function $\psi : \mathbf{R} \to \mathbf{R}$ such that $\psi(0) = 0$, and $|\psi(s)| \leq |s|^{p-1}$ for all s, we have*

$$\int \psi \circ u \, Lu \, dm \leq \lambda \cdot \int |u|^p \, dm \qquad \text{for all } u \in A.$$

Then $(L - \lambda, A)$ is dissipative on $L^p(E; m)$.

PROOF. Choose smooth increasing functions $\psi_n : \mathbf{R} \to \mathbf{R}$, $n \in \mathbf{N}$, such that $\psi_n(s) = 0$ if $|s| \leq n^{-1}$, $\psi_n(s) = |s|^{p-1} \cdot \text{sgn}(s)$ if $|s| \geq 2n^{-1}$, and $|\psi_n(s)| \leq |s|^{p-1}$ for all s. Then for any $u \in A$, $\psi_n \circ u$ is in $L^q(E; m)$, $\frac{1}{p} + \frac{1}{q} = 1$, and

$$\int \text{sgn}(u) |u|^{p-1} (L - \lambda) u \, dm = \lim_{n \to \infty} \int \psi_n \circ u \, Lu \, dm - \lambda \int |u|^p \, dm \leq 0$$

by dominated convergence. ∎

Now consider again the general situation described above, i.e., B is an arbitrary Banach space. If the operator (L, A) is dissipative, then it is closable. Let $\bar{L}$ denote the closure. We recall the Theorem of Lumer and Phillips, cf. e.g. [Pa 85, Ch. I, Thm. 4.3].

Theorem 1.1 (Lumer/Phillips) *Suppose (L, A) is dissipative. Then $\bar{L}$ is the generator of a C^0 semigroup of contractions on B if and only if the range of $(\lambda - L, A)$ is dense in B for some, or equivalently, all $\lambda > 0$.*

The next theorem shows that the Lumer–Phillips theorem is not only an existence, but also a uniqueness result.

Theorem 1.2 *Suppose there exists a C^0 semigroup $(T_t)_{t \geq 0}$ on B such that its generator extends the operator (L, A). Then the following assertions are equivalent:*

(i) *A is a core for the generator of $(T_t)_{t \geq 0}$.*

(ii) *The closure $\bar{L}$ of (L, A) is the generator of a C^0 semigroup.*

(iii) *$(T_t)_{t \geq 0}$ is the only C^0 semigroup on B which has a generator that extends (L, A).*

REMARK.   Suppose $(T_t)_{t \geq 0}$ as in the theorem is a contraction semigroup, and let $\varepsilon > 0$. Then the assertions in the theorem are also equivalent to:

(iii′) *$(T_t)_{t \geq 0}$ is the only C^0 semigroup on B such that its generator extends (L, A), and $\|T_t\| \leq e^{\varepsilon t}$.*

However, $(T_t)_{t \geq 0}$ can be the only C^0 *contraction* semigroup with a generator that extends (L, A), even if A is not a core for the generator of $(T_t)_{t \geq 0}$.

Proof of Theorem 1.2 and the remark. Obviously, (i) implies (ii). Suppose (ii) holds, and $\hat{L}$ is an extension of (L, A) which generates a C^0 semigroup. Then $\hat{L}$ is closed, and thus an extension of $\bar{L}$. Since both $\hat{L}$ and $\bar{L}$ generate C^0 semigroups, there exists $\lambda_0 \geq 0$ such that for all $\lambda \geq \lambda_0$, the operators $\lambda - \hat{L}$ and $\lambda - \bar{L}$ are bijections from their domains onto B, cf. e.g. [Pa 85]. But $\hat{L}$ extends $\bar{L}$, so both generators, and the corresponding semigroups, coincide. This proves that there is only one C^0 semigroup that has a generator which extends (L, A). Since $(T_t)_{t \geq 0}$ is such a C^0 semigroup, (iii) holds.
A proof of the more difficult implication (iii) $\Rightarrow$ (i) is given in [Ar 86, A-II, Thm. 1.33]. This completes the proof of the theorem.
If $(T_t)_{t \geq 0}$ is a contraction semigroup, then the proof in [Ar 86] actually implies the stronger implication (iii′) $\Rightarrow$ (i) for all $\varepsilon > 0$, whence we also obtain the assertion in the remark. ∎

The Lumer–Phillips theorem and Theorem 1.2 can be used to derive several criteria for existence and uniqueness of C^0 semigroups. For example, let L^* : $\mathrm{Dom}\,(L^*) \subseteq B^* \to B^*$ denote the adjoint of the operator (L, A).

Corollary 1.3 (The first adjoint criterion) *Suppose (L, A) is dissipative. Then $\bar{L}$ is the generator of a C^0 semigroup of contractions on B if and only if $\ker(\lambda - L^*) = \{0\}$ for some, or equivalently, all $\lambda > 0$.*

The corollary is an immediate consequence of the theorems of Lumer–Phillips and Hahn–Banach.

Corollary 1.4 (The second adjoint criterion) *Suppose both (L, A) and $(L^*, \mathrm{Dom}\,(L^*))$ are dissipative. Then $\bar{L}$ is the generator of a C^0 semigroup of contractions on B.*

PROOF.   Since L^* is dissipative, the equation $L^*u = u$ has only the trivial solution. Thus $\bar{L}$ generates a C^0 contraction semigroup by Corollary 1.3. ∎

Corollary 1.3 and 1.4 can often be used directly to prove existence and uniqueness. In some cases, however, the approximative criteria to be considered next are more powerful.

Let $f \in B$, and λ, $\varepsilon > 0$. An element $v \in B$ is called an ε–*approximative strong solution* of the equation

$$(1.13) \qquad\qquad \lambda v - Lv = f,$$

iff v is in the domain of $\bar{L}$, and

$$(1.14) \qquad\qquad \| \lambda v - \bar{L}v - f \| \le \varepsilon.$$

Similarly, for $T \in [0, \infty)$, a continuous function $u : [0, T] \to B$ is called an ε–*approximative strong solution* of the initial value problem

$$(1.15) \qquad\qquad \frac{d}{dt} u(t) = Lu(t), \qquad u(0) = f,$$

iff $u(t)$ is in the domain of $\bar{L}$ for all $t \in [0, T]$, the function $t \mapsto \bar{L}u(t)$ is Bochner–integrable on $[0, T]$, and

$$(1.16) \qquad \left\| u(t) - f - \int_0^t \bar{L}u(s)\,ds \right\| \le \varepsilon \qquad \text{for all } t \in [0, T].$$

We say that Equation (1.13) is *approximately solvable* iff there exist ε–approximative solutions v_ε for all $\varepsilon > 0$. Similarly, we say that (1.15) is *approximately solvable* iff there are ε–approximative solutions $u_{\varepsilon,T}$ on $[0, T]$ for all $\varepsilon > 0$ and $T \in (0, \infty)$, such that

$$\sup_{T>0} \| u_{\varepsilon,T}(T) \| < \infty \qquad \text{for each } \varepsilon > 0.$$

The following corollary to the Lumer–Phillips theorem is well–known. A Hilbert space version is given in Berezansky's book [Ber 86].

Corollary 1.5 (The approximative criterion) *Suppose that (L, A) is dissipative, and one of the following conditions holds for some dense subspace $D \subset B$:*

(i) There exists $\lambda > 0$ such that the equation (1.13) is approximately solvable for all $f \in D$.

(ii) The initial value problem (1.15) is approximately solvable for all $f \in D$.

Then $\bar{L}$ is the generator of a C^0 semigroup of contractions on B.

PROOF. The assertion that Condition (i) implies the claim follows immediately from the Lumer–Phillips theorem. We now show that (ii) implies (i).
Suppose (ii) holds, and fix $f \in D$ and $\varepsilon > 0$. Let $T \in (0, \infty)$ such that $e^{-T} \| u_{(\varepsilon/2),T}(T) \| \leq \varepsilon/2$, where $u_{(\varepsilon/2),T}$ is the approximative solution of the initial value problem for f. For brevity, we write u instead of $u_{(\varepsilon/2),T}$. Since $u(t)$ is continuous, and $\bar{L}u(t)$ is Bochner–integrable in B, the graph norm $\| u(t) \| + \| \bar{L}u(t) \|$ is integrable, i.e., u is Bochner integrable in the domain of $\bar{L}$ endowed with graph norm. Hence

$$v := \int_0^T e^{-t} u(t)\, dt$$

exists in the domain of $\bar{L}$, and $\bar{L}v = \int_0^T e^{-t} \bar{L}u(t)\, dt$. Integrating by parts, we have

$$\bar{L}v = e^{-T} \int_0^T \bar{L}u(s)\, ds + \int_0^T e^{-t} \int_0^t \bar{L}u(s)\, ds\, dt.$$

Hence

$$v - \bar{L}v - f = \int_0^T e^{-t} \left(u(t) - \int_0^t \bar{L}u(s)\, ds - f \right) dt$$
$$+ e^{-T} \left(u(T) - \int_0^T \bar{L}u(s)\, ds - f \right) - e^{-T} u(T).$$

Since u is an $\frac{\varepsilon}{2}$–approximative solution of (1.15), we obtain

$$\| v - \bar{L}v - f \| \leq \int_0^T e^{-t} \frac{\varepsilon}{2}\, dt + e^{-T} \frac{\varepsilon}{2} + e^{-T} \| u(T) \|$$
$$\leq (1 - e^{-T}) \frac{\varepsilon}{2} + e^{-T} \frac{\varepsilon}{2} + \frac{\varepsilon}{2} = \varepsilon,$$

i.e., v is an ε–approximative solution of Equation (1.13). Hence (1.13) is approximately solvable for $\lambda = 1$ and all $f \in D$. ∎

REMARK. *(Extension of the results to generators of non–contractive C^0 semigroups).* Suppose $(T_t)_{t \geq 0}$ is an arbitrary (not necessarily contractive) C^0 semigroup on B, and its generator extends the densely defined linear operator (L, A). Then there exist constants $M \in [1, \infty)$ and $\beta \in [0, \infty)$ such that

(1.17) $\| T_t \| \leq M \cdot e^{\beta t}$ for all $t \in [0, \infty)$,

cf. e.g. [Pa 85, Ch. 1, Thm. 2.2]. By considering $T_t^{(\beta)} := e^{-\beta t} T_t$ instead of T_t, one obtains a uniformly bounded C^0 semigroup. Its generator extends the operator $L^{(\beta)} := L - \beta$. By replacing the norm in B by an equivalent norm, $\left(T_t^{(\beta)} \right)_{t > 0}$

becomes a C^0 semigroup of contractions, cf. [Pa 85, Ch. I, Proof of Thm. 5.2]. Hence $L^{(\beta)}$ is dissipative w.r.t. this norm, and we may apply the above results to $L^{(\beta)}$. For example, we obtain the following uniqueness criterion:

Corollary 1.6 *Suppose $(T_t)_{t>0}$ is a C^0 semigroup, and let M and β be finite constants such that (1.17) holds. Assume that the generator of $(T_t)_{t\geq0}$ extends the densely defined linear operator (L, A). Then $(T_t)_{t\geq0}$ is the only C^0 semigroup with this property if and only if $\ker(\lambda - L^*) = \{0\}$ for some, or, equivalently, all $\lambda > \beta$.*

The proof follows by applying Corollary 1.3 and Theorem 1.2 to the operator $L^{(\beta)}$.

Finally, we recall another standard criterion that is often used to prove uniqueness of C^0 semigroups, provided existence is already known:

Theorem 1.3 *Suppose $(T_t)_{t>0}$ is a C^0 semigroup on B, and its generator extends the operator (L, A). If there exists a dense subspace D of B such that $T_t(D) \subseteq A$ for all $t \geq 0$, then $(T_t)_{t\geq0}$ is the only C^0 semigroup with a generator that extends (L, A).*

For the proof see e.g. [EthKur 86, Ch. 1, Prop. 3.3].

Appendix B   Some basic properties of diffusion operators on L^p spaces

Let $(E, \mathcal{B}, m)$ be a σ-finite measure space, and fix $p \in [1, \infty)$. We recall some basic facts about diffusion operators on $L^p(E; m)$.

Definition 1.5 *A densely defined linear operator (L, A) on $L^p(E; m)$ is called an **abstract diffusion operator** iff*

(i) *$\phi(u_1, u_2, \ldots, u_k)$ is in A for all $k \in \mathbf{N}$, $u_1, u_2, \ldots, u_k \in A$, and $\phi \in C^\infty(\mathbf{R}^k)$ such that $\phi(0) = 0$, and, in this case,*

$$L\,\phi(u_1, \ldots, u_k)$$
$$= \sum_{i=1}^{k} \frac{\partial\phi}{\partial x_i}(u_1, \ldots, u_k)\, L\,u_i + \sum_{i,j=1}^{k} \frac{\partial^2\phi}{\partial x_i \partial x_j}(u_1, \ldots, u_k)\, \Gamma(u_i, u_j)$$

where
$$\Gamma(u, v) := \frac{1}{2}\left(L(uv) - u\,Lv - v\,Lu\right), \quad u, v \in A.$$

(ii) *$\Gamma(u, u) \geq 0$ for all $u \in A$.*

*The bilinear operator $\Gamma : A \times A \to L(E; m)$ is called the **Carré du champ operator** of L.*

Here $L(E; m)$ denotes all m–classes of functions on E.

REMARK.   If (L, A) is an abstract diffusion operator, then the corresponding Carré du champ operator Γ satisfies a chain rule in each of its components, i.e.,

$$\Gamma\left(\phi(u_1,\ldots,u_k),\,v\right) \;=\; \sum_{i=1}^{k} \frac{\partial\phi}{\partial x_i}\,(u_1,\ldots,u_k)\,\Gamma\left(u_i,\,v\right)$$

for all $k \in \mathbf{N}$, $\phi \in C^\infty(\mathbf{R}^k)$ such that $\phi(0) = 0$, and $v, u_1, \ldots, u_k \in A$.

From now on, we fix an abstract diffusion operator (L, A) on $L^p(E; m)$, and $\alpha \geq 0$. We assume that m is a **sub–invariant measure** for the operator $(L - \alpha, A)$, i.e.,

(A 1)   For all $f \in A$ such that $f \geq 0$ m–a.e., f and Lf are in $L^1(E; m)$, and

$$\int Lf\,dm \;\leq\; \alpha \int f\,dm.$$

Lemma 1.8 *Suppose (A 1) holds. Then* $(L - \frac{\alpha}{p}, A)$ *is dissipative on* $L^p(E; m)$.

PROOF.   Let $\psi : \mathbf{R} \to \mathbf{R}$ be a smooth increasing function such that $|\psi(s)| \leq |s|^{p-1}$ for all s, and let $\Psi := \int_0^{\bullet} \psi(s)\,ds$. Then $0 \leq \Psi(t) \leq |t|^p/p$ for all $t \in \mathbf{R}$, and

$$L\left(\Psi \circ f\right) \;=\; \psi \circ f\, Lf + \psi' \circ f\, \Gamma(f,f) \;\geq\; \psi \circ f\, Lf \quad m\text{–a.e.}$$

for all $f \in A$. Hence, by the sub–invariance,

$$\int \psi \circ f\, Lf\,dm \;\leq\; \int L\left(\Psi \circ f\right)dm$$

$$\leq\; \alpha \int \Psi \circ f\,dm \;\leq\; \frac{\alpha}{p}\int |f|^p\,dm.$$

for all $f \in A$. This implies the dissipativity, cf. Lemma 1.7 in Appendix A. ∎

Now suppose that (A 1) holds, and the closure $\bar{L}$ of (L, A) generates a C^0 semigroup $(T_t)_{t \geq 0}$. Then, by Lemma 1.8, for all $\lambda > \alpha/p$, the resolvent operator $(\lambda - \bar{L})^{-1}$ exists and is a bounded linear operator on $L^p(E; m)$. Moreover,

$$\left\| (\lambda - \bar{L})^{-1} \right\| \;\leq\; \left(\lambda - \tfrac{\alpha}{p}\right)^{-1} \quad \text{for all } \lambda \geq \alpha/p, \qquad \text{and}$$

$$\|T_t\| \;\leq\; e^{\alpha t/p} \quad \text{for all } t \geq 0.$$

In fact, by the Lumer–Phillips theorem and Theorem 1.2 in Appendix A, $\left(e^{-\alpha t/p}\, T_t\right)_{t \geq 0}$ is a C^0 semigroup of *contractions* on $L^p(E; m)$, which implies the assertions above. Moreover, since (L, A) is a diffusion operator, we have:

Lemma 1.9 *Suppose (A 1) holds, and the closure $\bar{L}$ of (L, A) generates a C^0 semigroup $(T_t)_{t\geq 0}$. Then $(T_t)_{t\geq 0}$ is* **sub–Markov.**

The proof of the lemma will be given below.

REMARK.   The diffusion property on test–functions alone is not enough to guarantee that any semigroup generated by an extension of (L, A) is sub–Markov. In fact, if (L, A) is not $L^p(E; m)$ unique then there usually exist extensions that generate non–sub–Markovian C^0 semigroups. However, in the symmetric case, the assumption that A is a core for the generator of $(T_t)_{t\geq 0}$ can be replaced by the weaker condition that A is a form core for the quadratic form corresponding to $(T_t)_{t\geq 0}$:

Lemma 1.10 *Suppose that $p = 2$, and (L, A) is a densely defined, negative definite symmetric diffusion operator on $L^2(E; m)$, such that m is an invariant measure for (L, A). Let $L^{(0)}$ be the* **Friedrichs extension** *of (L, A), i.e., $L^{(0)}$ is the self–adjoint operator corresponding to the closure of the symmetric bilinear form $(\mathcal{E}, A)$ defined by $\mathcal{E}(f, g) = - \int f\, Lg\, dm$. Then $L^{(0)}$ is the generator of a sub–Markovian symmetric C^0 contraction semigroup on $L^2(E; m)$.*

REMARK.   Usually, the fact that m is an invariant measure for (L, A) automatically follows from the diffusion property and the symmetry of (L, A). If, for example, the constant function 1 is in A, then we have

$$\int Lf\, dm \;=\; \int f\, L1\, dm \;=\; 0 \qquad \text{for all } f \in A.$$

PROOF OF LEMMA 1.10.   Since $L^{(0)}$ is a negative definite self–adjoint operator on $L^2(E; m)$, it generates a symmetric C^0 contraction semigroup. The semigroup is sub–Markov if and only if the corresponding quadratic form is a Dirichlet form, cf. [MaRö 92, Ch. I, Sect. 4]. On functions in A, this quadratic form is given by

$$\mathcal{E}(f,g) \;=\; -\int f\, Lg\, dm \;=\; -\frac{1}{2}\int (f\, Lg + g\, Lf - L(fg))\, dm \;=\; \int \Gamma(f,g)\, dm,$$

where we have used the symmetry of (L, A) and the invariance of the measure m. Let $\phi : \mathbf{R} \to \mathbf{R}$ be a smooth function such that $\phi(0) = 0$, and $|\phi'(x)| \leq 1$ for all x. Then, by the product rule for the Carré du champ operator, we obtain

$$\mathcal{E}(\phi(f), \phi(f)) \;=\; \int |\phi'(f)|^2\, \Gamma(f,f)\, dm \;\leq\; \int \Gamma(f,f)\, dm \;=\; \mathcal{E}(f,f)$$

for all $f \in A$. This implies that the closure of the form $(\mathcal{E}, A)$ is a Dirichlet form, cf. e.g. [MaRö 92, Ch. I, Prop. 4.10]. ∎

PROOF OF LEMMA 1.9.   We first show that $\bar{L}$ has the following *Dirichlet property:*

$$(1.18) \qquad \int_{\{f\geq 1\}} \bar{L}f\, (f-1)^{p-1}\, dm \;\leq\; \frac{\alpha}{p} \int_{\{f\geq 1\}} (f-1)^p\, dm$$

for all $f \in \text{Dom}(\bar{L})$. We then conclude that the resolvent is sub–Markov, and finally obtain the sub–Markov property for the semigroup.

Step 1: Proof of (1.18).   Fix $f \in A$, and let $\psi_n : \mathbf{R} \to \mathbf{R}$, $n \in \mathbf{N}$, be smooth increasing functions such that $\psi_n(s) = 0$ for $s \leq 1$, $\psi_n(s) = (s-1)^{p-1}$ for $s \geq 1 + \frac{1}{n}$, and $0 \leq \psi_n(s) \leq (s-1)^{p-1}$ for all $s \in \mathbf{R}$. Let $\Psi_n := \int_1^s \psi_n(s)\, ds$. Then $0 \leq \Psi_n(t) \leq ((t-1)^+)^p / p$ for all $t \in \mathbf{R}$ and $n \in \mathbf{N}$. In the same way as in the proof of Lemma 1.8, we obtain

$$\int \psi_n \circ f\, Lf\, dm \ \leq\ \int L(\Psi_n \circ f)\, dm \ \leq\ \alpha \int \Psi_n \circ f\, dm$$

$$\leq\ \frac{\alpha}{p} \int_{\{f \geq 1\}} (f-1)^p\, dm.$$

For $n \to \infty$ this implies

$$(1.19) \qquad \int_{\{f \geq 1\}} (f-1)^{p-1} Lf\, dm \ \leq\ \frac{\alpha}{p} \int_{\{f \geq 1\}} (f-1)^p\, dm.$$

Both sides of (1.19) are functionals of f, which are bounded w.r.t. the graph norm of L. In fact,

$$\int_{\{f \geq 1\}} (f-1)^{p-1} Lf\, dm \ \leq\ \left(\int_{\{f \geq 1\}} (f-1)^p\, dm \right)^{1/q} \cdot \left(\int |Lf|^p\, dm \right)^{1/p},$$

where $q \in (1, \infty]$ such that $\frac{1}{p} + \frac{1}{q} = 1$. Since (1.19) holds for all $f \in A$, we obtain (1.18).

Step 2: $\lambda(\lambda - \bar{L})^{-1}$ is sub–Markov for all $\lambda > \alpha/p$.   Fix $\lambda > \alpha/p$. As noted above, $(\lambda - \bar{L})^{-1}$ exists and is bounded. Let $g \in L^p(E; m)$ such that $g \leq 1$ m-a.e., and let $f := \lambda(\lambda - \bar{L})^{-1} g$. Then $\lambda f - \bar{L} f = \lambda g$, whence, by (1.18),

$$\lambda \int_{\{f \geq 1\}} f\,(f-1)^{p-1}\, dm$$

$$= \ \lambda \int_{\{f \geq 1\}} g\,(f-1)^{p-1}\, dm \ + \ \lambda \int_{\{f \geq 1\}} \bar{L} f\,(f-1)^{p-1}\, dm$$

$$\leq\ \lambda \int_{\{f \geq 1\}} (f-1)^{p-1}\, dm \ + \ \frac{\alpha}{p} \int_{\{f \geq 1\}} (f-1)^p\, dm.$$

Thus

$$\left(\lambda - \frac{\alpha}{p}\right) \int_{\{f \geq 1\}} (f-1)^p\, dm \ \leq\ 0,$$

which implies $f \leq 1$ m-a.e.

Now suppose $0 \leq g \leq 1$ m-a.e. Then $-ng \leq 1$ for all $n \in \mathbf{N}$. By applying the above considerations both to g and to $-ng$, $n \in \mathbf{N}$, we obtain

$$-\frac{1}{n} \ \leq\ \lambda(\lambda - \bar{L})^{-1} g \ \leq\ 1 \qquad \text{for all } n \in \mathbf{N},$$

whence $0 \leq \lambda(\lambda - \bar{L})^{-1}g \leq 1$. This shows that $\lambda(\lambda - \bar{L})^{-1}$ is a sub–Markovian operator.

Step 3: $(T_t)_{t \geq 0}$ **is sub–Markov.** Fix $t \geq 0$. It is well-known that for $f \in \mathrm{Dom}\,(\bar{L})$, $T_t f$ can be represented as

$$
\begin{aligned}
T_t f &= \lim_{\lambda \to \infty} \exp\left(t\bar{L}\,(1 - \lambda^{-1}\bar{L})^{-1} \right)\, f \\
&= \lim_{\lambda \to \infty} \exp\left(t\lambda\,(\lambda\,(\lambda - \bar{L})^{-1} - 1) \right)\, f,
\end{aligned}
$$

see e.g. [Yo 80, IX.7, (7)]. Hence, by Step 2, $0 \leq T_t f \leq 1$ m–a.e. for all $f \in \mathrm{Dom}\,(\bar{L})$ such that $0 \leq f \leq 1$ m–a.e.
For an arbitrary function $f \in L^p(E;m)$ such that $0 \leq f \leq 1$ m–a.e., and $\lambda > \alpha/p$, we have $\lambda(\lambda - \bar{L})^{-1}f \in \mathrm{Dom}\,(\bar{L})$, and $0 \leq \lambda(\lambda - \bar{L})^{-1}f \leq 1$, whence

(1.20) $$0 \leq T_t\,\lambda\,(\lambda - \bar{L})^{-1}f \leq 1 \qquad m\text{–a.e.}$$

As $\lambda \uparrow \infty$, $\lambda(\lambda - \bar{L})^{-1}f$ converges to f in $L^p(E;m)$, see e.g. [Yo 80, IX.4, (6)]. Since T_t is bounded, (1.20) implies again $0 \leq T_t f \leq 1$ m–a.e. Hence T_t is sub–Markov. ∎

We finally remark that a sub–Markovian C^0 semigroup on $L^p(E;m)$ induces C^0 semigroups on $L^r(E;m)$ for all $r \in [p, \infty)$:

Lemma 1.11 (i) *Let* $(T_t)_{t \geq 0}$ *be a sub–Markovian C^0 semigroup on $L^p(E;m)$, and let $r \in [p, \infty)$. Then the restrictions of the linear operators T_t, $t \geq 0$, to $L^r(E;m) \cap L^p(E;m)$ are bounded w.r.t. the norm on $L^r(E;m)$. The unique continuous extensions $T_t^{(r)}$, $t \geq 0$, of these operators to $L^r(E;m)$ form a sub–Markovian C^0 semigroup on $L^r(E;m)$. Moreover, suppose f is a function in the domain of the generator L of $(T_t)_{t \geq 0}$, such that f and Lf are in $L^r(E;m)$. Then f is also in in the domain of the generator $L^{(r)}$ of $\left(T_t^{(r)} \right)_{t \geq 0}$, and $L^{(r)}f = Lf$.*
(ii) *If $(T_t)_{t \geq 0}$ is a sub–Markovian C^0 <u>contraction</u> semigroup on $L^p(E;m)$, then $\left(T_t^{(r)} \right)_{t \geq 0}$, as defined in (i), is a sub–Markovian C^0 <u>contraction</u> semigroup on $L^r(E;m)$ for all $r \in [p, \infty)$.*

Proof. Since $(T_t)_{t \geq 0}$ is sub–Markov, the restriction of T_t to $L^\infty(E;m) \cap L^p(E;m)$ is a contraction w.r.t. the L^∞ norm for all $t \geq 0$. By the Riesz–Thorin interpolation theorem, we can conclude that the restrictions of the operators T_t, $t \geq 0$, to $L^r(E;m) \cap L^p(E;m)$ are bounded w.r.t. the $L^r(E;m)$ norm, and

(1.21) $$\sup_{0 \leq t \leq t_0} \|T_t\|_{r \to r} < \infty \qquad \text{for all } t_0 \geq 0.$$

Let $T_t^{(r)}$, $t \geq 0$, denote the unique continuous extensions to $L^r(E;m)$. For $f \in L^\infty(E;m) \cap L^p(E;m)$ we have

$$
\begin{aligned}
\int \left| T_t^{(r)} f - f \right|^r dm &\leq \int |T_t f - f|^p\, dm \cdot \left(\|T_t f\|_{L^\infty(E;m)} + \|f\|_{L^\infty(E;m)} \right)^{r-p} \\
&\leq \int |T_t f - f|^p\, dm \cdot \left(2\,\|f\|_{L^\infty(E;m)} \right)^{r-p}.
\end{aligned}
$$

Since $(T_t)_{t\geq 0}$ is strongly continuous on $L^p(E\,;m)$, $T_t^{(r)}f \to f$ in $L^r(E\,;m)$ as $t \downarrow 0$. Since $L^\infty(E\,;m) \cap L^r(E\,;m)$ is dense in $L^r(E\,;m)$, the strong continuity of $\left(T_t^{(r)}\right)_{t\geq 0}$ on $L^r(E\,;m)$ follows by an $\varepsilon/3$–argument using (1.21). Obviously, $\left(T_t^{(r)}\right)_{t\geq 0}$ is sub–Markov. Moreover, if $(T_t)_{t\geq 0}$ is a contraction semigroup on $L^p(E\,;m)$, then the Riesz–Thorin theorem implies that $\left(T_t^{(r)}\right)_{t\geq 0}$ is a contraction semigroup on $L^r(E\,;m)$.

Now fix $f \in \mathrm{Dom}\,(L)$ such that f and Lf are in $L^r(E;m)$. We have

$$(1.22) \qquad T_t f - f \;=\; \int_0^t T_s \, L f \, ds \qquad \text{for all } t \geq 0,$$

where the integral is a Bochner integral in $L^p(E\,;m)$. Since Lf is in $L^r(E\,;m)$, the function $s \mapsto T_s Lf \;(= T_s^{(r)}Lf)$ is Bochner–integrable in $L^r(E;m)$ as well, and the L^r and L^p Bochner integrals coincide. By (1.22) and the strong continuity of $\left(T_t^{(r)}\right)_{t\geq 0}$, we then obtain

$$\frac{1}{t}\left(T_t^{(r)}f - f\right) \;=\; \frac{1}{t}\left(T_t f - f\right) \;=\; \int_0^t T_s^{(r)}Lf\, ds \;\to\; Lf$$

in $L^r(E;m)$ as $t \downarrow 0$, whence f is in the domain of $L^{(r)}$, and $L^{(r)}f = Lf$. ∎

By a similar interpolation argument, and an additional duality consideration, one can show:

Lemma 1.12 *Suppose that $(T_t)_{t\geq 0}$ is a $\underline{symmetric}$ C^0 semigroup on $L^2(E\,;m)$. Then $(T_t)_{t\geq 0}$ is a contraction semigroup, and the assertions of Lemma 1.11 hold for all $r \in [1,\infty)$.*

PROOF. See [Dav 89, Thm. 1.4.1]. ∎

Chapter 2

L^p uniqueness in finite dimensions

In this chapter we discuss uniqueness of diffusion operators with domain $C_0^\infty(U)$, where U is $\mathbf{R}^n$ or an open subset, on weighted L^p spaces, i.e., we ask under which conditions the closure of the operator generates a C^0 semigroup.

We first consider diffusion operators with regular coefficients on an interval in $\mathbf{R}^1$, cf. Section a). Here, the uniqueness problem for symmetric operators on L^2 has been solved completely in the classical Sturm–Liouville theory, cf. e.g. [JöRel 76], [Wei 87]. We partially extend the classical results to not necessarily symmetric diffusion operators on weighted L^p spaces, where we obtain sufficient conditions for uniqueness that are necessary in many cases (though not in all).

After demonstrating essential differences between diffusion operators on $\mathbf{R}^1$ and $\mathbf{R}^n$, $n \geq 2$, in Section b), we consider elliptic diffusion operators with regular coefficients on $\mathbf{R}^n$, cf. Section c). Here, a rather optimal condition for L^p uniqueness can be derived by using standard regularity theory and a localization argument, which can be considered as an improved and generalized version of Gaffney's classical argument showing essential self-adjointness of the Laplace–Beltrami operator on a complete Riemannian manifold [Ga 51].

We then turn to the singular case which is our main interest. The examples in Section d) show that here additional considerations are necessary, since operators for which one could expect L^p uniqueness by the results in the regular case turn out to be not necessarily unique if the coefficients are singular or degenerate.

We first study $L^p(I; \rho\,dx)$ uniqueness for Sturm–Liouville operators of type $\mathcal{L} = \frac{1}{\rho}\frac{d}{dx}(\alpha\frac{d}{dx}\cdot)$, where I is an interval, and α and ρ are continuous functions that are, however, now allowed to have zeros. Note that in non–divergence form, $\mathcal{L} = \frac{\alpha}{\rho}\frac{d^2}{dx^2} + \frac{\alpha'}{\rho}\frac{d}{dx}$, which means that zeros of α and ρ may cause very bad singularities of the coefficients of the non–divergence form operator. In Section e) we give a necessary and sufficient condition for $L^p(I; \rho\,dx)$ uniqueness of such operators, which depends on the behaviour of α and ρ both at the boundaries and at the zeros of α.

Finally, we present two new uniqueness results for non–symmetric diffusion operators of type $\Delta + \beta \cdot \nabla$ with singular drift β on weighted L^p spaces over $\mathbf{R}^n$, see Section f). The first result which applies for $p \in [1, 2]$, is in some sense rather optimal, whereas in some other sense it is still not completely satisfying, cf. Theorem 2.5 and the remarks below. However, the proof is surprisingly elementary, and the result seems to be the best known so far. In particular, the obtained criterion does not depend on the dimension, which is crucial for applications to infinite dimensional uniqueness problems. Using more advanced regularity results, we prove a second uniqueness criterion, cf. Theorem 2.6. This criterion is slightly better than the first one in $\mathbf{R}^2$, and it is also applicable for $p > 2$. However, the obtained condition has the disadvantage of being highly dimension dependent.

The concluding Section g) points out some remaining and newly arosen questions.

a)   The regular one–dimensional case

Let (x_0, y_0) , $-\infty \leq x_0 < y_0 \leq \infty$, be an interval, and $\rho : (x_0, y_0) \to \mathbf{R}$ an absolutely continuous function such that $\rho(x) > 0$ for all x. There is a well–known characterization of the essentially self-adjoint (i.e. L^2 unique) second order differential operators (*Sturm–Liouville operators*) on $L^2(x_0, y_0 \,;\, \rho\, dx)$ due to H. Weyl, cf. e.g. [JöRel 76], [Wei 87]. For Weyl' s characterization it is crucial that ρ and the second order coefficient of the differential operator have no zeros inside the interval, cf. the counterexamples in Section d), 1), below. This is what we call the *regular* case. In this section we prove counterparts to parts of Weyl' s result for not necessarily symmetric one–dimensional diffusion operators on $L^p(x_0, y_0 \,;\, \rho\, dx)$, $1 \leq p < \infty$.

Fix $p \in [1, \infty)$, and ρ as above. Let $(\,\mathcal{L}\,,\, C_0^\infty(x_0, y_0)\,)$ be the diffusion operator on $L^p(x_0, y_0 \,;\, \rho\, dx)$ given by

$$\mathcal{L}f = af'' + \beta f' \,,$$

where β is a function in $L^p_{\mathrm{loc}}(x_0, y_0 \,;\, \rho\, dx)$, and a is a locally Lipschitz continuous function on (x_0, y_0) such that $a(x) > 0$ for all x. For our purposes it is convenient to rewrite $\mathcal{L}$ in divergence form:

$$\mathcal{L}f = \tfrac{1}{\rho} \cdot (\rho a f')' + b \cdot f' \quad dx\text{–a.e. for all } f \in C_0^\infty(x_0, y_0),$$

where $b = \beta - \tfrac{1}{\rho}(\rho a)'$, which is again in $L^p_{\mathrm{loc}}(x_0, y_0; \rho\, dx)$. We assume:

(**A 1**)   There exists $\alpha \geq 0$ such that

$$\int bf' \, \rho\, dx \leq \alpha \int f \, \rho\, dx \quad \text{for all positive functions } f \in C_0^\infty(x_0, y_0),$$

i.e., $(\rho b)' \geq -\alpha\rho$ in the distributional sense. Note that (A 1) holds if and only if $\rho\,dx$ is a sub-invariant measure for the operator $(\mathcal{L} - \alpha,\ C_0^\infty(x_0, y_0))$, i.e.,

$$\int \mathcal{L}f\,\rho\,dx \leq \alpha \int f\,\rho\,dx \quad \text{for all positive functions } f \in C_0^\infty(x_0, y_0).$$

In particular, $(\mathcal{L} - \frac{\alpha}{p},\ C_0^\infty(x_0, y_0))$ is dissipative on $L^p(x_0, y_0\,;\,\rho\,dx)$ in this case, cf. Lemma 1.8 in Appendix B.

Since ρ and a are strictly positive, $\frac{b}{a}$ is locally dx-integrable. Fix $z_0 \in (x_0, y_0)$, and let $\pi : (x_0, y_0) \to \mathbf{R}$ be the absolutely continuous, strictly positive fuction given by

$$\pi := \rho \cdot \exp\left(-\int_{z_0}^{\bullet} \frac{b}{a}\,dx\right).$$

Note that up to a constant factor, the definition of π is independent of the choice of z_0.

REMARK.   Informally, the adjoint $\mathcal{L}^*$ of $\mathcal{L}$ w.r.t. the measure $\rho\,dx$ is given by

$$(2.1) \qquad \mathcal{L}^*f \ = \ \frac{1}{\rho}(\rho a f')' - \frac{1}{\rho}(\rho b f)' \ = \ \frac{1}{\rho}(\rho a f')' - bf' - \frac{(\rho b)'}{\rho}f$$
$$= \ \frac{1}{\pi}(\pi a f')' - \frac{(\rho b)'}{\rho}f \,.$$

Note, however, that in general we don't assume b to be absolutely continuous.

Let $q \in (1, \infty]$ such that $\frac{1}{q} + \frac{1}{p} = 1$. The main result of this section is:

Theorem 2.1 *Let $c \in (x_0, y_0)$. Suppose that the positive locally bounded function*

$$(2.2) \qquad y \mapsto \int_c^y \frac{1}{\pi(z)a(z)} \int_c^z \pi(x)\,dx\,dz \quad \left(= \int_y^c \frac{1}{\pi(z)a(z)} \int_z^c \pi(x)\,dx\,dz\right)$$

is neither in $L^q(c, y_0\,;\,\rho\,dx)$ nor in $L^q(x_0, c\,;\,\rho\,dx)$. Then the closure of $(\mathcal{L},\ C_0^\infty(x_0, y_0))$ on $L^p(x_0, y_0\,;\,\rho\,dx)$ generates a C^0 semigroup. In particular, $(\mathcal{L},\ C_0^\infty(x_0, y_0))$ is $L^p(x_0, y_0\,;\,\rho\,dx)$ unique.

The proof of Theorem 2.1 will be given below.

REMARK.   If the assumption in Theorem 2.1 holds for some $c \in (x_0, y_0)$ then it holds for all such c. In other words: The assumption is not fulfilled if and only if for some $c \in (x_0, y_0)$, the fuction F_c defined by (2.2) is in $L^q(c, y_0\,;\,\rho\,dx)$ or in $L^q(x_0, c\,;\,\rho\,dx)$. In fact, suppose first $p > 1$, and assume that, for some c, F_c is, for example, in $L^q(c, y_0\,;\,\rho\,dx)$. Since π and $\frac{1}{\pi a}$ are strictly positive, we obtain

$$(2.3) \qquad \int_c^{\bullet} \frac{1}{\pi a}\,dz \ \in \ L^q(c, y_0\,;\,\rho\,dx), \qquad \text{and, in particular,}$$

$$(2.4) \qquad \int_c^{y_0} \rho\,dy \ < \ \infty.$$

Now fix $\tilde{c} \in (x_0, y_0)$. Because ρ and $\frac{1}{\pi a}$ are locally bounded, (2.3) and (2.4) still hold, if c is replaced by $\tilde{c}$. From this and $F_c \in L^q(c, y_0; \rho dx)$ one easily concludes that $F_{\tilde{c}}$ is in $L^q(\tilde{c}, y_0; \rho dx)$ as well. If $p = 1$, i.e., $q = \infty$, we may argue similarly, but now (2.3) alone suffices to conclude $F_{\tilde{c}} \in L^q(\tilde{c}, y_0; \rho dx)$, whereas (2.4) does not necessarily hold.

In particular, we have the following remarkable consequence of Theorem 2.1:

Corollary 2.1 *Suppose $p > 1$, and the measure ρdx is infinite at both boundaries, i.e.,*

$$\int_{x_0}^c \rho\, dx = \int_c^{y_0} \rho\, dx = \infty \quad \text{for some (resp. all) } c \in (x_0, y_0).$$

Then the closure of the operator $(\frac{1}{\rho}\frac{d}{dx}(\rho a \frac{d}{dx}\cdot) + b\frac{d}{dx}, C_0^\infty(x_0, y_0))$ on $L^p(x_0, y_0; \rho dx)$ is the generator of a C^0 semigroup for every absolutely continuous function a on (x_0, y_0) such that $a(x) > 0$ for all x, and every function $b \in L^p_{loc}(x_0, y_0; \rho dx)$ satisfying (A 1).

The corollary is a direct consequence of Theorem 2.1, cf. (2.4) in the remark above. It is in sharp contrast to the situation in $\mathbf{R}^n$ for $n \geq 2$. Here the corresponding assertion is false even if all coefficients are smooth, cf. the counterexamples in Section b), 1), below.

There is also a non–uniqueness result which shows that Theorem 2.1 is sharp in special situations. We first give a general and not very smooth formulation of this result, which will be a very useful source of counterexamples. Afterwards we look at the consequences in special cases.

Theorem 2.2 *Suppose b is absolutely continuous, and $|(\rho b)'| \leq \alpha \rho$ dx–a.e. for some $\alpha \geq 0$. Moreover, let $c \in (x_0, y_0)$, and assume that*

(i) *The function*

$$y \mapsto \int_y^c \frac{1}{\pi(z)a(z)} \left(\int_z^c \left(\frac{\pi(x)}{\rho(x)}\right)^p \rho(x)\, dx \right)^{1/p} dz$$

is in $L^q(x_0, c; \rho dx)$.

(ii) *The measure ρdx is finite on (c, y_0), or $\limsup_{y \uparrow y_0} \int_c^y \frac{b}{a}\, dx < \infty$, or the function*

$$y \mapsto \int_c^y \frac{1}{\pi(z)a(z)} \left(\int_c^z \left(\frac{\pi(x)}{\rho(x)}\right)^p \rho(x)\, dx \right)^{1/p} dz$$

is in $L^q(c, y_0; \rho dx)$.

Then the operator $(\mathcal{L} - \frac{\alpha}{p}, C_0^\infty(x_0, y_0))$ is dissipative on $L^p(x_0, y_0; \rho dx)$, but the closure of $(\mathcal{L}, C_0^\infty(x_0, y_0))$ does not generate a C^0 semigroup.

The corresponding result with rôles of x_0 and y_0 interchanged is, of course, true as well. The proofs of both theorems will be given below.

REMARK. The condition that b is absolutely continuous and $|(\rho b)'| \le \alpha \rho$ holds if and only if $\rho\,dx$ is both a sub–invariant measure for $(\mathcal{L} - \alpha,\ C_0^\infty(x_0, y_0))$ and a super–invariant measure for $(\mathcal{L} + \alpha,\ C_0^\infty(x_0, y_0))$. In particular, if $\rho\,dx$ is an invariant measure for $(\mathcal{L},\ C_0^\infty(x_0, y_0))$ then $(\rho b)' = 0$.

We explicitly point out two consequences of the Theorems 2.1 and 2.2:

Corollary 2.2 (Small perturbations of the symmetric case)
Suppose b is absolutely continuous, $|(\rho b)'| \le \alpha \rho\ dx$–a.e. for some $\alpha \ge 0$, and $\frac{b}{a}$ is dx–integrable on (x_0, y_0). Let $c \in (x_0, y_0)$.

(i) *If $p > 1$ then the closure of $(\mathcal{L},\ C_0^\infty(x_0, y_0))$ on $L^p(x_0, y_0\,;\,\rho\,dx)$ generates a C^0 semigroup if and only if the function $\displaystyle\int_c^\bullet \frac{1}{\rho a}\,dx$ is neither in $L^q(c, y_0\,;\,\rho\,dx)$ nor in $L^q(x_0, c\,;\,\rho\,dx)$.*

(ii) *The closure of $(\mathcal{L},\ C_0^\infty(x_0, y_0))$ on $L^1(x_0, y_0\,;\,\rho\,dx)$ generates a C^0 semigroup if and only if the function $\displaystyle\frac{1}{\rho a}\int_c^\bullet \rho\,dx$ is neither dx–integrable at x_0 nor at y_0.*

PROOF. The integrability of $\frac{b}{a}$ implies that $\frac{\pi}{\pi}$ and $\frac{\varrho}{\pi}$ are bounded. Hence, if the non–integrability conditions in the corollary hold, then the closure of $(\mathcal{L},\ C_0^\infty(x_0, y_0))$ is a generator by Theorem 2.1. Conversely, if $p > 1$, and $\int_c^\bullet \frac{1}{\rho a}\,dx$ is, for example, in $L^q(x_0, c\,;\,\rho\,dx)$, then $\int_{x_0}^c \rho\,dx < \infty$, and thus Condition (i) in Theorem 2.2 is satisfied. By assumption, Condition (ii) also holds. Hence the closure of $(\mathcal{L},\ C_0^\infty(x_0, y_0))$ on $L^p(x_0, y_0\,;\,\rho\,dx)$ is not a generator. If $\frac{1}{\rho a}\int_c^\bullet \rho\,dx$ is dx–integrable at x_0 or y_0, then Theorem 2.2 immediately implies that the closure of $(\mathcal{L},\ C_0^\infty(x_0, y_0))$ on $L^1(x_0, y_0\,;\,\rho\,dx)$ generates a C^0 semigroup. ∎

REMARK. The corollary in particular applies in the *symmetric case*, i.e., $b \equiv 0$. For $p = 2$ it recovers the classical *limit point criterion* for essential self–adjointness of the Sturm–Liouville operator $(\frac{1}{\rho}\frac{d}{dx}(\rho a \frac{d}{dx}\cdot),\ C_0^\infty(x_0, y_0))$ on $L^2(x_0, y_0\,;\,\rho\,dx)$, cf. e.g. [Wei 87, Thm. 6.3 and 5.8]. It shows, moreover, that if the non–symmetric part of the operator is small in the approriate sense, then L^p uniqueness holds if and only if it holds for the symmetric part of the operator.

In Corollary 1 we have shown that if the measure $\rho\,dx$ is infinite at both boundaries then L^p uniqueness always holds for $p > 1$. If $\rho\,dx$ is finite we have the following condition for L^p uniqueness:

Corollary 2.3 *Suppose the measure $\rho\,dx$ is finite, b is absolutely continuous, $|(\rho b)'| \le \alpha \rho\ dx$–a.e. for some $\alpha \ge 0$, and the function $\exp(-\int_{x_0}^\bullet (b/a)\,dx)$ is in $L^p(x_0, y_0\,;\,\rho\,dx)$. Let $c \in (x_0, y_0)$. Then the closure of the operator*

$(\mathcal{L}, C_0^\infty(x_0, y_0))$ on $L^p(x_0, y_0 \,;\, \rho\, dx)$ is the generator of a C^0 semigroup if and only if the function

$$ y \longmapsto \int_c^y \frac{1}{\rho a} \exp\left(\int_c^\bullet \frac{b}{a}\, dx \right) dz $$

is neither in $L^q(c, y_0 \,;\, \rho\, dx)$ nor in $L^q(x_0, c \,;\, \rho\, dx)$.

PROOF.   By the assumptions, $\rho\, dx$ is finite and π/ρ is in $L^p(x_0, y_0 \,;\, \rho\, dx)$. Theorems 2.1 and 2.2 now imply that the closure generates if and only if $\int_c^\bullet (\pi a)^{-1}\, dz$ is neither in $L^q(c, y_0 \,;\, \rho\, dx)$ nor in $L^q(x_0, c \,;\, \rho\, dx)$. ∎

In Section b) below we will look at some examples, and show which differences may occur in higher dimensions. We now conclude this section by proving the theorems.

The following proof of Theorem 2.1 is a little intricate, because we don't assume b to be absolutely continuous. If b is absolutely continuous, it can be simplified considerably.

PROOF OF THEOREM 2.1.   Fix $\alpha \geq 0$ such that the condition in (A 1) holds, and let $\lambda := \alpha + 1$. Since $(\mathcal{L} - \frac{\alpha}{p}, C_0^\infty(x_0, y_0))$ is dissipative, and $\lambda > \frac{\alpha}{p}$, it suffices to show that there is no non-trivial solution $u \in L^q(x_0, y_0 \,;\, \rho\, dx)$ of the equation $L^* u = \lambda u$. Hence suppose there exists a non-trivial solution u.
Step 1: Regularity. Note that $L^* u = \lambda u$ means that u solves the ODE

$$ (2.5) \qquad\qquad (\rho a u' - \rho b u)' = \lambda\, \rho u $$

in the distributional sense. By a regularity result for ordinary differential equations, u is a "strong" solution of (2.5) in the following sense: u has an absolutely continuous dx-version, $\rho a u' - \rho b u$ has a C^1 version w, and $w' = \lambda \rho u$, cf. Theorem 2.7 in Appendix C. Here and from now on, u will always denote the absolutely continuous dx-version.
Step 2: Sturm–Liouville form of the adjoint ODE. Let (x_1, y_1) be a non–empty subinterval of (x_0, y_0). Suppose that $u \geq 0$ on (x_1, y_1). We show that

$$ (2.6) \qquad\qquad (\pi a u')(y) - (\pi a u')(x) \;\geq\; \int_x^y \pi u\, dx $$

for $dx\,dy$–a.e. $(x, y) \in (x_1, y_1) \times (x_1, y_1)$ such that $x \leq y$. Note that heuristically, (2.6) immediately follows from (2.1) and the assumption $(\rho b)' \geq -\alpha\rho$. We now give a rigorous proof. Fix a positive function $g \in C_0^\infty(x_1, y_1)$. Note that $(\frac{\pi}{\rho})' = -\frac{b}{a} \cdot \frac{\pi}{\rho}$ implies

$$ \left(\frac{\pi}{\rho} u \right)' \;=\; -\frac{\pi}{\rho}\frac{b}{a} u + \frac{\pi}{\rho} u' \;=\; \frac{\pi}{\rho^2 a} w \;\in\; L_{\mathrm{loc}}^\infty(x_1, y_1 \,;\, dx), \text{ and} $$

$$ \rho b \left(g \frac{\pi}{\rho} u \right)' \;=\; g' \pi b u + g \frac{\pi}{\rho}\frac{b}{a} w \;=\; g' \pi b u - g \left(\frac{\pi}{\rho} \right)' w. $$

Hence if (A 1) holds then

$$
\begin{aligned}
-\int_{x_1}^{y_1} g' \, \pi a u' \, dt \;&=\; -\int_{x_1}^{y_1} g' \frac{\pi}{\rho} w \, dt - \int_{x_1}^{y_1} g' \, \pi b u \, dt \\
&=\; \int_{x_1}^{y_1} g \frac{\pi}{\rho} w' \, dt + \int_{x_1}^{y_1} \left(g \left(\frac{\pi}{\rho}\right)' w - g' \, \pi b u \right) dt \\
&=\; \lambda \int_{x_1}^{y_1} g \pi u \, dt - \int_{x_1}^{y_1} \rho b \left(g \frac{\pi}{\rho} u \right)' dt \\
&\geq\; (\lambda - \alpha) \int_{x_1}^{y_1} g \pi u \, dt \;=\; \int_{x_1}^{y_1} g \pi u \, dt.
\end{aligned}
$$

It is now standard to conclude that (2.6) holds.

Step 3: Growth of u at the boundary. We finally show that by the inequality (2.6) derived in Step 2, the function u grows at one of the boundaries x_0 and y_0 so strongly that u cannot be in $L^q(x_0, y_0 \, ; \, \rho \, dx)$ provided the assumption in Theorem 2.1 holds. Hence in this case there is only the trivial solution of $L^* u = \lambda u$.

Since u is continuous and not identically zero, there exists an interval $[x_1, y_1]$, $x_0 < x_1 < y_1 < y_0$, such that either $u(x) > 0$ for all $x \in [x_1, y_1]$, or $-u(x) > 0$ for all $x \in [x_1, y_1]$. We may assume $u > 0$ on $[x_1, y_1]$, otherwise we consider the function $-u$ instead of u. From now on we will denote by u' a *fixed dx–version* of the derivative. By (2.6), we can find $z_1 \in (x_1, y_1)$ such that

$$
(2.7) \quad (\pi a u')(z) \;\geq\; (\pi a u')(z_1) + \int_{z_1}^{z} \pi u \, dx \qquad \text{for } dx\text{–a.e. } z \in (z_1, y_1),
$$

and

$$
(2.8) \quad (\pi a u')(z) \;\leq\; (\pi a u')(z_1) - \int_{z_1}^{z} \pi u \, dx \qquad \text{for } dx\text{–a.e. } z \in (x_1, z_1).
$$

We first consider the case where $u'(z_1) \geq 0$. I claim that then u is increasing on (z_1, y_0). In fact, let

$$
\bar{y} := \sup \{ y \in (z_1, y_0) \, ; \, u' \geq 0 \; dx\text{–a.e. on } (z_1, y) \}.
$$

By (2.7), $\bar{y} \geq y_1$. Since u is increasing on $(z_1, \bar{y})$, we have

$$
\inf \{ u(x); \; z_1 < x < \bar{y} \} = u(z_1) > 0.
$$

Now suppose $\bar{y} < y_0$. Then there exists $\varepsilon > 0$ such that $u \geq 0$ on $(z_1, \bar{y} + \varepsilon)$. Hence by (2.6) and (2.7),

$$
(\pi a u')(z) \;\geq\; (\pi a u')(z_1) + \int_{z_1}^{z} \pi u \, dx \;\geq\; 0
$$

for dx–a.e. $z \in (z_1, \bar{y} + \varepsilon)$, which is a contradiction to the definition of $\bar{y}$. Thus $\bar{y} = y_0$, and, in particular, $\inf \{u(x);\, z_1 < x < y_0\} = u(z_1) > 0$. By (2.7) we obtain

$$(\pi a u')(z) \geq (\pi a u')(z_1) + \int_{z_1}^{z} \pi u \, dx \geq u(z_1) \cdot \int_{z_1}^{z} \pi \, dx$$

for dx–a.e. $z \in (z_1, y_0)$. Hence

$$(2.9) \qquad u(y) \geq u(z_1) + u(z_1) \cdot \int_{z_1}^{y} (\pi(z)\, a(z))^{-1} \int_{z_1}^{z} \pi(x) \, dx \, dz$$

for all $y \in [z_1, y_0)$. By the assumption in Theorem 2.1 and the remark below Theorem 2.1, the positive function on the right hand side is not in $L^q(z_1, y_0;\, \rho\, dx)$, and so u is neither.

If $u'(z_1) \leq 0$, then by (2.6) and (2.8) we similarly obtain

$$(\pi a u')(z) \leq -u(z_1) \cdot \int_{z}^{z_1} \pi \, dx \qquad \text{for } dx\text{–a.e. } z \in (x_0, z_1),$$

whence

$$(2.10) \qquad u(y) \geq u(z_1) + u(z_1) \cdot \int_{y}^{z_1} (\pi(z)\, a(z))^{-1} \int_{z}^{z_1} \pi(x) \, dx \, dz$$

for all $y \in (x_0, z_1]$, which implies that u is not in $L^q(x_0, z_1;\, \rho\, dx)$ by the assumption. ∎

PROOF OF THEOREM 2.2.   The dissipativity of $(\mathcal{L} - \frac{\alpha}{p},\, C_0^\infty(x_0, y_0))$ follows as above. We show that there exists a non–trivial solution $u \in L^q(x_0, y_0;\, \rho\, dx)$ of the equation $L^* u = \lambda u$ where $\lambda := \alpha + 1$. This proves the assertion by Corollary 1.3 in Appendix A.

Step 1: Existence of a solution in $L^q(c, y_0;\, \rho\, dx)$.   We show that, by Assumption (ii), there exists a strictly positive function $u \in C^1(x_0, y_0)$ such that $\int_c^{y_0} |u|^q \rho\, dx < \infty$, u' is absolutely continuous, and

$$(2.11) \qquad (\rho a u' - \rho b u)' = \lambda \rho u.$$

In particular,

$$(2.12) \quad \int_{x_0}^{y_0} \mathcal{L} f \, u \rho \, dx \;=\; \int_{x_0}^{y_0} ((\rho a f')' + \rho b f')\, u \, dx$$

$$= -\int_{x_0}^{y_0} f'\, (\rho a u' - \rho b u) \, dx \;=\; \lambda \int_{x_0}^{y_0} f \, u \rho \, dx$$

for any $f \in C_0^\infty(x_0, y_0)$. For the proof we consider the corresponding system of first order linear ordinary differential equations:

$$(2.13) \qquad \begin{aligned} u' &= a^{-1} b\, u + a^{-1} \rho^{-1} w, \\ w' &= \lambda \rho\, u. \end{aligned}$$

The functions $a^{-1}b$, $a^{-1}\rho^{-1}$ and $\lambda\rho$ are continuous by assumption. Hence for any $x \in (x_0, y_0)$ and any $u_0, w_0 \in \mathbf{R}$ there exists a unique continuously differentiable solution (u, w) of (2.13) such that $u(x) = u_0$ and $w(x) = w_0$. Now let (u, w) be an arbitrary C^1 solution of (2.13). Then u' is absolutely continuous because of the assumptions on a, ρ and b, and u solves (2.11). In particular,

$$(2.14) \qquad \frac{1}{\pi}(\pi a u')' = \frac{1}{\rho}(\rho a u')' - bu'$$

$$= \frac{1}{\rho}(\rho a u' - \rho b u)' + \frac{(\rho b)'}{\rho}u = \left(\lambda + \frac{(\rho b)'}{\rho}\right)u$$

dx–a.e., whence

$$(2.15) \qquad (\pi a u u')' = \pi a (u')^2 + \left(\lambda + \frac{(\rho b)'}{\rho}\right)\pi u^2$$

$$\geq \pi a (u')^2 + \pi u^2 \geq 0 \quad dx\text{–a.e.}$$

From (2.15) we can conclude that there exists a C^1 solution (u, w) of (2.13) such that $u(x) > 0$ and $u'(x) < 0$ for all $x \in (x_0, y_0)$. In fact, consider the sets

$$K(x) := \{(u, w) \in C^1((x_0, y_0); \mathbf{R}^2);\ (u, w) \text{ solves } (2.13), \text{ and } u(x) \cdot u'(x) < 0\},$$

$x_0 < x < y_0$. By (2.15), K is decreasing, i.e.,

$$K(x) \supseteq K(y) \qquad \text{if } x \leq y.$$

Moreover, for $x \in (x_0, y_0)$, the map $(u, w) \mapsto (u(x), u'(x))$ is an isomorphism between the space of solutions of (2.13) and $\mathbf{R}^2$, which maps $K(x)$ to the open cone $\{(y, z) \in \mathbf{R}^2;\ y \cdot z < 0\}$. Hence $K(x)$ is a non–empty open cone itself, whence the intersection $\bigcap\{K(x);\ x_0 < x < y_0\}$ contains at least a half–ray R starting at 0, and the corresponding opposite half–ray $-R$. Now fix $(u, w) \in \bigcap\{K(x);\ x_0 < x < y_0\}$. Then $u \cdot u' < 0$ on (x_0, y_0). By continuity, we have either $u < 0$ and $u' > 0$ on (x_0, y_0), or $u > 0$ and $u' < 0$ on (x_0, y_0). By replacing (u, w) by $(-u, -w)$ if necessary, we may assume $u < 0$ and $u' > 0$ on (x_0, y_0). Since u is positive and decreasing, it is bounded on (c, y_0). By Assumption (ii) we can now conclude that u is in $L^q(c, y_0;\ \rho\,dx)$, which completes the proof of Step 1. In fact, if the measure $\rho\,dx$ is finite, then the boundedness of u directly implies $u \in L^q(c, y_0;\ \rho\,dx)$. If $\limsup_{y\uparrow y_0} \int_c^y \frac{b}{a}\,dx < \infty$ then $\frac{\pi}{\rho}$ is bounded from below on (c, y_0) by a strictly positive constant δ. Hence, by (2.14),

$$(\pi a u')' \geq \pi u \geq \delta\,\rho u \qquad dx\text{–a.e. on } (c, y_0),$$

and thus

$$\int_c^y |u|\,\rho\,dx = \int_c^y u\,\rho\,dx \leq \delta^{-1} \cdot ((\pi a u')(y) - (\pi a u')(c)) < -\delta^{-1}(\pi a u')(c)$$

for all $y \in (c, y_0)$. Therefore u is in $L^1(c, y_0;\ \rho\,dx)$. Since it is bounded, it is also in $L^q(c, y_0;\ \rho\,dx)$. Finally, if the function

$$y \mapsto \int_c^y \frac{1}{\pi(z)a(z)} \left(\int_c^z \left(\frac{\pi(x)}{\rho(x)}\right)^p \rho(x)\,dx\right)^{1/p} dz$$

is in $L^q(c, y_0 ; \rho\,dx)$, then it can be shown by a similar argumentation as we will use in the next step that u is in $L^q(c, y_0 ; \rho\,dx)$.

Step 2: Integrability of the solution at x_0. We show that, by Assumption (i), the function u constructed in Step 1 is also in $L^q(x_0, c ; \rho\,dx)$. Hence u is in $L^q(x_0, y_0 ; \rho\,dx)$, and, by (2.12), $L^*u = \lambda u$. This completes the proof of Theorem 2.2. To see that u is in $L^q(x_0, c ; \rho\,dx)$, note that by (2.14),

$$(\pi a u')' \leq (\lambda + \alpha)\, \pi u \qquad dx\text{–a.e. on } (x_0, y_0).$$

Hence for all $z \in (x_0, c)$,

$$(\pi a u')(z) \geq (\pi a u')(c) - (\lambda + \alpha) \int_z^c \pi u\, dx$$

$$\geq (\pi a u')(c) - (\lambda + \alpha) \left(\int_z^c (\pi/\rho)^p \rho\,dx \right)^{1/p} \|u\|_{L^q(z, c ; \rho\,dx)}.$$

Let $c_0 \in (x_0, c)$. Since u, π and ρ are strictly positive on $[c_0, c]$, there exists a finite constant K_1 such that for all $z \in (x_0, c_0)$,

$$(\pi a u')(z) \geq -K_1 \cdot \|u\|_{L^q(z, c ; \rho\,dx)} \cdot \left(\int_z^c (\pi/\rho)^p \rho\,dx \right)^{1/p}.$$

For $y \in (x_0, c_0)$ we obtain

$$(2.16)\quad u(y) = u(c) - \int_y^c u'(z)\, dz$$

$$\leq K_2 \cdot \int_y^c \frac{1}{\pi(z) a(z)} \left(\int_z^c (\pi/\rho)^p \rho\,dx \right)^{1/p} \|u\|_{L^q(z, c ; \rho\,dx)}\, dz,$$

where K_2 is a finite constant that does not depend on y.

Now suppose first $p > 1$, and let $f(x) := \int_x^c u^q \rho\,dx$. Then (2.16) implies

$$f \leq \int_{c_0}^c u^q \rho\,dx + K_2^q \cdot \int_{\bullet}^{c_0} \left(\int_y^c \frac{1}{\pi(z) a(z)} \left(\int_z^c (\pi/\rho)^p \rho\,dx \right)^{1/p} dz \right)^q \rho(y)\, f(y)\, dy$$

on (x_0, c_0). Hence, by the lemma of Gronwall and Assumption (i), f is bounded on (x_0, c_0), i.e., u is in $L^q(x_0, c ; \rho\,dx)$.

In the case $p = 1$ we can argue similarly. Here (2.16) implies

$$\|u\|_{L^\infty(y, c ; \rho\,dx)} \leq K_2 \cdot \int_y^c \frac{1}{\pi(z) a(z)} \int_z^c \pi\, dx\, \|u\|_{L^\infty(z, c ; \rho\,dx)}\, dz$$

for all $y \in (x_0, c_0)$. Hence, by Gronwall's lemma and Assumption (i), u is in $L^\infty(x_0, c ; \rho\,dx)$. ∎

b) Examples and counterexamples I: Regular diffusion operators on $\mathbf{R}^1$ and $\mathbf{R}^n$

We now look at some examples for the results in Section a). Moreover, we point out differences between the one–dimensional and the multi–dimensional case.

1) Diffusion operators on $L^p(\mathbf{R}^n; dx)$

By Corollary 2.1, the closure of *any* diffusion operator of type

$$\left(\frac{d}{dx} \left(a \frac{d}{dx} \cdot \right) + b \frac{d}{dx} \,,\; C_0^\infty(\mathbf{R}^1) \right)$$

on $L^p(\mathbf{R}^1; dx)$, $p > 1$, generates a C^0 semigroup provided a is absolutely continuous and strictly positive, b is in $L^p_{\text{loc}}(\mathbf{R}^1; dx)$, and b' is bounded from below in the distributional sense.

In $\mathbf{R}^n$, $n \geq 2$, the situation is completely different. This is demonstrated by the following example, which is a generalization of an example in [Dav 85, 3.5]:

EXAMPLE. Fix a large positive constant γ and $p \in [1, \infty)$. Let $\mathbf{R}^2_+ := \{x \in \mathbf{R}^2 ; x_1 > 0\}$. Suppose $F : (0, \infty) \to \mathbf{R}$ is a smooth bijection such that $F'(x) > 0$ for all x, $F(x) = -\exp(x^{-\gamma})$ if $x \leq 1$, and $F(x) = x$ if $x \geq 2$. We consider the bijection $\theta : \mathbf{R}^2_+ \to \mathbf{R}^2$ given by

$$\theta(x_1, x_2) = (F(x_1), x_2/F'(x_1)).$$

Since θ has unit Jacobian, it induces an isometry

$$\Phi : L^p(\mathbf{R}^2_+; dx) \longrightarrow L^p(\mathbf{R}^2; dx), \qquad (\Phi f)(x) = f(\theta^{-1}(x)).$$

For a diffusion operator $(\mathcal{L}, C_0^\infty(\mathbf{R}^2_+))$ on $L^p(\mathbf{R}^2_+; dx)$ of type

$$\mathcal{L}f = \sum_{j=1}^2 \left(\frac{\partial}{\partial x_j} \left(a \frac{\partial f}{\partial x_j} \right) + b_j \frac{\partial f}{\partial x_j} \right),$$

with coefficients a, b_1 $b_2 \in C^\infty(\mathbf{R}^2_+)$, the corresponding image diffusion operator $(\mathcal{L}^\Phi, C_0^\infty(\mathbf{R}^2))$ on $L^p(\mathbf{R}^2; dx)$ is given by

$$\mathcal{L}^\Phi f = \sum_{i,j=1}^2 \left(\frac{\partial}{\partial x_i} \left(a_{ij}^\Phi \frac{\partial f}{\partial x_j} \right) + b_j^\Phi \frac{\partial f}{\partial x_j} \right),$$

where

$$\left(a_{ij}^\Phi \left(\theta(x) \right) \right)_{i,j=1,2} = a(x) \cdot \begin{pmatrix} (F'(x_1))^2 & -\frac{x_2 \cdot F''(x_1)}{F'(x_1)} \\ -\frac{x_2 \cdot F''(x_1)}{F'(x_1)} & \left(\frac{x_2 F''(x_1)}{(F'(x_1))^2} \right)^2 + \frac{1}{(F'(x_1))^2} \end{pmatrix},$$

$$b_1^{\Phi}\left(\theta(x)\right) \;=\; b_1(x)\cdot F'(x_1), \qquad \text{and}$$

$$b_2^{\Phi}\left(\theta(x)\right) \;=\; -b_1(x)\cdot\frac{x_2\cdot F''(x_1)}{(F'(x_1))^2} \;+\; b_2(x)\cdot\frac{1}{F'(x_1)}.$$

We consider two cases:

(i) Suppose $\mathcal{L} = \Delta$, i.e., $a(x) = 1$ and $b(x) = 0$ for all x. Then

$$\mathcal{L}^{\Phi} \;=\; \sum_{i,j=1}^{2}\frac{\partial}{\partial x_i}\left(a_{ij}^{\Phi}\frac{\partial}{\partial x_j}\cdot\right)$$

with a_{ij}^{Φ}, $1 \le i, j \le 2$, as above. Since $(\Delta, C_0^{\infty}(\mathbf{R}_+^2))$ is not $L^p(\mathbf{R}_+^2\,;\,dx)$ unique, $(\mathcal{L}^{\Phi}, C_0^{\infty}(\mathbf{R}^2))$ is neither.

For later purposes, we determine the growth rate of $a^{\Phi}(x)$ as $|x| \to \infty$. An easy calculation shows

$$\limsup_{|x|\to\infty}\frac{\|a^{\Phi}(x)\|}{|x|^2(\log|x|)^{2+(2/\gamma)}} \;<\; \infty,$$

cf. also [Dav 85, 3.5], where this case has already been considered. Here $\|a^{\Phi}(x)\|$ denotes the norm of the coefficient matrix. Conversely, the results in Section c) below show that $L^p(\mathbf{R}^n\,;\,dx)$ uniqueness holds for operators of type

$$\cdot\;\sum_{i,j=1}^{n}\frac{\partial}{\partial x_i}\left(a_{ij}\frac{\partial}{\partial x_j}\cdot\right), \qquad n \ge 2,$$

provided the coefficient matrix $a(x) = (a_{ij}(x))$ is symmetric, Lipschitz continuous, locally bounded and locally strictly elliptic, and

$$\limsup_{|x|\to\infty}\frac{a_r(x)}{|x|^2(\log|x|)^2} \;<\; \infty, \qquad \text{where}\;\; a_r(x) \;=\; \frac{x\cdot a(x)\,x}{|x|^2}.$$

(ii) Suppose $a(x) = x_1^2$, $b_1(x) = 1$, and $b_2(x) = 0$ for all $x \in \mathbf{R}_+^2$, i.e.,

$$\mathcal{L} \;=\; \frac{\partial}{\partial x_1}\left(x_1^2\frac{\partial}{\partial x_1}\cdot\right) \;+\; \frac{\partial}{\partial x_1} \;+\; \frac{\partial^2}{\partial x_2^2}.$$

Obviously, $\operatorname{div} b = 0$, whence Lebesgue measuer is an invariant measure for $(\mathcal{L}, C_0^{\infty}(\mathbf{R}_+^2))$. It can be shown similarly to the proof of Theorem 2.2 that the ODE

$$(x^2 u')' \;-\; u' \;=\; u$$

has a non–trivial bounded solution $u \in C^2(0,\infty)$ that is dx–integrable at ∞. Thus u is in $L^q(0,\infty\,;\,dx)$, $p^{-1}+q^{-1}=1$, whence the operator

$$\left(\frac{d}{dx}x^2\frac{d}{dx}\cdot\;+\frac{d}{dx}\,,\;C_0^{\infty}(0,\infty)\right)$$

is not $L^p(0, \infty \,;\, dx)$ unique. It is not difficult to conclude that $(\mathcal{L}, C_0^\infty(\mathbf{R}_+^2))$ is not $L^p(\mathbf{R}_+^2 \,;\, dx)$ unique either. Therefore the closure of the image diffusion operator $(\mathcal{L}^\Phi, C_0^\infty(\mathbf{R}^2))$ on $L^p(\mathbf{R}^2 \,;\, dx)$ does *not* generate a C^0 semigroup. An easy calculation shows:

$$\limsup_{|x| \to \infty} \frac{\|a^\Phi(x)\|}{|x|^2(\log|x|)^2} < \infty, \quad \text{and} \quad \limsup_{|x| \to \infty} \frac{|b^\Phi(x)|}{|x|\,(\log|x|)^{1+(1/\gamma)}} < \infty.$$

Conversely, we will show in Section c) below, that $L^p(\mathbf{R}^n \,;\, dx)$ uniqueness holds for operators of type

$$\sum_{i,j=1}^n \frac{\partial}{\partial x_i}\left(a_{ij}\frac{\partial}{\partial x_j}\cdot\right) + \sum_{j=1}^n b_j \frac{\partial}{\partial x_j}, \qquad n \geq 2,$$

provided the coefficients a_{ij} and b_j satisfy regularity conditions, and

$$\limsup_{|x| \to \infty} \frac{a_r(x)}{|x|^2(\log|x|)^2} < \infty, \quad \text{and} \quad \liminf_{|x| \to \infty} \frac{b_r(x)}{|x|\log|x|} > -\infty,$$

where $b_r(x) := x \cdot b(x)/|x|$.

REMARK. In dimension 1, we may consider similarly transformations $\theta :$ $(0, \infty) \to \mathbf{R}$ which carry non $L^p(0, \infty \,;\, dx)$ unique operators defined on $C_0^\infty(0, \infty)$ onto operators defined on $C_0^\infty(\mathbf{R}^1)$. However, these transformations never preserve Lebesgue measure, but they map Lebesgue measure on $(0, \infty)$ to a measure on $\mathbf{R}$ which is finite either on $(-\infty, 0)$ or on $(0, \infty)$.

2) Operators on $L^p(\mathbf{R}^1 \,;\, e^{-x^2/2}\,dx)$

Let $p \in [1, \infty)$. Consider a diffusion operator on $C_0^\infty(\mathbf{R}^1)$ of type

$$\mathcal{L} = e^{x^2/2}\frac{d}{dx}\left(e^{-x^2/2}a\frac{d}{dx}\cdot\right) + b\cdot\frac{d}{dx} = \frac{d}{dx}\left(a\frac{d}{dx}\cdot\right) + (b - x\cdot a)\frac{d}{dx}$$

where a is absolutely continuous and strictly positive, and b is a function in $L^p(\mathbf{R}^1 \,;\, e^{-x^2/2}\,dx)$ satisfying (A 1) with $\rho(x) = e^{-x^2/2}$. This is the case if $b' - xb$ is bounded from below (in the distributional sense). Suppose that

$$\liminf_{x \to \infty} \frac{b(x)}{x\,a(x)} = 0, \quad \limsup_{x \to -\infty} \frac{b(x)}{|x|\,a(x)} = 0,$$

and $a(x)$ grows at most exponentially fast as $|x| \to \infty$. Then the function $\pi(x) = \exp\left(-(x^2/2) - \int_0^x (b/a)\,dx\right)$ decreases of order $\exp\left(-x^2/2\right)$ as $|x| \to \infty$, and $\pi(x)\,a(x)$ decreases at least of order $\exp\left(-x^2/2\right)$. Hence Theorem 2.1 shows that the closure of $(\mathcal{L}, C_0^\infty(\mathbf{R}^1))$ on $L^p(\mathbf{R}^1 \,;\, e^{-x^2/2}\,dx)$ generates a C^0 semigroup.

On the other hand, in contrast to the case of infinite measures, we cannot expect $L^p(\mathbf{R}^1 \,;\, e^{-x^2/2}\,dx)$ uniqueness for arbitrary functions a, b satisfying (A 1).

In fact, there is a diffeomorphism $\theta : \mathbf{R}^1 \to (0, \sqrt{2\pi})$ which maps Gauss measure to Lebesgue measure. Hence any non $L^p(0, \sqrt{2\pi} ; dx)$ unique diffusion operator defined on $C_0^\infty(0, \sqrt{2\pi})$ induces a non $L^p(\mathbf{R}^1 ; e^{-x^2/2} dx)$ unique diffusion operator on $C_0^\infty(\mathbf{R}^1)$.

3) One–dimensional generalized Schrödinger operators and small perturbations

Let (x_0, y_0) be an interval, and fix $p \in (1, \infty)$. Consider an operator of type

$$\mathcal{L} = \frac{1}{\rho} \frac{d}{dx} \left(\rho \frac{d}{dx} \cdot \right) + b \frac{d}{dx} = \frac{d^2}{dx^2} + \left(\frac{\rho'}{\rho} + b \right) \frac{d}{dx}$$

on $L^p(x_0, y_0 ; \rho \, dx)$, where ρ, $b : (x_0, y_0) \to \mathbf{R}$ are absolutely continuous, ρ is strictly positive, $|(\rho b)'| \leq \alpha \rho$ for some $\alpha \geq 0$, and b is dx–integrable on (x_0, y_0). In the case $b \equiv 0$, $\mathcal{L}$ is a one–dimensional generalized Schrödinger operator. The essential self–adjointness, or, equivalently, $L^2(x_0, y_0 ; \rho \, dx)$ uniqueness of these operators has been studied in [Wie 85]. Fix $c \in (x_0, y_0)$. By Corollary 2.2, $(\mathcal{L}, C_0^\infty(x_0, y_0))$ is $L^p(x_0, y_0 ; \rho \, dx)$ unique if and only if $\int_c^\bullet (1/\rho) \, dx$ is neither in $L^q(x_0, c ; \rho \, dx)$ nor in $L^q(c, y_0 ; \rho \, dx)$, $p^{-1} + q^{-1} = 1$. In particular, we obtain:

(i) If $(x_0, y_0) = \mathbf{R}^1$ then $(\mathcal{L}, C_0^\infty(x_0, y_0))$ is *always* $L^p(x_0, y_0 ; \rho \, dx)$ unique for *every* $p \in (1, \infty)$.

(ii) If $(x_0, y_0) = (0, \infty)$ and $\rho(x) \sim x^\gamma$ for small x, where $\gamma \geq 1$, then $(\mathcal{L}, C_0^\infty(x_0, y_0))$ is $L^p(x_0, y_0 ; \rho \, dx)$ unique if and only if $p \leq (\gamma + 1)/2$.

PROOF.   We only show

(2.17) $$\int_c^\infty \left(\int_c^y \frac{1}{\rho(z)} \, dz \right)^q \rho(x) \, dx = \infty$$

for any $q \in (1, \infty)$, $c \in \mathbf{R}$, and any continuous function $\rho : (c, \infty) \to (0, \infty)$ — it is then easy to see that (i) and (ii) hold. Note that (2.17) trivially holds if ρ is not dx–integrable at ∞. Otherwise, $\int_c^\infty (1/\rho) \, dx = \infty$. Let $f(x) := \int_c^x (1/\rho) \, dx$. We have

$$\int_c^\infty \left(\int_c^y \frac{1}{\rho(z)} \, dz \right)^q \rho(x) \, dx = \int_c^\infty f^q \cdot \frac{1}{f'} \, dx = (q-1) \cdot \int_c^\infty \frac{1}{(-f^{1-q})'} \, dx,$$

where $(-f^{1-q})' \geq 0$ since f is positive and incresing. If the right hand side would be finite, then $(-f^{1-q})'$ would not be dx–integrable on (c, ∞), i.e., $f^{1-q}(x)$ would become negative for large x. This is a contradiction. ∎

We will show in the next section that in the regular case (i.e., ρ strictly bounded away from zero) generalized Schrödinger operators on $\mathbf{R}^n$ are $L^p(\mathbf{R}^n ; \rho \, dx)$ unique for all $p \in (1, 2]$ as well.

c) Regular diffusion operators on $\mathbf{R}^n$

In $\mathbf{R}^n$, $n \geq 2$, precise characterizations of L^p uniqueness as in the one–dimensional case are not known. However, under approriate regularity and growth assumptions on the coefficients, L^p uniqueness of elliptic diffusion operators defined on all of $\mathbf{R}^n$ can be shown by using regularity results for the corresponding PDE. This will be demonstrated here for diffusion operators with locally Lipschitz continuous second order, and locally bounded first order coefficients, and $p \in (1, 2]$.

Let $n \in \mathbf{N}$. Fix a locally Lipschitz continuous function $\rho : \mathbf{R}^n \to \mathbf{R}$ such that $\rho(x) > 0$ for all x. Let $(\mathcal{L}, C_0^\infty(\mathbf{R}^n))$ be the diffusion operator on $L^p(\mathbf{R}^n ; \rho\, dx)$ given by

$$\mathcal{L}f = \sum_{i,j=1}^n a_{ij} \frac{\partial^2 f}{\partial x_i \partial x_j} + \sum_{j=1}^n \beta_j \frac{\partial f}{\partial x_j} \; ,$$

where β_j, $1 \leq j \leq n$, are locally bounded, and a_{ij}, $1 \leq i, j \leq n$, are Lipschitz continuous functions on $\mathbf{R}^n$. We suppose that $a = (a_{ij})$ is locally strictly elliptic, i.e., for every compact subset $K \subset \mathbf{R}^n$ there exists a constant $c_K > 0$ such that

$$\sum_{i,j=1}^n a_{ij}(x)\, \xi_i\, \xi_j \geq c_K\, |\xi|^2 \qquad \text{for all } \xi \in \mathbf{R}^n \text{ and } x \in K.$$

Without loss of generality, we may assume that a is symmetric, i.e., $a_{ij} = a_{ji}$ for all $1 \leq i, j \leq n$. As in the one–dimensional case, we rewrite $\mathcal{L}$ in divergence form:

$$\mathcal{L}f = \frac{1}{\rho} \sum_{i,j=1}^n \frac{\partial}{\partial x_i} \left(\rho\, a_{ij} \frac{\partial f}{\partial x_j} \right) + \sum_{j=1}^n b_j \frac{\partial f}{\partial x_j} \; ,$$

where

$$b_j = \beta_j - \frac{1}{\rho} \sum_{i=1}^n \frac{\partial}{\partial x_i} (\rho\, a_{ij}),$$

which is in $L^\infty_{\text{loc}}(\mathbf{R}^n ; dx)$. Throughout this section we assume:

(A 1$'$) There exists $\alpha \geq 0$ such that

$$\int b \cdot \nabla f\, \rho\, dx \leq \alpha \int f\, \rho\, dx \quad \text{for all positive functions } f \in C_0^\infty(\mathbf{R}^n),$$

i.e., $\text{div}\,(\rho b) \geq -\alpha\rho$ in the distributional sense.

REMARK. Note that Condition (A 1$'$) means that the $\rho\,dx$–**divergence** $\text{div}_{\rho\,dx}\, b = \frac{1}{\rho}\,\text{div}\,(\rho b)$ is **bounded from below** (in the distributional sense). In particular, (A 1$'$) holds if and only if $\rho\,dx$ is a sub–invariant measure for the operator $(\mathcal{L} - \alpha, C_0^\infty(\mathbf{R}^n))$. In this case, $(\mathcal{L} - \frac{\alpha}{p}, C_0^\infty(\mathbf{R}^n))$ is dissipative on $L^p(\mathbf{R}^n ; \rho\,dx)$ for any $p \in [1, \infty)$, cf. Lemma 1.8 in Appendix B.

We use the notations $a_r := e_r \cdot a e_r$, and $b_r := e_r \cdot b$, where $e_r(x) := x/|x|$, $x \in \mathbf{R}^n \setminus \{0\}$. The main result of this section is:

Theorem 2.3 *Suppose that*

$$(2.18) \qquad \limsup_{|x| \to \infty} \frac{a_r(x)}{|x|^2 (\log |x|)^2} < \infty , \;\; and \;\; \liminf_{|x| \to \infty} \frac{b_r(x)}{|x| \log |x|} > -\infty.$$

Then the closure of $(\mathcal{L}, C_0^\infty(\mathbf{R}^n))$ on $L^p(\mathbf{R}^n \, ; \, \rho \, dx)$ generates a C^0 semigroup for all $p \in (1, 2]$. In particular, $(\mathcal{L}, C_0^\infty(\mathbf{R}^n))$ is $L^p(\mathbf{R}^n \, ; \, \rho \, dx)$ unique for $p \in (1, 2]$.

The proof of Theorem 2.3 will be given below. It is a generalization of a proof of essential self–adjointness of generalized Schrödinger operators by N. Wielens [Wie 85]. The key ingredients are elliptic regularity results, and a localization technique which has been used in modified forms by many authors starting from [Ga 51]. The main reason why we give a detailed proof of Theorem 2.3 here is that the uniqueness results in the singular case presented in Section f) below are based on the same key ingredients, but in a refined form.

REMARKS. (o) Suppose $\mathcal{L}$ is *symmetric* w.r.t. the measure $\rho \, dx$, i.e., $b(x) = 0$ for all x. Then $L^p(\mathbf{R}^n \, ; \, \rho \, dx)$ uniqueness holds for all $p \in (1, 2]$ provided a_r satisfies the growth restriction (2.18). In particular, *generalized Schrödinger operators*, i.e., operators of type $(\Delta + \frac{\nabla \rho}{\rho} \cdot \nabla , \; C_0^\infty(\mathbf{R}^n))$ are $L^p(\mathbf{R}^n \, ; \, \rho \, dx)$ unique for all $p \in (1, 2]$, and any strictly positive function $\rho \in H_{\mathrm{loc}}^{1, \infty}(\mathbf{R}^n; dx)$, although the drift $\frac{\nabla \rho}{\rho}$ may grow very rapidly at infinity. This remarkable fact has first been proven by N. Wielens [Wie 85] in the case $p = 2$.

(i) *Optimality of growth conditions:* The function $|x| \log |x|$ in (2.18) may be replaced by any other function of type $f(|x|)$ where f is an increasing positive function on some interval (x_0, ∞), $x_0 > 0$, such that $1/f$ is not dx–integrable near ∞. In particular, we may slightly weaken the assumption (2.18) by replacing $|x| \log |x|$ for example by $|x| \log |x| \log \log |x|$. Nevertheless, the condition (2.18) is fairly sharp, as the example in Section b), 1), demonstrates.

(ii) $L^1(\mathbf{R}^n \, ; \, \rho \, dx)$ *uniqueness:* One might suspect that under condition (2.18), the closure of $(\mathcal{L}, C_0^\infty(\mathbf{R}^n))$ on $L^1(\mathbf{R}^n \, ; \, \rho \, dx)$ also generates a C^0 semi-group. This is false in general. In fact, E. B. Davies showed that for every $\gamma > 1$, and for every strictly positive function $a \in C^\infty(\mathbf{R}^n)$ such that $a(x) \sim |x|^2 (\log |x|)^\gamma$ as $|x| \to \infty$, and $\frac{\partial a}{\partial r}$ behaves approriately, the closure of the operator

$$\left(\sum_{i=1}^n \frac{\partial}{\partial x_i} \left(\rho \, a \, \frac{\partial}{\partial x_i} \cdot \right) , \; C_0^\infty(\mathbf{R}^n) \right)$$

on $L^1(\mathbf{R}^n \, ; \, dx)$ does *not* generate a C^0 semigroup, cf. [Dav 85, Note 6.6 and Thm. 2.2]. For symmetric diffusion operators, L^1 uniqueness turns out to be equivalent to conservativeness of the semigroup generated by the Friedrichs extension, cf. [Dav 85, Section 2]. In contrast to this, L^p uniqueness for $p > 1$ may hold even though the corresponding semigroup is not conservative. A detailed discussion of L^1 uniqueness will be given in the forthcoming paper [St 97].

(iii) $L^p(\mathbf{R}^n\,;\,\rho\,dx)$ *uniqueness for* $p > 2$: The restriction to $p \le 2$ in Theorem 2.3 is needed for the simple proof given below. Using more advanced regularity results, one can also consider the case $p > 2$, and obtain similar uniqueness results, cf. Theorem 2.6 below.

PROOF OF THEOREM 2.3. Fix $\alpha \ge 0$ such that the condition in (A 1') holds, and $p \in (1, 2]$. Since $(\mathcal{L} - \frac{\alpha}{p}, C_0^\infty(\mathbf{R}^n))$ is dissipative on $L^p(\mathbf{R}^n\,;\,\rho\,dx)$, we only have to show that the range of $\mathcal{L} - \frac{\alpha}{p} - 1$ is dense in $L^p(\mathbf{R}^n\,;\,\rho\,dx)$, or, equivalently, that there is no non–trivial solution of $L^* u = (\frac{\alpha}{p} + 1)\,u$. Hence fix a function $u \in L^q(\mathbf{R}^n\,;\,\rho\,dx)$, $\frac{1}{p} + \frac{1}{q} = 1$, such that

$$(2.19) \qquad \int \mathcal{L}f\,u\,\rho\,dx \;=\; (\tfrac{\alpha}{p} + 1)\int f\,u\,\rho\,dx$$

for all $f \in C_0^\infty(\mathbf{R}^n)$.

Step 1: u **is in** $H^{1,2}_{\mathrm{loc}}(\mathbf{R}^n\,;\,dx)$**.** This follows from elliptic regularity results. In fact, by Frehse [Fr 77, Lemma 2.1], (2.19) implies $u\rho \in H^{1,2}_{\mathrm{loc}}(\mathbf{R}^n\,;\,dx)$. Frehse only proves this in the case $\beta = 0$, but his arguments go through immediately for arbitrary locally bounded β (–Note that only the term A in Frehse's estimates changes. Here $\partial_k a_{ik}$ has to be replaced by $\beta_i + \partial_k a_{ik}$ which is locally bounded as well). Since ρ is locally Lipschitz continuous and strictly positive, u is in $H^{1,2}_{\mathrm{loc}}(\mathbf{R}^n\,;\,dx)$ as well. In particular, integration by parts in (2.19) yields

$$(2.20) \qquad \int (-\nabla f \cdot a\nabla u + b \cdot \nabla f\,u)\,\rho\,dx \;=\; (\tfrac{\alpha}{p} + 1)\int f\,u\,\rho\,dx$$

for all $f \in C_0^\infty(\mathbf{R}^n)$, where $a = (a_{ij})$ and $b = (b_j)$.

Step 2: $u \cdot |u|^{q-2}$ **is in** $H^{1,2}_{\mathrm{loc}}(\mathbf{R}^n\,;\,dx)$**.** Note that for $p = q = 2$, there is nothing to prove, hence we don't need the standard but non–trivial arguments that are necessary for $q \ge 2$. In general, by (i) we have $u \cdot |u|^{q-2} \in H^{1,2}_{\mathrm{loc}}(\mathbf{R}^n\,;\,dx)$, and

$$\nabla (u \cdot |u|^{q-2}) \;=\; (q-1) \cdot |u|^{q-2} \cdot \nabla u$$

provided u is in $L^\infty_{\mathrm{loc}}(\mathbf{R}^n\,;\,dx)$. Since, by (2.20), u is a weak solution of an elliptic PDE, the local boundedness can be shown by the Moser technique, cf. e.g. [GiTru 83, Thm. 8.17].

Step 3: u **vanishes.** Since a, b and ρ are locally bounded, and u is in $H^{1,2}_{\mathrm{loc}}(\mathbf{R}^n\,;\,dx)$, the equation (2.20) extends to all compactly supported functions $f \in H^{1,2}(\mathbf{R}^n\,;\,dx)$. Fix a positive function $\xi \in H^{1,\infty}(\mathbf{R}^n\,;\,dx)$ with compact support. Choosing $f = \xi \cdot u \cdot |u|^{q-2}$ in (2.20), we have

$$(\tfrac{\alpha}{p} + 1)\int \xi\,|u|^q\,\rho\,dx$$

$$= -(q-1)\int \xi\,\nabla u \cdot a\nabla u\,|u|^{q-2}\,\rho\,dx + \tfrac{q-1}{q}\int b \cdot \nabla(\xi|u|^q)\,\rho\,dx$$

$$- \int \nabla\xi \cdot a\nabla u\,u\,|u|^{q-2}\,\rho\,dx + \tfrac{1}{q}\int b \cdot \nabla\xi\,|u|^q\,\rho\,dx.$$

Here we used the equation

$$u \nabla (\xi u |u|^{q-2}) \;=\; \tfrac{q-1}{q} \nabla (\xi |u|^q) + \tfrac{1}{q} |u|^q \nabla \xi .$$

Since $\tfrac{q-1}{q} = \tfrac{1}{p}$, (A 1$'$) implies

$$\tfrac{\alpha}{p} \int \xi |u|^q \, \rho \, dx \;\geq\; \tfrac{q-1}{q} \int b \cdot \nabla (\xi |u|^q) \, \rho \, dx .$$

Choosing $\xi = \phi^2$, where ϕ is a compactly supported function in $H^{1,\infty}(\mathbf{R}^n \,;\, dx)$ such that $0 \leq \phi \leq 1$, we obtain

$$\int \phi^2 |u|^q \rho \, dx + (q-1) \cdot \int \phi^2 \nabla u \cdot a \nabla u \, |u|^{q-2} \, \rho \, dx$$

$$\leq \;\; -2 \int \phi \nabla \phi \cdot a \nabla u \, u |u|^{q-2} \, \rho \, dx + \tfrac{1}{q} \int \phi \, b \cdot \nabla \phi \, |u|^q \, \rho \, dx$$

$$\leq \;\; 2 \cdot \|\nabla \phi \cdot a \nabla \phi\|^{1/2}_{L^\infty(\mathbf{R}^n;dx)} \cdot \|u\|^{q/2}_{L^q(\mathbf{R}^n;\,\rho\,dx)} \left(\int \phi^2 \, \nabla u \cdot a \nabla u \, |u|^{q-2} \, \rho \, dx \right)^{1/2}$$

$$+ \; \tfrac{1}{q} \cdot \|(b \cdot \nabla \phi)^+\|_{L^\infty(\mathbf{R}^n;dx)} \cdot \|u\|^{q}_{L^q(\mathbf{R}^n;\,\rho\,dx)} .$$

Hence

$$\int \phi^2 |u|^q \, \rho \, dx \;\leq\; C \cdot \left(\|\nabla \phi \cdot a \nabla \phi\|_{L^\infty(\mathbf{R}^n;dx)} + \|(b \cdot \nabla \phi)^+\|_{L^\infty(\mathbf{R}^n;dx)} \right) ,$$

where C is a finite constant that does not depend on ϕ. Now fix $k \in \mathbf{N}$. Let

$$A_k \;:=\; \{x \in \mathbf{R}^n \,;\, \log \log^+ \log^+ |x| \leq k\},$$

and consider the function $\phi_k(x) := \min \left((k - \log \log^+ \log^+ |x|)^+, 1 \right)$, i.e., $\phi_k(x) = 1$ if $x \in A_{k-1}$, $\phi_k(x) = 0$ if $x \in \mathbf{R}^n \setminus A_k$, and $\phi_k(x) = k - \log \log \log |x|$ else. ϕ_k is a compactly supported function in $H^{1,\infty}(\mathbf{R}^n \,;\, dx)$, and

$$\nabla \phi_k \;=\; \frac{-e_r}{|x| \, \log |x| \, \log \log |x|} \cdot \chi_{A_k \setminus A_{k-1}} \qquad dx\text{--a.e.}$$

Hence

$$\int_{A_{k-1}} |u|^q \, \rho \, dx \;\leq\; \int \phi_k^2 \, |u|^q \, \rho \, dx$$

$$\leq \; C \cdot \left(\sup_{x \in A_k \setminus A_{k-1}} \frac{a_r(x)}{(|x| \, \log |x| \, \log \log |x|)^2} + \sup_{x \in A_k \setminus A_{k-1}} \frac{b_r^-(x)}{|x| \, \log |x| \, \log \log |x|} \right)$$

By Assumption (2.18), we finally obtain

$$\int_{\mathbf{R}^n} |u|^q \, \rho \, dx \;=\; \lim_{k \to \infty} \int_{A_{k-1}} |u|^q \, \rho \, dx \;\leq\; 0 ,$$

whence $u = 0$ dx-a.e. ∎

d) Examples and counterexamples II: Differences between the singular and the regular case

We have already seen that non–uniqueness of diffusion operators on $\mathbf{R}^n$ may originate from boundaries or too strong growth of the coefficients at infinity. Roughly speaking, the strong coefficient growth at infinity produces a "boundary at infinity". In this sense, all examples of non–uniqueness given so far arise from generalized boundaries. Is this the only possible source of non–uniqueness ? The theorems above show that in the regular case the answer is "Yes". In singular cases, however, the situation is different. In this section we demonstrate:

- Zeros of the density ρ may cause non–uniqueness of diffusion operators on $L^p(\mathbf{R}^n\,;\,\rho\,dx)$ — even in $\mathbf{R}^1$, and even if ρ vanishes only at one point.

- Local degeneracy of the diffusion matrix a of an operator may cause non–uniqueness.

Conversely, we show that certain rotationally invariant generalized Schrödinger operators on $\mathbf{R}^n$, $n \geq 2$, are always essentially self–adjoint, although they have a singularity at 0. For $n = 1$, the corresponding assertion is false.

1) Non–uniqueness of singular generalized Schrödinger operators on $\mathbf{R}^1$

Fix $p \in (1, \infty)$, and let $\rho : \mathbf{R} \to [0, \infty)$ be an absolutely continuous function such that $\rho(0) = 0$, $\rho(x) > 0$ for all $x \neq 0$, and ρ'/ρ is in $L^p(\mathbf{R}\,;\,\rho\,dx)$. Then the generalized Schrödinger operator

$$\mathcal{L} \;=\; \frac{1}{\rho}\,\frac{d}{dx}\left(\rho\,\frac{d}{dx}\,\bullet\right) \;=\; \frac{d^2}{dx^2} \;+\; \frac{\rho'}{\rho}\,\frac{d}{dx}$$

defined on $C_0^\infty(\mathbf{R})$ is a dissipative operator on $L^p(\mathbf{R}\,;\,\rho\,dx)$. Let $q \in (1, \infty]$ such that $p^{-1} + q^{-1} = 1$.

Lemma *Suppose that* $\int_{\bullet}^{1}(1/\rho)\,dy$ *is not in* $L^q(0, 1\,;\,\rho\,dx)$, *and* $\int_{-1}^{\bullet}(1/\rho)\,dy$ *is not in* $L^q(-1, 0\,;\,\rho\,dx)$. *Then the generalized Schrödinger operator* $(\mathcal{L}, C_0^\infty(\mathbf{R}))$ *is not* $L^p(\mathbf{R}\,;\,\rho\,dx)$ *unique.*

EXAMPLE. Let $\rho(x) = |x|^\gamma$, $\gamma > p - 1$. Then ρ'/ρ is in $L^p(\mathbf{R}\,;\,\rho\,dx)$, i.e., the corresponding generalized Schrödinger operator $\frac{d^2}{dx^2} + \frac{\gamma}{x}\frac{d}{dx}$ is a dissipative operator on $L^p(\mathbf{R}\,;\,\rho\,dx)$. However, if $p > (\gamma+1)/2$ (i.e., $\gamma < 2p-1$), then the conditions in the lemma are satisfied, whence the operator is *not* $L^p(\mathbf{R}\,;\,\rho\,dx)$ *unique*. In particular, for $1 < \gamma < 3$, the operator $\left(\frac{d^2}{dx^2} + \frac{\gamma}{x}\frac{d}{dx}, C_0^\infty(\mathbf{R})\right)$ is a symmetric but *not essentially self–adjoint* operator on $L^2(\mathbf{R}\,;\,|x|^\gamma dx)$.

Examples for non–uniqueness in *higher dimensions* can be constructed easily from the example above. For example, the generalized Schrödinger operator $\left(\Delta + \frac{x}{x_1}\frac{\partial}{\partial x_1}, C_0^\infty(\mathbf{R}^n)\right)$, $n \geq 2$, is $L^p(\mathbf{R}^n; |x_1|^\gamma dx)$ unique if and only if the one–dimensional projection $\left(\frac{d^2}{dx^2} + \frac{\gamma}{x}\frac{d}{dx}, C_0^\infty(\mathbf{R})\right)$ is $L^p(\mathbf{R}; |x|^\gamma dx)$ unique.

PROOF OF THE LEMMA. Note that by the non–integrability assumptions on $\int_{\bullet}^1 (1/\rho)\,dy$ respectively $\int_{-1}^{\bullet}(1/\rho)\,dy$ at 0, the operators $(\mathcal{L}, C_0^\infty(0,\infty))$ and $(\mathcal{L}, C_0^\infty(-\infty,0))$ are not $L^p(0,\infty; \rho\,dx)$ unique respectively $L^p(-\infty,0; \rho\,dx)$ unique, cf. Corollary 2.2 in Section a). Similarly to the proof of Theorem 2.2, we can show that there exists a strictly positive function $u_1 \in C^1(0,\infty)$ such that $\int_0^\infty |u_1|^q \rho\,dx < \infty$, u_1' is absolutely continuous, and $(\rho u_1')' = \rho u_1$, as well as a strictly negative function $u_2 \in C^1(-\infty,0)$ such that $\int_{-\infty}^0 |u_2|^q \rho\,dx < \infty$, u_2' is absolutely continuous, and $(\rho u_2')' = \rho u_2$. In particular, the limits of $\rho(x)\,u_1'(x)$ as $x \downarrow 0$ and of $\rho(x)\,u_2'(x)$ as $x \uparrow 0$ exist, since the derivatives of the functions are dx–integrable at 0. Let $u \in L^q(\mathbf{R}; \rho\,dx)$ be a non–trivial function such that $u = \lambda_1 u_1$ on $(0,\infty)$ and $u = \lambda_2 u_2$ on $(-\infty,0)$ for some factors $\lambda_1, \lambda_2 \in \mathbf{R}$, and

$$(2.21) \qquad \lim_{x\uparrow 0} \rho(x)\,u'(x) \;=\; \lim_{x\downarrow 0} \rho(x)\,u'(x).$$

I claim that $L^* u = u$. Fix $f \in C_0^\infty(\mathbf{R})$. Then

$$
\begin{aligned}
\int_\varepsilon^\infty u\,\mathcal{L}f\,\rho\,dx &= \int_\varepsilon^\infty u\,(\rho f')'\,dx \\
&= -\int_\varepsilon^\infty u'\,\rho f'\,dx - (u\rho f')(\varepsilon) \\
(2.22) \qquad &= \int_\varepsilon^\infty (\rho u')'\,f\,dx + (u'\rho f)(\varepsilon) - (u\rho f')(\varepsilon) \\
&= \int_\varepsilon^\infty u\,f\,\rho\,dx + (u'\rho f)(\varepsilon) - (u\rho f')(\varepsilon)
\end{aligned}
$$

for any $\varepsilon > 0$. Similarly,

$$(2.23) \qquad \int_{-\infty}^{-\varepsilon} u\,\mathcal{L}f\,\rho\,dx = \int_{-\infty}^{-\varepsilon} u\,f\,\rho\,dx - (u'\rho f)(-\varepsilon) + (u\rho f')(-\varepsilon)$$

for any $\varepsilon > 0$. Since the limits as $x \downarrow 0$ resp. $x \uparrow 0$ of $\rho(x)\,u'(x)$ are finite, $u(x)$ grows at most of order $\int_x^1 (1/\rho)\,dy$ as $x \downarrow 0$ resp. of order $\int_{-1}^x (1/\rho)\,dy$ as $x \uparrow 0$. Hence there exist finite constants $K_1, K_2 > 0$ such that

$$\rho|u| \;\leq\; K_1\rho\left(1 + \int_{\bullet}^1 (1/\rho)\,dy\right) \;=\; K_1\rho - K_1\cdot\left[\left(\log\int_{\bullet}^1 (1/\rho)\,dy\right)'\right]^{-1}$$

on $(0,1)$, and

$$\rho|u| \;\leq\; K_2\rho\left(1 + \int_{-1}^{\bullet} (1/\rho)\,dy\right) \;=\; K_2\rho + K_2\cdot\left[\left(\log\int_{-1}^{\bullet} (1/\rho)\,dy\right)'\right]^{-1}$$

on $(-1, 0)$. If $1/\rho$ is dx-integrable at $0+$ respectively $0-$, then $\rho(0) = 0$ immediately implies

$$\lim_{x \downarrow 0} \left(\rho(x) \, u(x) \right) = 0 \qquad \text{resp.} \qquad \lim_{x \uparrow 0} \left(\rho(x) \, u(x) \right) = 0.$$

Otherwise, $\log \int_{\bullet}^{1} (1/\rho) \, dy$ resp. $\log \int_{-1}^{\bullet} (1/\rho) \, dy$ is unbounded at $0+$ resp. $0-$, whence

$$\lim_{n \to \infty} \left(\log \int_{\bullet}^{1} (1/\rho) \, dy \right)' (\varepsilon_n) = -\infty, \qquad \text{resp.}$$

$$\lim_{n \to \infty} \left(\log \int_{-1}^{\bullet} (1/\rho) \, dy \right)' (\varepsilon_n) = \infty$$

for some sequence $\varepsilon_n \downarrow 0$ resp. $\varepsilon_n \uparrow 0$. Thus we obtain

$$\lim_{n \to \infty} \left(\rho(\varepsilon_n) \, u(\varepsilon_n) \right) = 0$$

in this case. In any case, (2.22), (2.23), and (2.21) imply

$$\int_{-\infty}^{\infty} u \, \mathcal{L}f \, \rho \, dx = \int_{-\infty}^{\infty} u \, f \, \rho \, dx \qquad \text{for all } f \in C_0^{\infty}(\mathbf{R}).$$

Hence $L^* u = u$, i.e., $(\mathcal{L}, C_0^{\infty}(\mathbf{R}))$ is not $L^p(\mathbf{R}; \rho \, dx)$ unique. ∎

2) Essential self–adjointness of rotationally invariant generalized Schrödinger operators on $\mathbf{R}^n$, $n \geq 2$, with a singularity at 0

Fix $n \geq 2$. We now give an example which is quite contrary to the examples in 1). We consider a generalized Schrödinger operator

$$\mathcal{L} = \frac{1}{\rho} \operatorname{div} (\rho \nabla \cdot) = \Delta + \frac{\nabla \rho}{\rho} \cdot \nabla$$

with domain $C_0^{\infty}(\mathbf{R}^n)$ on $L^2(\mathbf{R}^n; \rho \, dx)$. Here $\rho(x) = |x|^{\gamma}$ for some $\gamma > 2 - n$, or $\rho(x) = \exp\left(\frac{\gamma}{k} |x|^k\right)$ for some $\gamma \in \mathbf{R}$ and $k > 0$, or $\rho(x) = \exp\left(-\frac{\gamma}{k} |x|^{-k}\right)$ for some $\gamma \geq 0$ and $k > 0$, whence

(2.24) $$\frac{\nabla \rho}{\rho}(x) = \gamma \cdot |x|^{-1} \cdot e_r(x), \quad \text{resp.} \quad \gamma \cdot |x|^{-1 \pm k} \cdot e_r(x),$$

where $e_r(x) = x/|x|$. Note that the conditions $\gamma > 2 - n$ in the first case resp. $\gamma \geq 0$ in the third case are necessary to guarantee that ρ is locally dx-integrable and $\nabla \rho / \rho$ is in $L^2(\mathbf{R}^n; \rho \, dx)$, i.e., both the domain and the range of $(\mathcal{L}, C_0^{\infty}(\mathbf{R}^n))$ are in $L^2(\mathbf{R}^n; \rho \, dx)$. In the case $\rho(x) = \exp\left(\frac{\gamma}{k} |x|^k\right)$ for $\gamma \in \mathbf{R}$ and $k > 0$, we assume for simplicity $n \geq 3$.

Lemma *Under the assumptions above, the operator $(\mathcal{L}, C_0^\infty(\mathbf{R}^n))$ is essentially self–adjoint on $L^2(\mathbf{R}^n\,;\,\rho\,dx)$.*

Hence for rotationally invariant generalized Schrödinger operators on $\mathbf{R}^n$, $n \geq 2$ (respectively ≥ 3), with polynomially growing drift, essential self–adjointness always holds, in spite of the singularity at 0. Recall that in contrast to this, we have shown in the example in 1), that for $n = 1$ essential self–adjointness does not hold if $\rho(x) = |x|^\gamma$, $1 < \gamma < 3$.

PROOF OF THE LEMMA. Suppose that u is a solution of $L^* u = u$, where L^* is the adjoint of $(\mathcal{L}, C_0^\infty(\mathbf{R}^n))$ on $L^2(\mathbf{R}^n\,;\,\rho\,dx)$. We have to show that u vanishes.

Step 1: Reduction to the one–dimensional case. Let ψ be the smooth function on $(0,\infty)$ given by $\psi(r) = r^{\gamma+n-1}$, resp. $\psi(r) = \exp(\frac{\gamma}{k} r^k) \cdot r^{n-1}$, resp. $\psi(r) = \exp(-\frac{\gamma}{k} r^{-k}) \cdot r^{n-1}$. Let dy denote the surface measure on the unit sphere S^{n-1}. Note that the measure $\rho\,dx$ transformed in polar coordinates is given by $\psi(r)\,dr\,dy$. Now fix an orthonormal basis e_i, $i \in \mathbf{N} \cup \{0\}$, of $L^2(S^{n-1}; dy)$ consisting of smooth eigenfunctions of the Laplace–Beltrami operator $\Delta_{S^{n-1}}$. We may assume that e_0 is constant. Consider the functions $u_i \in L^2(0,\infty\,;\,\psi\,dr)$, $i \geq 0$, given by

$$u_i(r) = \int_{S^{n-1}} u(r \cdot y)\, e_i(y)\, dy.$$

Obviously, it is enough to show $u_i = 0$ dr–a.e. on $(0,\infty)$ for all $i \geq 0$. Now fix $i \geq 0$, and a smooth function g on $(0,\infty)$. If $i > 0$ then we assume $g \in C_0^\infty(0,\infty)$, if $i = 0$ we only assume that g is constant in a neighbourhood of 0, and $g(x)$ vanishes for large x. In both cases, there exists a function $f \in C_0^\infty(\mathbf{R}^n)$ such that $f(r \cdot y) = g(r) \cdot e_i(y)$ for all $r > 0$ and $y \in S^{n-1}$. We have

$$(\mathcal{L}f)(r \cdot y) = \frac{1}{\psi(r)} \frac{\partial}{\partial r}\left(\psi(r)\frac{\partial}{\partial r}(g(r) \cdot e_i(y))\right) + \frac{1}{r^2} g(r)\, \Delta_{S^{n-1}} e_i(y)$$

$$= \frac{1}{\psi(r)}(\psi \cdot g')'(r) \cdot e_i(y) + \frac{\lambda_i}{r^2} g(r)\, e_i(y),$$

where λ_i denotes the eigenvalue of $\Delta_{S^{n-1}}$ corresponding to e_i. Since f is in $C_0^\infty(\mathbf{R}^n)$, the equation $L^* u = u$ implies

$$\int_0^\infty u_i\,(\psi g')'\, dr = \int_0^\infty u_i \frac{1}{\psi}(\psi g')'\,\psi\,dr$$

$$= \int_{\mathbf{R}^n} u\,(\mathcal{L}f - \lambda_i |x|^{-2} f)\,\rho\,dx$$

(2.25)
$$= \int_{\mathbf{R}^n} u\,f\,(1 - \lambda_i |x|^{-2})\,\rho\,dx$$

$$= \int_0^\infty u_i\, g\,(1 - \lambda_i r^{-2})\,\psi\,dr.$$

Note that all the integrals exist, because $\lambda_0 = 0$, whereas for $i > 0$, g is in $C_0^\infty(0,\infty)$, and f is in $C_0^\infty(\mathbf{R}^n \setminus \{0\})$.

Step 2: Regularity and growth of u_i. Since ψ is strictly positive and absolutely continuous, and $\lambda_i \leq 0$, we can now use similar arguments as in the regular one–dimensional case. The equation (2.25) holds for all $g \in C_0^\infty(0, \infty)$. By a slight modification of the proof of Theorem 2.7 resp. Corollary 2.4 in Appendix C, we can conclude that u_i has a C^1 version (which we again denote by u_i), u_i' is absolutely continuous, and

$$(2.26) \qquad (\psi\, u_i')' = (1 - \lambda_i\, r^{-2})\, \psi\, u_i \qquad dr\text{–a.e. on } (0, \infty).$$

Now suppose u_i does not vanish, and fix $z_1 \in (0, \infty)$ such that $u_i(z_1) \neq 0$. We may assume $u_i(z_1) > 0$, otherwise we consider $-u_i$ instead of u_i. Since $\lambda_i \leq 0$, a similar argument as we have used in the proof of Theorem 2.1 shows that u_i is increasing on $[z_1, \infty)$ or decreasing on $(0, z_1]$, depending on whether $u_i'(z_1) \geq 0$ or $u_i'(z_1) \leq 0$. In the first case,

$$\psi\, u_i' = (\psi\, u_i')(z_1) + \int_{z_1}^{\bullet} (1 - \lambda_i\, r^{-2})\, \psi\, u_i\, dr \geq u_i(z_1) \cdot \int_{z_1}^{\bullet} \psi\, dr$$

on (z_1, ∞). This leads to a contradiction to $u_i \in L^q(0, \infty; \psi\, dr)$, because it implies that $u_i'(r)$ grows at least of order $1/\psi(r)$ as $r \to \infty$, whereas

$$\int_c^\infty \left(\int_c^s (1/\psi(r))\, dr \right)^2 \psi(s)\, ds = \infty$$

for any $c \in (0, \infty)$ (and any continuous strictly positive function ψ on $(0, \infty)$), cf. (2.17).

Now suppose that u_i is decreasing on $(0, z_1]$. We argue differently depending on whether $i = 0$ or $i \geq 1$.

Case (i): $i = 0$. Since $\lambda_0 = 0$, the equation (2.26) implies

$$\psi\, u_0' = (\psi\, u_0')(z_1) - \int_{\bullet}^{z_1} \psi\, u_0\, dr \leq -u_0(z_1) \cdot \int_{\bullet}^{z_1} \psi\, dr.$$

Hence

$$(2.27) \qquad\qquad \limsup_{r\downarrow 0} (\psi\, u_0')(r) < 0.$$

Now fix $g \in C^\infty(0, \infty)$ such that $g(r) = 1$ for small r, and g vanishes for large r. Since $i = 0$, the equation (2.25) holds for g, although g does not vanish at 0 (— Here we use that we consider the operator $\mathcal{L}$ with domain $C_0^\infty(\mathbf{R}^n)$, and not only with domain $C_0^\infty(\mathbf{R}^n \setminus \{0\})$. In the latter case, essential self–adjointness does not necessarily hold, e.g. $(\Delta, C_0^\infty(\mathbf{R}^n \setminus \{0\}))$ is essentially self-adjoint on $L^2(\mathbf{R}^n; dx)$ if and only if $n \geq 4$). By (2.25) and (2.26) we obtain

$$\int_0^\infty u_0\, g\, \psi\, dr = \int_0^\infty u_0\, (\psi g')'\, dr = \lim_{\varepsilon\downarrow 0} \int_\varepsilon^\infty u_0\, (\psi g')'\, dr$$

$$= \lim_{\varepsilon\downarrow 0} \left(\int_\varepsilon^\infty (\psi u_0')'\, g\, dr - (u_0\psi g')(\varepsilon) + (u_0'\psi g)(\varepsilon) \right)$$

$$(2.28) \qquad\qquad = \lim_{\varepsilon\downarrow 0} \left(\int_\varepsilon^\infty \psi\, u_0\, g\, dr + (u_0'\psi)(\varepsilon) \right),$$

because $g'(r)$ vanishes for small r. (2.28) is clearly a contradiction to (2.27). Hence u_0 vanishes.

Case (ii): $i \geq 1$. It is well–known that the largest non–zero eigenvalue of $\Delta_{S^{n-1}}$ is $-(n-1)$. Thus $\lambda_i \leq -(n-1)$. Since u_i is decreasing on $(0, z_1)$, and thus positive, the equation (2.26) implies

$$(2.29) \qquad (\psi u_i')' \geq \frac{n-1}{r^2} \, \psi \, u_i \qquad dr\text{–a.e. on } (0, z_1).$$

We first consider the case $\rho(x) = |x|^\gamma$ for some $\gamma > 2 - n$. Let v be a solution of the ordinary differential equation

$$(2.30) \qquad (\psi v')' = \frac{n-1}{r^2} \, \psi \, v \qquad \text{on } (0, \infty)$$

such that $0 < v(z_1) < u_i(z_1)$ and $v'(z_1) = 0$. Since $u_i'(z_1) \leq 0$, an easy comparison argument shows that $0 \leq v(r) < u_i(r)$ for all $r \in (0, z_1]$. Hence to obtain a contradiction to $u_i \in L^2(0, \infty \, ; \, \psi \, dr)$, it is enough to show that v is not in $L^2(0, z_1 \, ; \, \psi \, dr)$. Since $\psi(r) = r^{\gamma+n-1}$, (2.30) can be written as

$$v'' = -\frac{\psi'}{\psi} v' + \frac{n-1}{r^2} v = \frac{1-\gamma-n}{r} v' + \frac{n-1}{r^2} v.$$

A fundamental system is $r^{\alpha_+}, r^{\alpha_-}$, where α_+ and α_- are the roots of the equation $\alpha(\alpha - 1) = (1 - \gamma - n)\alpha + n - 1$, i.e.,

$$\alpha_\pm = 1 - \frac{\gamma+n}{2} \pm \sqrt{n - 1 + \left(1 - \frac{\gamma+n}{2}\right)^2}.$$

Since $v(z_1) > 0$ and $v'(z_1) = 0$, v is not a multiple of r^{α_+}, whence v increases of order r^{α_-} as $r \downarrow 0$. We have

$$\int_0^{z_1} r^{2 \cdot \alpha_-} \, \psi(r) \, dr = \int_0^{z_1} r^{2 \cdot \alpha_- + \gamma + n - 1} \, dr = \infty, \qquad \text{since}$$

$$2 \cdot \alpha_- + \gamma + n - 1 = 1 - 2\sqrt{n - 1 + (1 - (\gamma + n)/2)^2} \leq 1 - 2\sqrt{n-1} \leq -1.$$

Hence v is not in $L^2(0, z_1 \, ; \, \psi \, dr)$, which is a contradiction to $u_i \in L^2(0, \infty \, ; \, \psi \, dr)$. In the case $\rho(x) = \exp\left(\frac{\gamma}{k} |x|^k\right)$, $\gamma \in \mathbb{R}$, $k > 0$, we can argue similarly. Fix $\varepsilon > 0$. We have

$$\frac{\psi'(r)}{\psi(r)} = (\log \psi)'(r) = \gamma r^{k-1} + (n - 1) \cdot r^{-1} \geq (1 - \varepsilon) \cdot (n - 1) \cdot r^{-1}$$

on some interval $(0, \delta)$, $0 < \delta \leq z_1$. Since $-u_i' \geq 0$ on $(0, z_1)$, (2.29) implies

$$u_i'' \geq -\frac{\psi'}{\psi} u_i' + \frac{n-1}{r^2} u_i \geq -(1 - \varepsilon) \frac{n-1}{r} u_i' + \frac{n-1}{r^2} u_i \quad \text{on } (0, \delta).$$

Now, as above, comparison with the ODE

$$v'' = (1 - \varepsilon) \frac{1-n}{r} v' + \frac{n-1}{r^2} v$$

shows that u_i increases at least of order $r^{\alpha-}$ as $r \downarrow 0$, where

$$\alpha_- = -\frac{(1-\varepsilon)(n-1)-1}{2} - \sqrt{n-1+\left(\frac{(1-\varepsilon)(n-1)-1}{2}\right)^2}.$$

Since we have assumed $n \geq 3$ in the case under consideration, we obtain

$$2\alpha_- + n - 1 \leq \varepsilon(n-1) + 1 - 2\sqrt{n-1} \leq -1,$$

provided ε is chosen small enough. Hence $\int_0^\infty u_i^2 \psi \, dr = \infty$, which is a contradiction.

Finally, in the case $\rho(x) = \exp\left(-\frac{\gamma}{k}|x|^{-k}\right)$, $\gamma > 0$ and $k > 0$,

$$\frac{\psi'(r)}{\psi(r)} = \frac{\gamma}{r^{1+k}} + \frac{n-1}{r} \geq \frac{\gamma}{r^{1+k}}.$$

Hence, by (2.29),

$$u_i'' \geq -\frac{\psi'}{\psi} u_i' + \frac{n-1}{r^2} u_i \geq -\frac{\gamma}{r^{1+k}} u_i' \qquad \text{on } (0, z_1).$$

Comparison with the ODE $v'' = -\gamma r^{-1-k} v'$ shows that $u_i(r)$ grows at least of order $\int_\bullet^{z_1} \exp\left(\frac{\gamma}{k} t^{-k}\right) dt$ as $r \downarrow 0$. This is again a contradiction to $u_i \in L^2(0, \infty; \psi \, dr)$. Hence u_i vanishes in any case. ∎

3) Degeneracy of second order coefficients

As an extreme example of a degenerate second order diffusion operator we consider the operator $\mathcal{L} = \mathrm{sgn}(x) \cdot \frac{d}{dx}$ on $C_0^\infty(\mathbf{R})$. Here the second order part vanishes completely. Since sgn is an increasing function, Lebesgue measure is a sub–invariant measure for $\mathcal{L}$, hence $(\mathcal{L}, C_0^\infty(\mathbf{R}))$ is dissipative on $L^p(\mathbf{R}; dx)$ for all $p \in [1, \infty)$. However, this operator is *not* $L^p(\mathbf{R}; dx)$ *unique* for any p. In fact, the function $u(x) = \mathrm{sgn}(x) \exp(-|x|)$ is in $L^q(\mathbf{R}; dx)$ for all $q \in (1, \infty]$, and solves $\mathcal{L}^* u = u$, since

$$\int_{-\infty}^\infty \mathrm{sgn}(x) f'(x) \, \mathrm{sgn}(x) \, e^{-|x|} \, dx = \int_0^\infty f'(x) e^{-x} \, dx + \int_{-\infty}^0 f'(x) e^x \, dx$$

$$= \int_0^\infty f(x) e^{-x} \, dx - f(0) - \int_{-\infty}^0 f(x) e^x \, dx + f(0)$$

$$= \int_{-\infty}^\infty f(x) \, \mathrm{sgn}(x) \, e^{-|x|} \, dx$$

for all $f \in C_0^\infty(\mathbf{R})$. The example shows that in contrast to locally strictly elliptic diffusion operators, one cannot in general expect L^p uniqueness of degenerate diffusion operators if the first order coefficients are not continuous.

REMARK. Note that there is a parallel between the examples 1) and 3) given above. In both cases, the first order coefficients are singular in comparison to the second order coefficients of the operator. In the first example this is due to a singularity of the first order coefficients while the second order coefficients are constant, whereas in 3) it is due to the vanishing of the second order coefficients while the first order coefficients remain bounded.

e)   The singular one–dimensional case

As the examples in Section d) show, we cannot expect uniqueness of singular diffusion operators in general – even if the state space is $\mathbf{R}^1$ and the coefficients do not grow too strongly as $|x| \to \infty$. We will now prove a sharp uniqueness result similar to Corollary 2.2 in Section a) for the singular case. For simplicity, we will restrict ourselves to the symmetric case, i.e., with the notations from Section a), $b \equiv 0$. Fix $p \in (1, \infty)$, and an interval (x_0, y_0), $-\infty \leq x_0 \leq y_0 \leq \infty$. We consider a divergence form operator of type

$$\mathcal{L} = \frac{1}{\rho} \frac{d}{dx} \left(\alpha \frac{d}{dx} \cdot \right)$$

with domain $C_0^\infty(x_0, y_0)$ on $L^P(x_0, y_0 \,;\, \rho\, dx)$. In non–divergence form,

$$\mathcal{L} = a \frac{d^2}{dx^2} + \beta \frac{d}{dx} \,, \qquad \text{where } a = \frac{\alpha}{\rho} \text{ and } \beta = \frac{\alpha'}{\rho}.$$

Here we assume that ρ is a continuous function on (x_0, y_0), α is absolutely continuous, $\rho > 0$ and $\alpha > 0$ dx–a.e., and α/ρ and α'/ρ are in $L^p_{\text{loc}}(x_0, y_0 \,;\, \rho\, dx)$. Hence ρ and α **may have zeros**, but only on a measure zero set. In particular, the dx–classes and the $\rho\, dx$–classes of functions are the same. The integrability assumptions on α/ρ and α'/ρ are needed to ensure that $\mathcal{L}f$ is in $L^P(x_0, y_0 \,;\, \rho\, dx)$ for any $f \in C_0^\infty(x_0, y_0)$.

In the sequel we call points $s \in (x_0, y_0)$ such that $\alpha(s) = 0$ "singularities". Note that if $\mathcal{L}$ is given in the form $\frac{1}{\rho} \frac{d}{dx}(\rho a \frac{d}{dx} \cdot)$ as in Section a), then $\alpha = \rho \cdot a$, i.e., both zeros of ρ and a are singularities in our sense.

REMARKS.   (i) At first glance, our definition of "singularity" may seem strange. In particular, one may wonder, why we don' t consider "singularities" where α becomes infinite. Note, however, that already the condition that the test–functions are in $L^P(x_0, y_0 \,;\, \rho\, dx)$ forces ρ to be locally dx–integrable. The unavoidable assumption $(\alpha'/\rho) \in L^p_{\text{loc}}(x_0, y_0 \,;\, \rho\, dx)$ then implies that α'/ρ is $\rho\, dx$–integrable, i.e., α' is dx–integrable. Hence α is *automatically locally bounded* whenever the $L^P(x_0, y_0 \,;\, \rho\, dx)$ uniqueness problem on $C_0^\infty(x_0, y_0)$, which we look at here, makes sense. On the other hand, *zeros* of α and ρ can produce very *singular drift and diffusion coefficients*, if the operator is written in non–divergence form, cf. the examples in the last section, and the example after Definition 2.1 below.

(ii) The assumption that the first order coefficient b of the operator $\mathcal{L}$ in divergence form vanishes, is, in the one–dimensional case, not as restrictive as it might seem. In fact, every operator of type

$$\frac{1}{\rho}\frac{d}{dx}\left(\rho a \frac{d}{dx}\cdot\right) + b\frac{d}{dx}$$

on $C_0^\infty(x_0, y_0)$, where b/a is locally dx–integrable, can be written in the form

$$\frac{1}{\tilde{\rho}}\frac{d}{dx}\left(\bar{\alpha}\frac{d}{dx}\cdot\right),$$

where $\tilde{\rho} := \exp(\int_{z_0}^{\bullet}(b/a)\,dx)\cdot\rho$, $z_0 \in (x_0, y_0)$ fixed, and $\bar{\alpha} := \tilde{\rho}a$. Hence if we consider the operator on $L^p(x_0, y_0\,;\,\tilde{\rho}\,dx)$ instead of $L^p(x_0, y_0\,;\,\rho\,dx)$, the results below apply.

Since $\rho\,dx$ is an invariant (even reversible) measure for $(\mathcal{L}, C_0^\infty(x_0, y_0))$, the operator is dissipative on $L^p(x_0, y_0\,;\,\rho\,dx)$.

Let $S := \{s \in (x_0, y_0); \alpha(s) = 0\}$ denote the singularity set. For simplicity, we make the following additional assumption:

(S 1)   Any point $s \in S$ is either an isolated point in S, or both $s-$ and $s+$ are accumulation points of S, i.e., each of the intervals $(s - \varepsilon, s)$ and $(s, s+\varepsilon)$, $\varepsilon > 0$, contains infinitely many zeros of α.

Let $q \in (1, \infty)$ such that $p^{-1} + q^{-1} = 1$.

Definition 2.1 *We say that the $L^p(\rho\,dx)$ limit point case holds at x_0 (resp. y_0) iff there does not exist $c \in (x_0, y_0)$ such that the function $\int_{\bullet}^{c}(1/\alpha)\,dx$ is in*

$L^q(x_0, c;\, \rho\,dx)$ *(resp.* $\int_{c}^{\bullet}(1/\alpha)\,dx$ *is in $L^q(c, y_0;\, \rho\,dx)$).*
Similarly, if s is an isolated point in S, then we say that the $L^p(\rho\,dx)$ limit point case holds at $s-$ (resp. $s+$) iff there does not exist $c \in (x_0, s)$ (resp. $c \in (s, y_0)$) such that the function $\int_{c}^{\bullet}(1/\alpha)\,dx$ is in $L^q(c, s;\, \rho\,dx)$ (resp.
$\int_{\bullet}^{c}(1/\alpha)\,dx$ *is in $L^q(s, c;\, \rho\,dx)$).*

Corollary 2.2 shows that in the regular case, i.e., when $S = \emptyset$, the operator $(\mathcal{L}, C_0^\infty(x_0, y_0))$ is $L^p(x_0, y_0\,;\,\rho\,dx)$ unique if and only if the $L^p(\rho\,dx)$ limit point case holds both at x_0 and y_0. In particular, for $p = 2$ our definition of the limit point case coincides with Weyl' s original definition, cf. the remark below Corollary 2.2.

EXAMPLE. Let $t \in [x_0, y_0)$, $\gamma \ge 0$, and $\delta > 0$. Suppose that

$$\rho(x) \sim (x - t)^\gamma \quad \text{and} \quad \alpha(x) \sim (x - t)^\delta \quad \text{at } t+.$$

Then the $L^p(\rho\,dx)$ limit point case holds at $t+$ if and only if

$$p \cdot (2 + \gamma - \delta) \leq 1 + \gamma.$$

Note that the diffusion coefficient $a = (\alpha/\rho)$ of the operator $\mathcal{L}$ in non–divergence form satisfies $a(x) \sim (x - t)^{\delta-\gamma}$ at $t+$. Hence if $a(x)$ decreases at least of order $(x-t)^2$ at $t+$, then the $L^p(\rho\,dx)$ limit point case holds for any function ρ which is bounded in a neighbourhood of t, and any $p \in (1,\infty)$. On the other hand, if $a(x)$ does not converge to zero at $t+$, or if even $a(x) \sim (x-t)^{-k}$ for some $k > 0$, then the $L^p(\rho\,dx)$ limit point case only holds for large γ, respectively for small p.

Suppose the $L^p(\rho\,dx)$ limit point case holds at x_0, y_0, and at $s-$ and $s+$ for any point $s \in S$. Then, by Theorem 2.2, the operator $(\mathcal{L}, C_0^\infty(x_1,y_1))$ is $L^p(x_1,y_1;\rho\,dx)$ unique for every connection component (x_1,y_1) of $(x_0,y_0) \setminus S$. It is not difficult to conclude that hence the operator $(\mathcal{L}, C_0^\infty((x_0,y_0) \setminus S))$ is $L^p(x_0,y_0;\rho\,dx)$ unique, and therefore $(\mathcal{L}, C_0^\infty(x_0,y_0))$ is $L^p(x_0,y_0;\rho\,dx)$ unique as well. The point is, however, that there are other cases where $(\mathcal{L}, C_0^\infty((x_0,y_0)\setminus S))$ is *not* $L^p(x_0,y_0;\rho\,dx)$ unique, but $(\mathcal{L}, C_0^\infty(x_0,y_0))$ *is* unique.

We now give a necessary and sufficient condition for $L^p(x_0,y_0;\rho\,dx)$ uniqueness in the singular case.

Theorem 2.4 *Suppose (S 1) holds.*
Then the closure of the operator $(\mathcal{L}, C_0^\infty(x_0,y_0))$ on $L^p(x_0,y_0;\rho\,dx)$ generates a C^0 semigroup if and only if the following conditions hold:

- *x_0 is an accumulation point of S, or the $L^p(\rho\,dx)$ limit point case holds at x_0.*

- *y_0 is an accumulation point of S, or the $L^p(\rho\,dx)$ limit point case holds at y_0.*

- *For any isolated point $s \in S$, the $L^p(\rho\,dx)$ limit point case holds at $s-$ or $s+$.*

REMARKS.   (i) The non–symmetric case can be treated similarly, but the conditions needed to prove uniqueness become more intricate, and are not sharp in general. The proof also becomes more involved. Since we just want to demonstrate the basic phenomena we restrict ourselves to the symmetric case. Similarly, Assumption (S 1) can be dropped but we do not obtain a sharp result in this case.

(ii) The case $p = 1$ can be treated similarly. However, the non–integrability conditions on $\int_c^\bullet (1/\alpha)\,dx$ in the definition of the $L^p(\rho\,dx)$ limit point case have to be replaced by the condition that $(1/\alpha)\int_c^\bullet \rho\,dx$ is not dx–integrable on the corresponding interval, cf. Corollary 2.2 in Section a).

The theorem contains (in particular) two important qualitative statements: Firstly, singularities may cause non–uniqueness. Secondly, "good" behaviour of the coefficients *at only one side* of the singularities and at the boundaries is already enough to avoid non–uniqueness.

EXAMPLE. (Generalized Schrödinger operators with a singularity at 0, cf. also the example in Section d), 2), above)
Fix γ_1, $\gamma_2 \in (p - 1, \infty)$. Suppose that $(x_0, y_0) = \mathbf{R}$, $\rho(x) = \alpha(x) = x^{\gamma_1}$ for $x \geq 0$, and $\rho(x) = \alpha(x) = |x|^{\gamma_2}$ for $x \leq 0$. Then

$$\mathcal{L} = \frac{d^2}{dx^2} + \frac{\gamma(x)}{x} \frac{d}{dx} \, ,$$

where $\gamma(x) = \gamma_1$ for $x > 0$, and $\gamma(x) = -\gamma_2$ for $x < 0$. The operator $(\mathcal{L}, C_0^\infty(\mathbf{R}))$ is dissipative on $L^p(\mathbf{R}; \rho \, dx)$. It is $L^p(\mathbf{R}; \rho \, dx)$ unique if and only if $p \leq (1 + \gamma_1)/2$ or $p \leq (1 + \gamma_2)/2$. In contrast to this, the operator $(\mathcal{L}, C_0^\infty(\mathbf{R} \setminus \{0\}))$ is $L^p(\mathbf{R}; \rho \, dx)$ unique if and only if $p \leq (1 + \gamma_1)/2$ *and* $p \leq (1 + \gamma_2)/2$.

In the remaining part of this section, we prove Theorem 2.4.

PROOF OF THE "IF"–PART. Let $u \in L^q(x_0, y_0; \rho \, dx)$ be a solution of $L^* u = u$. We have to show that $u = 0$ dx–a.e. The proof is similar to that of Theorem 2.1 but we will need an additional consideration concerning the continuity of $\alpha u'$ at the singularities, cf. the claim below. Since S has Lebesgue measure zero, it is enough to show that u vanishes dx–a.e. on each connection component of the open set $(x_0, y_0) \setminus S$. We fix a component (x_1, y_1), $x_0 \leq x_1 < y_1 \leq y_0$.
Step 1: Regularity. Obviously, α is strictly positive on (x_1, y_1). Hence by Corollary 2.4 in Appendix C, u has a continuously differentiable dx–version on (x_1, y_1), which we denote again by u, u' is absolutely continuous on (x_1, y_1), and $(\alpha u')' = \rho u$.
Step 2: Growth of u' at boundaries respectively singularities. Suppose u does not vanish on (x_1, y_1). Fix $z_1 \in (x_1, y_1)$ such that $u(z_1) \neq 0$. Without loss of generality, we assume $u(z_1) > 0$. By the equation $(\alpha u')' = \rho u$, we obtain

(2.31) $(\alpha u')(z) \geq u(z_1) \cdot \int_{z_1}^{z} \rho \, dx \quad$ for all $z \in (z_1, y_1)$,

or

(2.32) $(\alpha u')(z) \leq -u(z_1) \cdot \int_{z}^{z_1} \rho \, dx \quad$ for all $z \in (x_1, z_1)$,

depending on whether $u'(z_1) \geq 0$ or $u'(z_1) \leq 0$, cp. Step 3 in the proof of Theorem 2.1. In particular, u is positive and increasing on (z_1, y_1), respectively positive and decreasing on (x_1, z_1).
Step 3: Contradiction. Assume that (2.31) holds — in the other case we can argue similarly. There are three possibilities:
1) $y_1 = y_0$: Since $u(z_1) > 0$, the inequality (2.31) implies

$$\liminf_{z \uparrow y_1} (\alpha u')(z) > 0.$$

Hence $u'(z)$ grows at least of order $1/\alpha(z)$ as $z \uparrow y_1$, and therefore $u \geq c \cdot \int_{z_1}^{\bullet} (1/\alpha)\, dz$ on $[z_1, y_1)$ for some strictly positive constant c. On the other hand, since y_0 equals y_1, y_0 is not an accumulation point of S. Hence the $L^p(\rho\, dx)$ limit point case holds at y_0, i.e., $\int_{z_1}^{\bullet} (1/\alpha)\, dz$ is not in $L^q(z_1, y_1 ; \rho\, dx)$. This is a contradiction to $u \in L^q(x_0, y_0 ; \rho\, dx)$, whence u vanishes on (x_1, y_1).

2) y_1 is an isolated point in S, and the $L^p(\rho\, dx)$ limit point case holds at y_1-. Then we obtain a contradiction in the same way as in Case 1).

3) y_1 is an isolated point in S, and the $L^p(\rho\, dx)$ limit point case holds at y_1+. Since ρ is dx–integrable on a neighbourhood of y_1, $1/\alpha$ is not dx–integrable at y_1+, cf. the definition of the $L^p(\rho\, dx)$ limit point case . Note that u is C^1 both on the left and on the right of y_1. We will show:

Claim: $\quad \lim_{z \uparrow y_1} (\alpha u')(z) = \lim_{z \downarrow y_1} (\alpha u')(z)$.

Suppose the claim holds. Then, by (2.31), both limits are strictly positive. Hence $u'(z)$ grows at least of order $1/\alpha(z)$ as $z \downarrow y_1$. Since $1/\alpha$ is not dx–integrable at y_1+, we obtain $u \leq -c \cdot \int_{\bullet}^{z_2} (1/\alpha)\, dz$ on (y_1, z_2) for some $z_2 > y_1$ and some constant $c > 0$. Since the $L^p(\rho\, dx)$ limit point case holds at y_1+, this is a contradiction to $u \in L^q(x_0, y_0 ; \rho\, dx)$. Thus u vanishes on (x_1, y_1). It only remains to prove the claim.

Proof of the claim: Roughly speaking, the claim is a consequence of the equation $\mathcal{L}^* u = u$ "evaluated at the point y_1". Fix $y_2 \in (y_1, y_0)$ such that $(y_1, y_2) \cap S = \emptyset$. Let f be a function in $C_0^\infty(x_1, y_2)$ such that $f(x) = 1$ for all x in a neighbourhood of y_1. In particular, f' vanishes in this neighbourhood. We have

(2.33) $$ \int_{x_1}^{y_2} u\, \mathcal{L}f\, \rho\, dx = \int_{x_1}^{y_2} u f\, \rho\, dx. $$

On the other hand, u is C^1 both on (x_1, y_1) and (y_1, y_2), and $(\alpha u')' = \rho u$ on these intervals. Hence, for small $\varepsilon, \delta > 0$,

$$ \int_{x_1}^{y_1-\varepsilon} u\, \mathcal{L}f\, \rho\, dx = \int_{x_1}^{y_1-\varepsilon} u\, (\alpha f')'\, dx $$

$$ = -\int_{x_1}^{y_1-\varepsilon} \alpha u'\, f'\, dx + (\alpha u f')(y_1 - \varepsilon) $$

$$ = \int_{x_1}^{y_1-\varepsilon} (\alpha u')'\, f\, dx + (\alpha u f')(y_1 - \varepsilon) - (\alpha u' f)(y_1 - \varepsilon) $$

(2.34) $$ = \int_{x_1}^{y_1-\varepsilon} u f\, \rho\, dx - (\alpha u')(y_1 - \varepsilon), $$

and, similarly,

(2.35) $$ \int_{y_1+\delta}^{y_2} u\, \mathcal{L}f\, \rho\, dx = \int_{y_1+\delta}^{y_2} u f\, \rho\, dx - (\alpha u')(y_1 + \delta). $$

By (2.33), (2.34) and (2.35), and since u is in $L^q(x_0, y_0 ; \rho\, dx)$, we see that $(\alpha u')(y_1 + \delta) - (\alpha u')(y_1 - \varepsilon)$ converges to 0 as $\varepsilon, \delta \downarrow 0$. ∎

PROOF OF THE "ONLY IF"–PART.   We have to show non–uniqueness of the
operator $(\mathcal{L}, C_0^\infty(x_0, y_0))$ on $L^p(x_0, y_0 ; \rho\, dx)$, i.e., the existence of a non–trivial
solution of $\mathcal{L}^* u = u$, if one of the conditions in Theorem 2.4 does not hold. The
key ideas of the following non–uniqueness proof are already contained in the
proofs of Theorem 2.2 and the lemma in Section d).
Suppose that one of the conditions in Theorem 2.4 does not hold. There are
three possibilities:

  1) α is strictly positive in a neighbourhood of x_0, and the $L^p(\rho\, dx)$ limit point
     case does not hold at x_0.

  2) α is strictly positive in a neighbourhood of y_0, and the $L^p(\rho\, dx)$ limit point
     case does not hold at y_0.

  3) There exists $s \in (x_0, y_0)$ such that $\alpha(s) = 0$, α is strictly positive on
     $(s - \varepsilon, s)$ and $(s, s + \varepsilon)$ for some $\varepsilon > 0$, and the $L^p(\rho\, dx)$ limit point case
     holds neither at $s-$ nor at $s+$.

Obviously, Case 1) and 2) can be treated similarly, hence we may assume that
1) or 3) holds. If 1) holds then we set $y_1 := \inf\{z \in (x_0, y_0); \; \alpha(z) = 0\}$,
respectively $y_1 := y_0$ if there is no zero of α. If 3) holds then we set $y_1 :=$
$\inf\{z \in (s, y_0); \; \alpha(z) = 0\}$, respectively $y_1 := y_0$ if α is strictly positive on
(s, y_0), and $x_1 := \sup\{z \in (x_0, s); \; \alpha(z) = 0\}$, respectively $x_1 := x_0$ if $\alpha > 0$
on (x_0, s). We will construct a non–trivial solution of the equation $L^* u = u$
on the interval (x_0, y_1) in Case 1) respectively on (x_1, y_1) in Case 3), such that
$\alpha(z)\, u'(z)$ and $\alpha(z)\, u(z)$ converge to 0 as $z \uparrow y_1$ (and as $z \downarrow x_1$ in Case 3)).
This solution can then be trivially extended to a solution on the whole interval
(x_0, y_0).

**Step 1: Construction of a local solution of $(\alpha u')' = \rho u$ with "good"
behaviour at y_1 resp. x_1.**
Let $t := x_0$ if 1) holds respectively $t := s$ if 3) holds.

Claim: There exists a continuously differentiable function $u : (t, y_1) \to \mathbf{R}$ with
the following properties:
(i) u' is absolutely continuous, and $(\alpha u')' = \rho u$.
(ii) u is strictly positive and strictly decreasing.

(iii) If $y_1 \neq y_0$ then $\alpha(x)\, u'(x)$ converges to 0 as $x \uparrow y_1$.
(iv) u is in $L^q(x, y_1 ; \rho\, dx)$ for any $x \in (t, y_1)$.

Similarly, we can in Case 3) construct a strictly positive and strictly increasing
function $u \in C^1(x_1, s)$ such that (i) and (ii) hold, $\lim_{x \downarrow x_1} \alpha(x)\, u'(x) = 0$ if
$x_1 \neq x_0$, and u is in $L^q(x_1, x ; \rho\, dx)$ for any $x \in (x_1, s)$.

Proof of the claim: We first consider the case where $1/\alpha$ is not dx–integrable
at y_1-. Then the proof is similar to Step 1 in the proof of Theorem 2.2: Since
α is strictly positive, and α and ρ are continuous on (t, y_1), there exists a unique
C^1 solution (u, w) of the system

$$(2.36) \qquad\qquad u' = (1/\alpha)\, w, \qquad w' = \rho\, u,$$

on (t, y_1) for any initial value $u(x) = u_0$, $w(x) = w_0$, $x \in (t, y_1)$, $u_0, w_0 \in \mathbf{R}$. If (u, w) is such a solution then u' is absolutely continuous, and $(\alpha u')' = \rho u$. In particular,

$$(2.37) \qquad (\alpha u u')' = \alpha (u')^2 + \rho u^2 \geq 0 \qquad dx\text{–a.e. on } (t, y_1).$$

Hence the set–valued function

$$K(x) := \{(u, w) \in C^1((t, y_1); \mathbf{R}^2); \ (u, w) \text{ solves (2.36), and } u(x) \cdot u'(x) < 0 \},$$

$t < x < y_1$, is decreasing. Since $K(x)$ is a non–empty open cone for any x, the intersection $\bigcap \{K(x); \ t < x < y_1\}$ contains at least a half–ray R starting at 0, and the corresponding opposite half–ray $-R$. If (u, w) is an element in the intersection, then $u \cdot u' < 0$ on (t, y_1). Hence, by continuity, either $u < 0$ and $u' > 0$ on (t, y_1), or $u > 0$ and $u' < 0$ on (t, y_1). By replacing (u, w) by $(-u, -w)$ if necessary, we have found a function $u \in C^1(t, y_1)$ satisfying (i) and (ii).

Since $(\alpha u')' = \rho u$, $\alpha u'$ is increasing. On the other hand, it is negative. Hence the limit as $x \uparrow y_1$ of $(\alpha u')(x)$ exists and is negative. Suppose it is *strictly* negative. Then $u'(x)$ decreases of order $1/\alpha$ as $x \uparrow y_1$. This is a contradiction, since u is positive on (t, y_1), and we are considering the case where $1/\alpha$ is *not* dx–integrable. Hence $\lim_{x \uparrow y_1} (\alpha u')(x) = 0$, i.e., (iii) holds.

Finally, for $x \in (t, y_1)$ we have

$$\int_x^{y_1} |u| \, \rho \, dx = \lim_{y \uparrow y_1} \int_x^y u \, \rho \, dx$$
$$= \lim_{y \uparrow y_1} (\alpha u')(y) - (\alpha u')(x) = -(\alpha u'')' < \infty.$$

Hence u is in $L^1(x, y_1 ; \rho \, dx)$. Since u is decreasing and positive, it is bounded on (x, y_1). Thus u is in $L^q(x, y_1 ; \rho \, dx)$.

It remains to consider the case where $1/\alpha$ is dx–integrable at y_1-. If $y_1 = y_0$ then we may argue as above, only that now we do not obtain $\lim_{x \uparrow y_1} (\alpha u')(x) = 0$. Nevertheless, the limit exists in $(-\infty, 0]$, which suffices to show (iv), whereas (iii) holds automatically.

Now suppose $y_1 < y_0$. Then ρ is dx–integrable in a neighbourhood of y_1. Hence the coefficients of the system (2.36) are both dx–integrable at y_1-. In this case, it is well–known that (2.36) even has a solution $(u, w) \in C^1((t, y_1) \to \mathbf{R}^2) \cap C((t, y_1] \to \mathbf{R}^2)$ for any given initial values $u(y_1) = u_0$, $w(y_1) = w_0$, $u_0, w_0 \in \mathbf{R}$. We consider the solution satisfying $u(y_1) = 1$, $w(y_1) = 0$. Clearly, (iii) holds. Moreover, since $w' = \rho u$, we have $u > 0$ and $w < 0$ on $(y_1 - \varepsilon, y_1)$ for some $\varepsilon > 0$. Hence, by (2.37), u is strictly positive and strictly decreasing on $(t, y_1]$, i.e., (ii) holds. The validity of (iv) can now be shown in the same way as in the case where $1/\alpha$ is not dx–integrable at y_1-.

This proves the claim. In Case 3), the corresponding assertion with y_1 replaced by x_1 can be shown similarly.

Step 2: Integrability of u at the boundary resp. singularity where the limit point case does not hold. Let I denote the interval (x_0, y_1) in Case

1) respectively one of the intervals (x_1, s) or (s, y_1) in Case 3). We will show that the solution u of $(\alpha u')' = \rho u$ on I, that has been constructed in Step 1, is in $L^q(I; \rho dx)$. The proof is essentially the same as the proof of Step 2 in Theorem 2.2. We restrict ourselves to the case $I = (t, y_1)$ where $t = x_0$ resp. $t = s$ as in Step 1 — the case $I = (x_1, s)$ can be treated similarly. Fix $c \in I$. By Assertion (iv) in the claim above it only remains to show $u \in L^q(t, c; \rho dx)$. We know that at $t+$, the $L^p(\rho dx)$ limit point case does not hold. In particular, the measure ρdx is finite on (t, c). The equation $(\alpha u')' = \rho u$ implies

$$(\alpha u')(z) = (\alpha u')(c) - \int_z^c \rho u \, dx$$

$$\geq (\alpha u')(c) - \|u\|_{L^q(z,c; \rho dx)} \cdot \left(\int_t^c \rho \, dx \right)^{1/p}$$

for all $z \in (t, c)$. Fix $c_0 \in (t, c)$. Then there exists a finite constant K_1 such that

$$(\alpha u')(z) \geq -K_1 \cdot \|u\|_{L^q(z,c; \rho dx)} \qquad \text{for all } z \in (t, c_0).$$

Hence there also is a finite constant K_2 such that

$$u(y) = u(c) - \int_y^c u'(z) \, dz$$

$$\leq K_2 \cdot \int_y^c (1/\alpha(z)) \cdot \|u\|_{L^q(z,c; \rho dx)} \, dz \quad \text{for all } y \in (t, c_0).$$

Let $f(x) := \int_x^c u^q \, \rho dx$, $t < x < c$. Then

$$f \leq \int_{c_0}^c u^q \, \rho dx + K_2^q \cdot \int_{\bullet}^{c_0} \left(\int_y^c (1/\alpha(z)) \, dz \right)^q \rho(y) \, f(y) \, dy$$

on (t, c_0). Since the $L^p(\rho dx)$ limit point case does not hold at $t+$, the lemma of Gronwall implies that f is bounded on (t, c_0), i.e., u is in $L^q(t, c; \rho dx)$.

Step 3: Boundary values of αu and $\alpha u'$ at singularities. This step is only needed in Case 3). So suppose 3) holds, and let u be the solution of $(\alpha u')' = \rho u$ on (x_1, s) constructed above. We will show:

(i) The limit of $(\alpha u')(x)$ as $x \uparrow s$ exists and is finite.

(ii) There exists a sequence $x_n \uparrow x$ such that $\lim_{n \to \infty} (\alpha u)(x_n) = 0$.

The corresponding assertion on (s, y_1) is, of course, true as well, and can be proven similarly.

To prove (i) note that, by Step 2, u is in $L^q(x_1, s; \rho dx)$. Since the measure ρdx is finite in a neighbourhood of s, u is in particular ρdx–integrable at $s-$, i.e., ρu is dx–integrable at $s-$. The equation $(\alpha u')' = \rho u$ now implies (i).

To prove (ii), we fix $c \in (x_1, s)$. By (i), there exists a finite constant K such that

$$u \leq K \cdot \left(1 + \int_c^{\bullet} (1/\alpha) \, dz \right) \qquad \text{on } (c, s),$$

and thus

$$\alpha u \leq K \cdot \left(\alpha + \alpha \int_c^\bullet (1/\alpha)\, dz \right)$$

(2.38)
$$= K \cdot \alpha + K \cdot \left[\left(\log \int_c^\bullet (1/\alpha)\, dz \right)' \right]^{-1}.$$

Since u is positive on (x_1, s), and $\alpha(s) = 0$, we immediately obtain $\lim_{x \uparrow s} (\alpha u)(x) = 0$ if $\int_c^s (1/\alpha)\, dz$ is finite. Now suppose $\int_c^s (1/\alpha)\, dz = \infty$. Then $\log \int_c^\bullet (1/\alpha)\, dz$ is unbounded at $s-$, whence there exists a sequence $x_n \uparrow s$ such that

$$\lim_{n \to \infty} \left(\log \int_c^\bullet (1/\alpha)\, dz \right)' (x_n) = \infty.$$

Thus, by (2.38), (ii) holds as well.

Step 4: Construction of a global solution. We finally construct a solution $\bar{u}$ of $L^* \bar{u} = \bar{u}$ on (x_0, y_0). We first look at the case where 1) holds. Let $\bar{u} \in L^q(x_0, y_0 ; \rho\, dx)$ denote the function which on the interval (x_0, y_1) coincides with the solution of $(\alpha u')' = \rho u$ constructed in Step 1, and which vanishes on $[y_1, y_0)$. Clearly, $\bar{u}$ is non–trivial. Fix $f \in C_0^\infty(x_0, y_0)$. We have

$$\int_{x_0}^{y_0} \bar{u}\, \mathcal{L}f\, \rho\, dx = \int_{x_0}^{y_1} u\, (\alpha f')'\, dx = \lim_{y \uparrow y_1} \int_{x_0}^y u\, (\alpha f')'\, dx.$$

For $y \in (x_0, y_1)$,

$$\int_{x_0}^y u\, (\alpha f')'\, dx = - \int_{x_0}^y u'\, \alpha f'\, dx + (u\alpha f')\, (y)$$

(2.39)
$$= \int_{x_0}^y \rho u f\, dx + (\alpha u f')\, (y) - (\alpha u' f)\, (y).$$

If $y_1 = y_0$ then $f'(y)$ and $f(y)$ vanish for y close to y_1, and hence we immediately obtain

$$\int_{x_0}^{y_0} \bar{u}\, \mathcal{L}f\, \rho\, dx = \int_{x_0}^{y_1} u f\, \rho\, dx = \int_{x_0}^{y_0} \bar{u}\, f\, \rho\, dx.$$

Now suppose $y_1 < y_0$. Then $\alpha(y_1) = 0$. By the properties of u proven in Step 1, $u(y)$ has a finite limit as $y \uparrow y_1$, and $\lim_{y \uparrow y_1} \alpha(y)\, u'(y) = 0$. Thus the boundary terms on the right hand side of (2.39) converge to 0 as $y \uparrow y_1$, whence

$$\int_{x_0}^{y_0} \bar{u}\, \mathcal{L}f\, \rho\, dx = \lim_{y \uparrow y_1} \int_{x_0}^y u\, (\alpha f')'\, dx = \int_{x_0}^{y_1} u f\, \rho\, dx = \int_{x_0}^{y_0} \bar{u}\, f\, \rho\, dx.$$

for all $f \in C_0^\infty(x_0, y_0)$. Thus $\bar{u}$ is a non–trivial solution of $L^* \bar{u} = \bar{u}$.

Now suppose 3) holds. Let $u_1 \in C^1(x_1, s)$ and $u_2 \in C^1(s, y_1)$ be the non–trivial solutions of the ODE $(\alpha u')' = \rho u$ constructed in Step 1. By Step 3, the limits

$\lim_{x\uparrow s}(\alpha u_1')(x)$ and $\lim_{x\downarrow s}(\alpha u_2')(x)$ exist and are finite. We fix $\lambda_1, \lambda_2 \in \mathbf{R}$ such that not both λ_1 and λ_2 vanish, and

$$(2.40) \qquad \lambda_1 \cdot \lim_{x\uparrow s} (\alpha u_1')\,(x) \;=\; \lambda_2 \cdot \lim_{x\downarrow s} (\alpha u_2')\,(x).$$

Let $\bar{u}$ be the non–trivial function in $L^q(x_0, y_0\,;\, \rho\,dx)$ which coincides with $\lambda_1 u_1$ on (x_1, s), with $\lambda_2 u_2$ on (s, y_1), and which vanishes outside (x_1, y_1). Fix $f \in C_0^\infty(x_0, y_0)$. We have

$$(2.41) \qquad \int_{x_0}^{y_0} \bar{u}\, \mathcal{L}f\, \rho\,dx \;=\; \lambda_1 \int_{x_1}^{s} u_1\, (\alpha f')'\, dx \;+\; \lambda_2 \int_{s}^{y_1} u_2\, (\alpha f')'\, dx.$$

For $s < t < y < y_1$,

$$\int_{t}^{y} u_2\, (\alpha f')'\, dx \;=\; \int_{t}^{y} u_2\, f\, \rho\,dx \;+\; (\alpha u_2 f')|_t^y \;-\; (\alpha u_2' f)|_t^y.$$

As for Case 1) above, we can show that the boundary terms $(\alpha u_2 f')(y)$ and $(\alpha u_2' f)(y)$ converge to 0 as $y \uparrow y_1$. Moreover, by Step 3, the limit as $t \downarrow s$ of $(\alpha u_2')(t)$ exists and is finite, and $\lim_{n\to\infty} (\alpha u_2)(t_n) = 0$ for some sequence $t_n \downarrow s$. Hence for $y \uparrow y_1$ and $t \downarrow s$, we obtain

$$(2.42) \qquad \int_{s}^{y_1} u_2\, (\alpha f')'\, dx \;=\; \int_{s}^{y_1} u_2\, f\, \rho\,dx \;+\; f(s) \cdot \lim_{t\downarrow s} (\alpha u_2')\,(t).$$

Similarly, we can show

$$(2.43) \qquad \int_{x_1}^{s} u_1\, (\alpha f')'\, dx \;=\; \int_{x_1}^{s} u_1\, f\, \rho\,dx \;-\; f(s) \cdot \lim_{r\uparrow s} (\alpha u_1')\,(r).$$

By (2.41), (2.42), (2.43), and (2.40), we obtain

$$\int_{x_0}^{y_0} \bar{u}\, \mathcal{L}f\, \rho\,dx \;=\; \lambda_1 \int_{x_1}^{s} u_1\, f\, \rho\,dx \;+\; \lambda_2 \int_{s}^{y_1} u_2\, f\, \rho\,dx$$

$$+\; f(s) \cdot \left(\lambda_2 \lim_{t\downarrow s} (\alpha u_2')\,(t) \;-\; \lambda_1 \lim_{r\uparrow s} (\alpha u_1')\,(r) \right)$$

$$=\; \int_{x_0}^{y_0} \bar{u}\, f\, \rho\,dx$$

for all $f \in C_0^\infty(x_0, y_0)$. This implies $L^*\bar{u} = \bar{u}$. ∎

f)   Singular diffusion operators on $\mathbf{R}^n$

Let $n \in \mathbf{N}$, $n \geq 2$. We finally prove uniqueness results for singular diffusion operators on $\mathbf{R}^n$. These results are contained in the article [Eb 99]. Fix a positive Radon measure m on $\mathbf{R}^n$ such that the support of m is $\mathbf{R}^n$, or, more generally, the boundary of the support of m has measure 0. The latter condition is to ensure that the derivativion operators ∇ and Δ on $C_0^\infty(\mathbf{R}^n)$ respect m-classes, i.e., $\nabla f = \nabla g$ m–a.e. and $\Delta f = \Delta g$ m–a.e. for all f, $g \in C_0^\infty(\mathbf{R}^n)$ such that $f = g$ m–a.e.

We will prove two uniqueness results for singular diffusion operators of type $\Delta + \beta \cdot \nabla$, where β is a function in $L_{\mathrm{loc}}^p(\mathbf{R}^n \to \mathbf{R}^n\,;m)$. The first result gives a good *dimension–independent* condition for $L^p(\mathbf{R}^n\,;m)$ uniqueness for small p up to $p = 2$, provided m is absolutely continuous w.r.t. Lebesgue measure. The proof is surprisingly elementary in the sense that no advanced elliptic regularity theory w.r.t. L^r–norms for $r \neq 2$ is required. We only use standard techniques on L^2 respectively $H^{1,2}$ spaces w.r.t. Lebesgue measure. In contrast to this, the second result gives a slightly better (and in some sense optimal) condition for $L^p(\mathbf{R}^n\,;m)$ uniqueness in dimension $n = 2$, and it also works for $p > 2$, and for not absolutely continuous measures m. However, the conditions for L^p uniqueness in the second result are dimension dependent, and far from the optimal condition in high dimensions. Moreover, the proof of this result is based on a highly non–trivial regularity result by Bogachev, Krylov and Röckner [BoKryRö 96].

To state the results, we fix $p \in [1, \infty)$, and a function $\beta \in L_{\mathrm{loc}}^p(\mathbf{R}^n \to \mathbf{R}^n\,;m)$. Let $(\mathcal{L}, C_0^\infty(\mathbf{R}^n))$ be the densely defined operator on $L^p(\mathbf{R}^n\,;m)$ given by

$$\mathcal{L}f \;=\; \Delta f + \beta \cdot \nabla f.$$

Note that $\mathcal{L}$ is well–defined, since Δ and ∇ respect m–classes. We assume that $(\mathcal{L} - \lambda, C_0^\infty(\mathbf{R}^n))$ is dissipative on $L^p(\mathbf{R}^n\,;m)$ for some $\lambda \geq 0$. This is always the case if m is a sub–invariant measure for the operator $(\mathcal{L} - p\lambda, C_0^\infty(\mathbf{R}^n))$, cf. Lemma 1.8 in Appendix B.

Let $\beta_r := e_r \cdot \beta$, where $e_r(x) = x/|x|$. For $r > 0$, the open ball of radius r around 0 will be denoted by B_r. Let $f^- := -\min(f, 0)$ denote the negative part of a function f. For both results we need the following assumption on the growth of $\beta_r^-(x)$ for large x:

(G 1) There exists a decomposition $\beta = \beta^{\mathrm{sing}} + \beta^{\mathrm{reg}}$, β^{sing}, $\beta^{\mathrm{reg}} : \mathbf{R}^n \to \mathbf{R}^n$, such that the "singular" part β^{sing} satisfies

(2.44) $$\lim_{r \to \infty} \frac{1}{r} \|(\beta_r^{\mathrm{sing}})^-\|_{L^p(B_r;m)} \;=\; 0,$$

and the "regular" part β^{reg} is locally bounded, and satisfies

(2.45) $$\limsup_{r \to \infty} \frac{m(B_r)}{r^k} \;<\; \infty \quad \text{for some } k > 0, \text{ and}$$

$$\limsup_{|x| \to \infty} \frac{(\beta_r^{\mathrm{reg}})^-(x)}{|x|} \;<\; \infty,$$

   or

(2.46)   The measure m is finite, and

$$\limsup_{|x| \to \infty} \frac{(\beta_r^{\mathrm{reg}})^-(x)}{|x| \log |x|} \;<\; \infty.$$

REMARKS. (i) The restrictions on the volume growth are essential for our proofs because we use lower estimates for L^1 norms w.r.t. the measure m, from which we derive lower estimates for the corresponding L^q-norms. For the same reason, we don't recover completely the optimal growth condition for $(\beta_r^{\mathrm{reg}})^-$ $(\sim |x| \log |x|)$, if m is not finite. Other methods of proof might lead to slightly better conditions here.

(ii) The condition (2.44) is a somehow restrictive assumption on the decay of singularities if the measure m is infinite. Note, however, that if m has a density which decreases exponentially fast as $|x| \to \infty$, then (2.44) is a very weak assumption on β^{sing}. Moreover, (2.44) is of course always satisfied with $\beta^{\mathrm{sing}} = \beta$, if β is globally in $L^p(\mathbf{R}^n\, ; m)$.

We set $p/(2-p) := \infty$ if $p = 2$. We now state the two results:

Theorem 2.5 *Let $p \in [1,2]$. Suppose that $m = \rho\, dx$ with $\rho \in L_{\mathrm{loc}}^{p/(2-p)}(\mathbf{R}^n, dx)$, β is in $L_{\mathrm{loc}}^{2p}(\mathbf{R}^n \to \mathbf{R}^n\, ; m)$, $\beta\rho^{1/2}$ is in $L_{\mathrm{loc}}^{2p/(2-p)}(\mathbf{R}^n \to \mathbf{R}^n\, ; m)$, and (G 1) holds.*
Then the closure of $(\mathcal{L}, C_0^\infty(\mathbf{R}^n))$ generates a C^0 semigroup on $L^p(\mathbf{R}^n\, ; m)$. In particular, $(\mathcal{L}, C_0^\infty(\mathbf{R}^n))$ is $L^p(\mathbf{R}^n\, ; m)$ unique.

Theorem 2.6 *Let $p \in [1,\infty)$. Suppose that β is in $L_{\mathrm{loc}}^{(1+\frac{n}{2})p+\varepsilon}(\mathbf{R}^n \to \mathbf{R}^n\, ; m)$ for some $\varepsilon > 0$, and (G 1) holds. Then the closure of $(\mathcal{L}, C_0^\infty(\mathbf{R}^n))$ generates a C^0 semigroup on $L^p(\mathbf{R}^n\, ; m)$.*

The proofs of the theorems will be given below.

REMARKS. (iii) The first local conditon $\beta \in L_{\mathrm{loc}}^{2p}(\mathbf{R}^n \to \mathbf{R}^n\, ; m)$ in Theorem 2.5 is sharp in the following sense:
Let $p \in [1,2]$. For every $\varepsilon > 0$, there exist ρ and β such that the operator $(\mathcal{L}, C_0^\infty(\mathbf{R}^n))$ is not $L^p(\mathbf{R}^n\, ; m)$ unique, although all the assumptions in Theorem 2.5 are satisfied, except that β is not in $L_{\mathrm{loc}}^{2p}(\mathbf{R}^n \to \mathbf{R}^n\, ; m)$ but only in $L_{\mathrm{loc}}^{2(p-\varepsilon)}(\mathbf{R}^n \to \mathbf{R}^n\, ; m)$. In fact, let $\rho(x) := |x_1|^{2p-1-\varepsilon}$, and $\beta(x) := \frac{\nabla\rho}{\rho}(x) = (2p - 1 - \varepsilon)x_1^{-1}e_1$. Then $(\mathcal{L}, C_0^\infty(\mathbf{R}^n))$ is a generalized Schrödinger operator which is not $L^p(\mathbf{R}^n\, ; \rho\, dx)$ unique, cf. the example and the remark below in Section d), 1). However,

$$\int |\beta|^{2(p-\varepsilon)}\, \rho\, dx \;=\; \mathrm{const.} \cdot \int |x_1|^{\varepsilon - 1}\, dx \;<\; \infty.$$

For $p = 2$, it has already been claimed in [LiSem 92] that a condition of type $\beta \in L_{\mathrm{loc}}^{2p}(\mathbf{R}^n \to \mathbf{R}^n\, ; m)$ would be optimal in the sense above. However, the counterexample given in [LiSem 92, p. 212] is not correct, since for rotationally invariant generalized Schrödinger operators in $\mathbf{R}^n$, $n \geq 2$, with polynomially growing ρ, essential self-adjointness always holds, cf. Section d), 2). The example above closes the gap in [LiSem 92].

(iv) Unfortunately, in Thm. 2.5 we need not only the rather optimal local condition $\beta \in L_{\mathrm{loc}}^{2p}(\mathbf{R}^n \to \mathbf{R}^n\, ; m)$, but also the condition $\beta\rho^{1/2} \in L_{\mathrm{loc}}^{2p/(2-p)}(\mathbf{R}^n \to$

$\mathbf{R}^n$; m). In many concrete cases, the second condition is weaker than the first one. For example, in the example in Section d), 1), our result implies $L^p(\mathbf{R} \, ; m)$ uniqueness for $\gamma > 2p - 1$, whereas it is known that $L^p(\mathbf{R} \, ; m)$ uniqueness holds for $\gamma \geq 2p - 1$. Thus we almost recover the optimal result.

(v) In Theorem 2.6 we only need the rather optimal local condition $\beta \in L_{\text{loc}}^{2p+\varepsilon}(\mathbf{R}^n \to \mathbf{R}^n \, ; m)$ for some $\varepsilon > 0$, if $n = 2$. However, in higher dimensions the assumed condition $\beta \in L_{\text{loc}}^{(1+\frac{n}{2})p+\varepsilon}(\mathbf{R}^n \to \mathbf{R}^n \, ; m)$ is far from optimal.

(vi) Although the condition $\beta \in L_{\text{loc}}^{2p}(\mathbf{R}^n \to \mathbf{R}^n \, ; m)$ seems to be rather optimal if nothing is known about the shape of the singularities, one may hope for much better results if one knows a priori that the singularities are located on a lower-dimensional submanifold, or if they are even isolated points in $\mathbf{R}^n$, $n \geq 2$, cf. the conjecture in Section g) below, and the examples in Section d), 2).

The advantages and disadvantages of the imposed global condition (G 1) have been discussed above.

Relations to previous results: In particular, the following results on uniqueness of singular diffusion operators of type $(\Delta + \beta \cdot \nabla, C_0^\infty(\mathbf{R}^n))$ on weighted L^p spaces are known:

- V. Liskevič and Y. Semenov [LiSem 92] have shown essential self-adjoint-ness for generalized Schrödinger operators $(\Delta + \frac{\nabla \rho}{\rho} \cdot \nabla, C_0^\infty(\mathbf{R}^n))$ on $L^2(\mathbf{R}^n \, ; \rho \, dx)$ under the condition $\frac{\nabla \rho}{\rho} \in L^4(\mathbf{R}^n \to \mathbf{R}^n \, ; \rho \, dx)$. In the special situation they considered, their condition is locally weaker than ours, but the assumed global integrability of $\left| \frac{\nabla \rho}{\rho} \right|^4$ is restrictive. The technique of proof is different from ours. It is based on the approximative criterion, cf. Corollary 1.5 in Appendix A, and an L^4 gradient estimate for solutions of parabolic PDE.
  In [Li 94], V. Liskevič extends the method from [LiSem 92] to prove $L^p(\mathbf{R}^n \, ; \rho \, dx)$ uniqueness of generalized Schrödinger operators for $p > 3/2$ under the global condition $\frac{\nabla \rho}{\rho} \in L^{2p}(\mathbf{R}^n \to \mathbf{R}^n \, ; \rho \, dx)$. Attempts are going on to prove a localized version of the Liskevič/Semenov result.

- A corresponding global criterion for essential self-adjointness of diffusion operators on $\mathbf{R}^n$ with strictly elliptic non-constant diffusion matrix and singular drift is proven in [LiTuv 93].

- A criterion for essential self-adjointness of generalized Schrödinger operators is also given in [BogKryRö 96]. Here it is only assumed that $\left| \frac{\nabla \rho}{\rho} \right|^\gamma$ is *locally* $\rho \, dx$-integrable for some $\gamma > n$, but ρ is assumed to be locally uniformly positive, which is restrictive. Note, however, that we apply a regularity result obtained in [BogKryRö 96] to prove the uniqueness result in Theorem 2.6, where ρ is allowed to have zeros.

- The easier problem of L^1 uniqueness of singular generalized Schrödinger operators is studied in [LiSem 96]. Here L^1 uniqueness is shown under the condition $\rho^{1/2} \in H^{1,2}(\mathbf{R}^n\,;\,dx)$, which implies $\frac{\nabla \rho}{\rho} \in L^2(\mathbf{R}^n \to \mathbf{R}^n\,;\,\rho\,dx)$. A complete treatment of L^1 uniqueness for not necessarily symmetric diffusion operators on $\mathbf{R}^n$ will be given in the forthcoming paper [St 97], cf. also [St 96] for first steps in this direction. This includes a localized version of the results in [LiSem 96] mentioned above.

Let $C^1(\,[0,\infty)\,)$ denote the space of all continuously differentiable functions on $[0,\infty)$, where the derivative is taken to the right at 0. For the proof of the theorems, we need the following comparison lemma:

Lemma 2.1 *Let* $A \in C(\,[0,\infty)\,)$, $B \in C^1(\,[0,\infty)\,)$, *and* $r_1 \in (0,\infty)$, *such that* $A + B' > 0$ *on* (r_1,∞). *Suppose* G *and* K *are functions in* $C^1(\,[0,\infty)\,)$ *such that* $G(0) = K(0) = 0$, *and the following inequalities hold:*

$$(2.47) \quad -G'(r) + \int_0^r A(s)\,G(s)\,ds \;\leq\; \int_0^r B(s)\,G'(s)\,ds \qquad and$$

$$-K'(r) + \int_0^r A(s)\,K(s)\,ds \;\geq\; \int_0^r B(s)\,K'(s)\,ds \quad \text{for all } r \geq r_1.$$

$$(2.48) \qquad G(r_1) \;>\; K(r_1) \quad and$$

$$\int_0^{r_1} (\,A(s) + B'(s)\,)\,G(s)\,ds \;\geq\; \int_0^{r_1} (\,A(s) + B'(s)\,)\,K(s)\,ds.$$

Then $G(r) > K(r)$ *for all* $r \in [r_1,\infty)$.

PROOF OF THE LEMMA. Partial integration yields

$$-G'(r) + \int_0^r (A + B')\,G\,ds \;\leq\; B(r)\,G(r), \quad and$$

$$-K'(r) + \int_0^r (A + B')\,K\,ds \;\geq\; B(r)\,K(r) \quad \text{for all } r \geq r_1.$$

Suppose that $G(r) \leq K(r)$ for some $r \geq r_1$, and let $u := \inf\{r \geq r_1\,;\,G(r) \leq K(r)\}$. Since $G(r_1) > K(r_1)$, u is in (r_1,∞). Obviously, $G(u) = K(u)$ and $G'(u) \leq K'(u)$. Hence

$$\int_0^u (A + B')\,(G - K)\,ds \;\leq\; G'(u) + B(u)\,G(u)\, -\, K'(u)\, -\, B(u)\,K(u) \;\leq\; 0.$$

This is a contradiction, because, on the other hand,

$$\int_0^u (A + B')\,(G - K)\,ds \;=\; \int_0^{r_1} (A + B')\,(G - K)\,ds + \int_{r_1}^u (A + B')\,(G - K)\,ds,$$

which is strictly positive, since $G > K$ and $A + B' > 0$ on (r_1,u), and (2.48) holds. ∎

PROOF OF THEOREM 2.5. Let $q \in [2, \infty]$ such that $\frac{1}{p} + \frac{1}{q} = 1$. We choose $\lambda \geq 0$ such that $(\mathcal{L} - \lambda, C_0^\infty(\mathbf{R}^n))$ is dissipative on $L^p(\mathbf{R}^n; m)$, and we fix a large constant $\gamma > \lambda$ to be specified below. We show in several steps that the existence of a non-trivial solution $u \in L^q(\mathbf{R}^n; m)$ of the equation $L^*u = \gamma u$ leads to a contradiction, if γ is chosen large enough. This proves the assertion, cf. Corollary 1.3 in Appendix A. Thus suppose u is a non-trivial solution of $L^*u = \gamma u$.

Step 1: Regularity. We will show that ρu is in $H_{\mathrm{loc}}^{1,2}(\mathbf{R}^n; dx)$, and

$$(2.49) \qquad \int \nabla f \cdot \nabla(\rho u)\, dx \;+\; \gamma \int f\, \rho u\, dx \;=\; \int \beta \cdot \nabla f\, \rho u\, dx$$

for all compactly supported functions $f \in H^{1,2}(\mathbf{R}^n; dx)$.

Note that the equation $L^*u = \gamma u$ can be rewritten as

$$(2.50) \quad \int_{\mathbf{R}^n} (\gamma - \Delta) f\, \rho u\, dx \;=\; \int_{\mathbf{R}^n} \beta \cdot \nabla f\, \rho u\, dx \quad \text{for all } f \in C_0^\infty(\mathbf{R}^n).$$

Let $\varphi \in C_0^\infty(\mathbf{R}^n)$ be a positive function such that $\int_{\mathbf{R}^n} \varphi\, dx = 1$ and $\varphi(x) = \varphi(-x)$ for all x, and let φ_ε, $\varepsilon > 0$, $\varphi_\varepsilon(x) = \varepsilon^{-n} \cdot \varphi(x/\varepsilon)$, denote the corresponding dirac sequence. Let $\mathcal{E}_\gamma$ be the bilinear form

$$\mathcal{E}_\gamma(f,g) \;=\; \int \nabla f \cdot \nabla g\, dx \;+\; \gamma \int f \cdot g\, dx$$

on $H^{1,2}(\mathbf{R}^n; dx)$. Fix a function η in $C_0^\infty(\mathbf{R}^n)$ such that $0 \leq \eta \leq 1$. Then for all $f \in C_0^\infty(\mathbf{R}^n)$, and $0 < \varepsilon \leq 1$,

$$\mathcal{E}_\gamma(f, (\eta\rho u) * \varphi_\varepsilon) \;=\; \int_{\mathbf{R}^n} (\gamma - \Delta) f\ (\eta\rho u) * \varphi_\varepsilon\, dx$$

$$= \int_{\mathbf{R}^n} (\gamma - \Delta)(f * \varphi_\varepsilon)\, \eta\, \rho u\, dx$$

$$= \int (\gamma - \Delta)((f * \varphi_\varepsilon)\eta)\, \rho u\, dx \;+\; \int (f * \varphi_\varepsilon\, \Delta\eta + 2\nabla(f*\varphi_\varepsilon)\cdot\nabla\eta)\, \rho u\, dx$$

$$= \int \beta \cdot \nabla(f * \varphi_\varepsilon)\,\eta\rho u\, dx \;+\; \int (f * \varphi_\varepsilon\,(\Delta\eta + \beta\nabla\eta) + 2\nabla(f*\varphi_\varepsilon)\nabla\eta)\, \rho u\, dx$$

$$\leq C \cdot \mathcal{E}_\gamma(f,f)^{1/2} \cdot \|(1 + |\beta|)\rho u\|_{L^2(\mathrm{supp}\,\eta;\, dx)}$$

for some finite constant C depending only on η. Choosing $f = (\eta\rho u) * \varphi_\varepsilon$, we obtain

$$(2.51) \quad \mathcal{E}_\gamma((\eta\rho u) * \varphi_\varepsilon, (\eta\rho u) * \varphi_\varepsilon)^{1/2}$$

$$\leq C \cdot \|(1 + |\beta|) \cdot \rho u\|_{L^2(\mathrm{supp}\,\eta;\, dx)} \;=\; C \cdot \|(1 + |\beta|)\rho^{1/2} u\|_{L^2(\mathrm{supp}\,\eta;\, m)}$$

$$\leq C \cdot \|(1 + |\beta|)\rho^{1/2}\|_{L^s(\mathrm{supp}\,\eta;\, m)} \cdot \|u\|_{L^q(\mathbf{R}^n;\, m)},$$

where $s = \frac{2p}{2-p}$, i.e., $\frac{1}{s} = \frac{1}{p} - \frac{1}{2} = \frac{1}{2} - \frac{1}{q}$. By assumption, $(1 + |\beta|)\rho^{1/2}$ is in $L_{\mathrm{loc}}^s(\mathbf{R}^n; m)$, whence the right hand side is finite. Hence the functions $(\eta\rho u)*\varphi_\varepsilon$,

$0 < \varepsilon \leq 1$, are uniformly bounded elements in $H^{1,2}(\mathbf{R}^n \,;\, dx)$. Therefore, $\eta \rho u$ is in $H^{1,2}(\mathbf{R}^n \,;\, dx)$ as well. Since η is an arbitrary function in $C_0^\infty(\mathbf{R}^n)$ such that $0 \leq \eta \leq 1$, we obtain $\rho u \in H^{1,2}_{\mathrm{loc}}(\mathbf{R}^n \,;\, dx)$. Integration by parts in (2.50) now yields (2.49) for all $f \in C_0^\infty(\mathbf{R}^n)$. Since, by the last estimate in (2.51), $\beta \rho u$ is in $L^2_{\mathrm{loc}}(\mathbf{R}^n \to \mathbf{R}^n \,;\, dx)$, (2.49) even holds for all compactly supported functions $f \in H^{1,2}(\mathbf{R}^n \,;\, dx)$.

Step 2: Inequality for $\rho |u|$.   Since ρu is in $H^{1,2}_{\mathrm{loc}}(\mathbf{R}^n \,;\, dx)$, $\rho |u|$ is in $H^{1,2}_{\mathrm{loc}}(\mathbf{R}^n \,;\, dx)$ as well. We will derive the following inequality for $\rho |u|$ from (2.49):

$$(2.52) \qquad \int \nabla \xi \cdot \nabla \left(\rho |u| \right) dx \; + \; \gamma \int \xi \, \rho |u| \, dx \; \leq \; \int \beta \cdot \nabla \xi \, \rho |u| \, dx$$

for all positive, compactly supported functions $\xi \in H^{1,2}(\mathbf{R}^n \,;\, dx)$.

For $\varepsilon > 0$, let $\psi_\varepsilon : \mathbf{R} \to \mathbf{R}$ be given by $\psi_\varepsilon(x) = \mathrm{sgn}\,(x)$ if $|x| \geq \varepsilon$, and $\psi_\varepsilon(x) = x/\varepsilon$ if $|x| \leq \varepsilon$. Obviously, ψ_ε is Lipschitz continuous, whence $\psi_\varepsilon(\rho u)$ is in $H^{1,2}_{\mathrm{loc}}(\mathbf{R}^n \,;\, dx)$, and

$$\nabla \psi_\varepsilon(\rho u) \;=\; \psi_\varepsilon'(\rho u) \cdot \nabla(\rho u) \;=\; \varepsilon^{-1} \cdot \chi_{\{\rho|u|\leq\varepsilon\}} \cdot \nabla(\rho u) \qquad dx\text{-a.e.}$$

Fix a positive function $\xi \in C_0^\infty(\mathbf{R}^n)$. Setting $f := \xi \cdot \psi_\varepsilon(\rho u)$ in (2.49), we obtain

$$
\begin{aligned}
& \int \nabla \xi \cdot \nabla(\rho u)\, \psi_\varepsilon(\rho u)\, dx \; + \; \gamma \int \xi \rho u\, \psi_\varepsilon(\rho u)\, dx \\
(2.53) \qquad &= \; \int \beta \cdot \nabla \xi \, \rho u \, \psi_\varepsilon(\rho u)\, dx \; + \; \varepsilon^{-1} \int_{\{\rho|u|\leq\varepsilon\}} \xi\, \beta \cdot \nabla(\rho u)\, \rho u\, dx \\
& \quad - \; \varepsilon^{-1} \int_{\{\rho|u|\leq\varepsilon\}} \xi\, |\nabla(\rho u)|^2\, dx .
\end{aligned}
$$

To estimate the right-hand side of (2.53) note that

$$\int_{\{\rho|u|\leq\varepsilon\}} \xi\, \beta \cdot \nabla(\rho u)\, \rho u\, dx \; \leq \; \left(\int_{\{\rho|u|\leq\varepsilon\}} \xi\, |\nabla(\rho u)|^2\, dx \right)^{1/2} \cdot C_\varepsilon^{1/2},$$

where

$$C_\varepsilon \; := \; \int_{\{\rho|u|\leq\varepsilon\}} \xi\, |\beta|^2\, \rho^2\, u^2\, dx \; \leq \; \varepsilon \cdot \| u \|_{L^q(\mathbf{R}^n\,;\,m)} \cdot \| \chi_{\{0<\rho|u|\leq\varepsilon\}} \xi |\beta|^2 \|_{L^p(\mathbf{R}^n\,;\,m)}.$$

Since β is in $L^{2p}_{\mathrm{loc}}(\mathbf{R}^n \to \mathbf{R}^n \,;\, m)$, $\varepsilon^{-1} C_\varepsilon$ converges to 0 as ε tends to 0. We obtain

$$
\begin{aligned}
& \varepsilon^{-1} \int_{\{\rho|u|\leq\varepsilon\}} \xi\, \beta \cdot \nabla(\rho u)\, \rho u\, dx \; - \; \varepsilon^{-1} \int_{\{\rho|u|\leq\varepsilon\}} \xi\, |\nabla(\rho u)|^2\, dx \\
& \leq \; \varepsilon^{-1} \left(\left(\int_{\{\rho|u|\leq\varepsilon\}} \xi\, |\nabla(\rho u)|^2\, dx \right)^{1/2} \cdot C_\varepsilon^{1/2} \; - \; \int_{\{\rho|u|\leq\varepsilon\}} \xi\, |\nabla(\rho u)|^2\, dx \right) \\
& \leq \; C_\varepsilon / 4\varepsilon \; \xrightarrow{\varepsilon\downarrow 0} \; 0.
\end{aligned}
$$

Now, by letting ε tend to 0 in (2.53), we obtain (2.52), because

$$\lim_{\varepsilon \downarrow 0} \int \nabla \xi \cdot \nabla(\rho u)\, \psi_\varepsilon(\rho u)\, dx \;=\; \int \nabla \xi \cdot \nabla(\rho u)\, \mathrm{sgn}\,(\rho u)\, dx$$

$$=\; \int \nabla \xi \cdot \nabla(\rho\, |u|)\, dx$$

by dominated convergence, and the other integrals in (2.53) converge similarly. Hence (2.52) holds for all positive functions $\xi \in C_0^\infty(\mathbf{R}^n)$. Since $|\beta| \rho u$ is in $L_{\mathrm{loc}}^2(\mathbf{R}^n\,;\, dx)$ by (2.51), (2.52) is true for all positive compactly supported functions $\xi \in H^{1,2}(\mathbf{R}^n\,;\, dx)$ as well.

Step 3: Localization.   We finally show that by (2.52) and Assumption (G 1), u cannot be globally in $L^q(\mathbf{R}^n\,;\, m)$ if γ has been chosen large enough, i.e., there is no non–trivial solution of $L^* u = \gamma u$ in this case.

Fix $r > 0$, and let $\xi(x) := (r - |x|)^+$. By (2.52), we obtain

$$(2.54) \qquad -\int_{B_r} e_r \cdot \nabla(\rho\, |u|)\, dx \;+\; \gamma \int_{B_r} (r - |x|)\, \rho\, |u|\, dx$$

$$\leq \; -\int_{B_r} \beta_r\, \rho\, |u|\, dx.$$

Note that $\mathrm{div}\, e_r(x) = (n-1)/|x| \geq 0$ dx–a.e., whence $\mathrm{div}\, e_r$ is in $L_{\mathrm{loc}}^{2-\varepsilon}(\mathbf{R}^n\,;\, dx)$ for every $\varepsilon > 0$. The function $\rho |u|$ is in $H_{\mathrm{loc}}^{1,2}(\mathbf{R}^n\,;\, dx)$, and hence, by the Sobolev embedding theorem, in $L_{\mathrm{loc}}^{2+\delta}(\mathbf{R}^n\,;\, dx)$ for some $\delta > 0$. Thus, $\mathrm{div}\,(\rho |u| e_r)$ is in $L_{\mathrm{loc}}^1(\mathbf{R}^n\,;\, dx)$. We have

$$(2.55) \quad -e_r \cdot \nabla(\rho\, |u|) \;=\; -\mathrm{div}\,(\rho |u| e_r) + \rho |u|\, \mathrm{div}\,(e_r) \;\geq\; -\mathrm{div}\,(\rho |u| e_r)$$

dx–a.e. For $s \geq 0$ let

$$g(s) \;:=\; \int_{B_s} \mathrm{div}\,(\rho |u| e_r)\, dx.$$

By multiplying with a test–function and integrating, one easily verifies that

$$g(r) \;=\; \int_{\partial B_r} \rho\, |u|\, dy \qquad \text{for a.e. } r \geq 0,$$

i.e., g is a continuous modification of the function $r \mapsto \int_{\partial B_r} \rho\, |u|\, dy$. Moreover,

$$\int_{B_r} (r - |x|)^+ \rho\, |u|\, dx \;=\; \int_0^r \int_{B_s} \rho\, |u|\, dx\, ds \;=\; \int_0^r \int_0^s g(t)\, dt\, ds.$$

Hence, by (2.54) and (2.55),

$$(2.56) \quad -g(r) + \gamma \int_0^r \int_0^s g(t)\, dt\, ds$$

$$\leq \; -\int_{B_r} \beta_r\, \rho\, |u|\, dx \;\leq\; \int_{B_r} \beta_r^-\, \rho\, |u|\, dx$$

$$\leq \; \|(\beta_r^{\mathrm{sing}})^-\|_{L^p(B_r;m)} \cdot \|u\|_{L^q(\mathbf{R}^n\,;\, m)} \;+\; \int_{B_r} (\beta_r^{\mathrm{reg}})^-\, \rho\, |u|\, dx,$$

where $\beta = \beta^{\text{sing}} + \beta^{\text{reg}}$ is a decomposition as in (G 1). We first consider the case where Assumption (2.45) holds. Then there exist finite positive constants α and C such that

(2.57) $\qquad \lim_{r \to \infty} r^{-\alpha p} \, m\,(B_r) \;=\; 0\,, \qquad$ and

(2.58) $\qquad (\beta_r^{\text{reg}}(x))^- \;\leq\; C \cdot |x| \quad$ for all x outside some ball around 0.

By changing the decomposition $\beta = \beta^{\text{sing}} + \beta^{\text{reg}}$, we may even assume that (2.58) holds for all $x \in \mathbf{R}^n$. Since u does not vanish, we have

$$\liminf_{r \to \infty} r^{-1} \int_0^r \int_0^s g(t)\,dt\,ds \;>\; 0, \qquad \text{whereas}$$

$$\lim_{r \to \infty} r^{-1} \left\| (\beta_r^{\text{sing}})^- \right\|_{L^p(B_r;m)} \;=\; 0$$

by Assumption (2.44). Hence there exists $r_0 > 0$ such that $\displaystyle\int_0^{r_0} g(t)\,dt > 0$, and

(2.59) $\qquad\displaystyle -g(r) + \frac{\gamma}{2} \int_0^r \int_0^s g(t)\,dt\,ds \;\leq\; \int_{B_r} (\beta_r^{\text{reg}})^- \, \rho\,|u|\,dx$

$$\leq\; C \int_{B_r} |x|\,\rho\,|u|\,dx \;=\; C \int_0^r s\,g(s)\,ds$$

for all $r \geq r_0$.

We can now apply the comparison lemma (Lemma 2.1). Let $G(r) := \displaystyle\int_0^r g(t)\,dt$, and $K(r) := \varepsilon \cdot r^\alpha$, where $\varepsilon > 0$ is a fixed constant. If γ has been chosen sufficiently large (i.e., $\gamma > 2C\alpha$) in the beginning, and $r_1 \in [r_0, \infty)$ is a sufficiently large constant (which does not depend on ε), then

$$-K'(r) + \frac{\gamma}{2} \int_0^r K(s)\,ds \;=\; \varepsilon\left(-\alpha r^{\alpha-1} + \frac{\gamma}{2(\alpha+1)} r^{\alpha+1} \right)$$

$$\geq\; \varepsilon \cdot C \cdot \frac{\alpha}{\alpha+1} r^{\alpha+1} \;=\; C \cdot \int_0^r s\,K'(s)\,ds$$

for all $r \geq r_1$, whereas G satisfies the opposite inequality by (2.59). Thus (2.47) holds with $A(s) = \gamma/2$ and $B(s) = Cs$. Since $r_1 \geq r_0$, we have $G(r_1) = \int_0^{r_1} g(t)\,dt > 0$, and $\int_0^{r_1} (A(s) + B'(s))\,G(s)\,ds > 0$, whence (2.48) holds if ε is chosen small enough. By Lemma 2.1, we then obtain

$$\varepsilon r^\alpha \;=\; K(r) \;<\; G(r) \;=\; \int_0^r g(s)\,ds \;=\; \int_{B_r} \rho\,|u|\,dx$$

$$\leq\; (m(B_r))^{1/p}\,\|u\|_{L^q(\mathbf{R}^n\,;\,m)} \qquad \text{for all } r \geq r_1.$$

This is a contradiction to the volume growth restriction (2.57), so there is no non–trivial solution u of $L^* u = \gamma u$ if (2.45) holds, and γ is large enough.

If (2.46) holds then we can argue similarly. We have

$$(2.60) \qquad (\beta_r^{\mathrm{reg}}(x))^- \leq C \left(1 + |x| \log^+ |x|\right) \leq C \left(|x| + e\right) \log(|x| + e)$$

for all $x \in \mathbf{R}^n$, where C is a finite constant. Hence, by (2.56) and the assumption on β^{sing}, there exists $r_0 > 0$ such that $\int_0^{r_0} g(t)\, dt > 0$, and

$$(2.61) \qquad -g(r) + \frac{\gamma}{2} \int_0^r \int_0^s g(t)\, dt\, ds \ \leq\ \int_{B_r} (\beta_r^{\mathrm{reg}})^- \, \rho\, |u|\, dx$$

$$\leq\ C \cdot \int_0^r (s + e)\, \log(s + e)\, g(s)\, ds \qquad \text{for all } r \geq r_0.$$

On the other hand, for $\varepsilon > 0$, the function $K(r) := \varepsilon \cdot \log\log(r + e)$ satisfies

$$-K'(r) + \frac{\gamma}{2} \int_0^r K(s)\, ds$$

$$= \ \varepsilon \left(-((r + e) \log(r + e))^{-1} + \frac{\gamma}{2} \int_0^r \log\log(s + e)\, ds \right)$$

$$\geq \ \varepsilon \cdot C \cdot r \ = \ C \cdot \int_0^r (s + e)\, \log(s + e)\, K'(s)\, ds$$

for all $r \geq r_1$, provided $r_1 \in [r_0, \infty)$ is a sufficiently large constant. Hence (2.47) holds with $G(r) = \int_0^r g(t)\, dt$, $A(r) := \gamma/2$, and $B(r) := C \cdot (r + e) \log(r + e)$. As above, we see that (2.48) also holds, if ε is chosen small enough. By Lemma 2.1 we then obtain

$$\varepsilon \log\log(r + e) \ = \ K(r) \ < \ G(r) \ = \ \int_0^r g(s)\, ds \ = \ \int_{B_r} \rho\, |u|\, dx$$

$$\leq \ (m(B_r))^{1/p}\, \|u\|_{L^q(\mathbf{R}^n\,;\, m)} \qquad \text{for all } r \geq r_1.$$

This is a contradiction to the assumed finiteness of the measure m, so there is no non–trivial solution u of $L^* u = \gamma u$ if (2.46) holds either. ∎

To prove Theorem 2.6 we need the following special case of a regularity result by Bogachev, Krylov and Röckner [BoKryRö 96, Thm. 1, (ii) and (iii)]:

Lemma 2.2 *Suppose $n \geq 2$. Let $\kappa \in [1, n]$ and $c \in \mathbf{R}$. Suppose μ is a signed Radon measure on $\mathbf{R}^n$ such that β is in $L_{\mathrm{loc}}^{\kappa + \varepsilon}(\mathbf{R}^n \to \mathbf{R}^n\,;\, \mu)$ for some $\varepsilon > 0$, and*

$$\int (\Delta f + \beta \cdot \nabla f + cf)\, d\mu \ = \ 0 \qquad \text{for all } f \in C_0^\infty(\mathbf{R}^n).$$

Then μ is absolutely continuous w.r.t. Lebesgue measure, and $\dfrac{d\mu}{dx}$ is in $H_{\mathrm{loc}}^{1, n/(n - \kappa + 1)}(\mathbf{R}^n\,;\, dx) \cap L_{\mathrm{loc}}^{n/(n - \kappa)}(\mathbf{R}^n\,;\, dx)$.

The highly non–trivial proof of the lemma is based on regularity results in fractional Sobolev spaces, cf. [BoKryRö 96].

Proof of Theorem 2.6. Let $u \in L^q(\mathbf{R}^n\,;\,m)$, $\frac{1}{q} + \frac{1}{p} = 1$, be a solution of $L^* u = \gamma u$ for some $\gamma \geq \lambda$. Note that, since $n \geq 2$, the assumption implies $\beta \in L^{2p}_{\text{loc}}(\mathbf{R}^n \to \mathbf{R}^n\,;\,m)$, which was one of the assumptions in Theorem 2.5. We will show moreover, that the signed measure $u \cdot m$ is absolutely continuous w.r.t. Lebesgue measure, the density $\frac{d(u \cdot m)}{dx}$ is in $H^{1,2}_{\text{loc}}(\mathbf{R}^n\,;\,dx)$, and $|\beta| \cdot \frac{d(u \cdot m)}{dx}$ is in $L^2_{\text{loc}}(\mathbf{R}^n\,;\,dx)$. Once we have shown this, the proof of Theorem 2.6 can be carried out in the same way as the proof of Theorem 2.5, starting from the end of Step 1. In fact, we only have to replace everywhere ρu by $\frac{d(u \cdot m)}{dx}$, and $\rho |u|$ by $\left| \frac{d(u \cdot m)}{dx} \right|$. Instead of $|\beta| \rho u \in L^2_{\text{loc}}(\mathbf{R}^n\,;\,dx)$ (for which the assumption $|\beta| \rho^{1/2} \in L^{2p/(2-p)}_{\text{loc}}(\mathbf{R}^n\,;\,m)$ in Theorem 2.5 was needed), we can use now that $|\beta| \cdot \frac{d(u \cdot m)}{dx}$ is in $L^2_{\text{loc}}(\mathbf{R}^n\,;\,dx)$.

By the assumption, there exists $\varepsilon > 0$ such that

$$\int_K |\beta|^{1 + \frac{n}{2} + \varepsilon} |u|\, dm \;\leq\; \|u\|_{L^q(\mathbf{R}^n\,;\,m)} \cdot \left(\int_K |\beta|^{(1 + \frac{n}{2} + \varepsilon)p}\, dm \right)^{1/p} \;<\; \infty$$

for any compact subset $K \subset \mathbf{R}^n$. Hence, by Lemma 2.2, the measure $u \cdot m$ is absolutely continuous w.r.t. Lebesgue measure, and $\frac{d(u \cdot m)}{dx} \in H^{1,2}_{\text{loc}}(\mathbf{R}^n\,;\,dx) \cap L^{2n/(n-2)}_{\text{loc}}(\mathbf{R}^n\,;\,dx)$, where $2n/(n-2) := \infty$ if $n = 2$. In particular,

$$\int_K |\beta|^2 \left(\frac{d(u \cdot m)}{dx} \right)^2 dx$$

$$\leq \left(\int_K |\beta|^{1 + \frac{n}{2}} \left| \frac{d(u \cdot m)}{dx} \right| dx \right)^{\frac{4}{n+2}} \cdot \left\| \left| \frac{d(u \cdot m)}{dx} \right|^{\frac{2n}{n+2}} \right\|_{L^{\frac{n+2}{n-2}}(K\,;\,dx)}$$

$$= \left(\int_K |\beta|^{1 + \frac{n}{2}} |u|\, dm \right)^{\frac{4}{n+2}} \cdot \left\| \frac{d(u \cdot m)}{dx} \right\|^{\frac{n+2}{2n}}_{L^{\frac{2n}{n-2}}(K\,;\,dx)} \;<\; \infty$$

for any compact subset $K \subset \mathbf{R}^n$. Thus $|\beta| \dfrac{d(u \cdot m)}{dx}$ is in $L^2_{\text{loc}}(\mathbf{R}^n\,;\,dx)$, which completes the proof. $\blacksquare$

Appendix C   Regularity of distributional solutions of ordinary differential equations

Fix $p \in [1, \infty)$, and let (x_0, y_0), $-\infty \leq x_0 < y_0 \leq \infty$, be an interval. Let α, β and ρ be functions on (x_0, y_0) such that

- ρ is continuous, and $\rho > 0$ dx–a.e.

- α is absolutely continuous, and $\alpha(x) > 0$ for all x.

- α/ρ, α'/ρ and β/ρ are in $L^p_{loc}(x_0, y_0; \rho\, dx)$.

We consider a diffusion operator $(\mathcal{L}, C_0^\infty(x_0, y_0))$ on $L^p(x_0, y_0; \rho\, dx)$ of type

$$\mathcal{L}f \;=\; \frac{1}{\rho}\Big((\alpha f')' + \beta f'\Big) \qquad \Big(= \frac{\alpha}{\rho}f'' + \frac{\alpha' + \beta}{\rho}f'\Big).$$

Theorem 2.7 *Let $u \in L^q(x_0, y_0\,;\, \rho\, dx)$, $\frac{1}{p} + \frac{1}{q} = 1$, be a solution of $L^*u = \lambda u$ for some $\lambda \in \mathbf{R}$. Then u solves the O.D.E.*

$$(\alpha u' - \beta u)' \;=\; \lambda\rho u$$

in the following sense: u has an absolutely continuous dx–version $\tilde{u}$, $\alpha\tilde{u}' - \beta u$ has a C^1 version w, and $w' = \lambda\rho\tilde{u}$.

If β is continuous, then $\alpha\tilde{u}' - \beta\tilde{u} = w$ implies the continuity of $\tilde{u}'$:

Corollary 2.4 *Suppose in addition, that β is continuous. Then any solution u of $L^*u = \lambda u$, $\lambda \in \mathbf{R}$, has a modification $\tilde{u} \in C^1(x_0, y_0)$, $\alpha\tilde{u}' - \beta\tilde{u}$ is in $C^1(x_0, y_0)$, and $(\alpha\tilde{u}' - \beta\tilde{u})' = \lambda\rho\tilde{u}$.*

PROOF OF THEOREM 2.7.   Let (x_1, y_1), $x_0 < x_1 < y_1 < y_0$, be a relatively compact subinterval of (x_0, y_0). On (x_1, y_1), ρ and α are bounded. The assumptions α/ρ, α'/ρ, $\beta/\rho \in L^p_{loc}(x_0, y_0; \rho\, dx)$ imply that $\alpha \cdot u$ $(= (\alpha/\rho)u\rho)$, $\alpha' \cdot u$, and $\beta \cdot u$ are dx–integrable on (x_1, y_1). Hence for $f \in C_0^\infty(x_1, y_1)$ we have

$$\left|\int_{x_1}^{y_1} u\,(\alpha f'' + (\alpha' + \beta)\,f')\,dx\right| \;=\; \left|\int_{x_0}^{y_0} u\,\mathcal{L}f\,\rho\, dx\right|$$

$$=\; |\lambda| \cdot \left|\int_{x_0}^{y_0} u\,f\,\rho\, dx\right|$$

$$\le\; |\lambda| \cdot |u|_{L^q(x_0, y_0\,;\,\rho\, dx)} \cdot \left(\int_{x_1}^{y_1} \rho\, dx\right)^{1/p} \cdot \|f\|_{L^\infty(x_1, y_1\,;\, dx)}$$

$$\le\; C \cdot \|f'\|_{L^1(x_1, y_1\,;\, dx)}\,,$$

where C is a finite constant that does not depend on f. Thus the linear functional

$$\ell(\eta) \;:=\; \int_{x_1}^{y_1} (u\,\alpha\,\eta' + u\,(\alpha' + \beta)\,\eta)\,dx$$

is continuous on $\{f'\,;\, f \in C_0^\infty(x_1, y_1)\}$ w.r.t. the norm in $L^1(x_1, y_1\,;\, dx)$, i.e., there exists $v \in L^\infty(x_1, y_1\,;\, dx)$ such that

$$\int_{x_1}^{y_1} (u\,\alpha\,\eta' + u\,(\alpha' + \beta)\,\eta)\,dx \;=\; \int_{x_1}^{y_1} v\,\eta\,dx, \qquad \text{respectively}$$

(2.62) $$\int_{x_1}^{y_1} u\,\alpha\,\eta'\,dx \;=\; \int_{x_1}^{y_1} (v - u\,(\alpha' + \beta))\,\eta\,dx$$

for all $\eta = f'$, $f \in C_0^\infty(x_1, y_1)$.

Now fix an arbitrary function $\eta_0 \in C_0^\infty(x_1, y_1)$ such that $\int_{x_1}^{y_1} \eta_0 \, dx \neq 0$. Note that

(2.63) $\qquad C_0^\infty(x_1, y_1) = \{f' + t \cdot \eta_0 \, ; \, f \in C_0^\infty(x_1, y_1), \, t \in \mathbf{R}\}.$

We choose $c \in \mathbf{R}$ such that

$$\int_{x_1}^{y_1} u \, \alpha \, \eta_0' \, dx = \int_{x_1}^{y_1} (v - u \, (\alpha' + \beta) + c) \, \eta_0 \, dx.$$

Then, by (2.63) and (2.62),

$$\int_{x_1}^{y_1} u \, \alpha \, \eta' \, dx = \int_{x_1}^{y_1} (v - u \, (\alpha' + \beta) + c) \, \eta \, dx$$

for all $\eta \in C_0^\infty(x_1, y_1)$. Hence $u \, \alpha$ has an absolutely continuous dx–version on (x_1, y_1). Since α is absolutely continuous and strictly positive, u also has an absolutely continuous dx–version $\tilde{u}$ on (x_1, y_1). Moreover,

$$\lambda \int_{x_1}^{y_1} u \, f \, \rho \, dx = \int_{x_1}^{y_1} u \, \mathcal{L}f \, \rho \, dx$$
$$= \int_{x_1}^{y_1} u \, (\alpha f')' \, dx + \int_{x_1}^{y_1} u \, \beta \, f' \, dx$$
$$= -\int_{x_1}^{y_1} (\alpha \tilde{u}' - \beta u) \, f' \, dx \qquad \text{for all } f \in C_0^\infty(x_1, y_1).$$

Hence $\alpha \tilde{u}' - \beta u$ also has an absolutely continuous dx–version w on (x_1, y_1), and $w' = \lambda \rho \tilde{u}$ dx–a.e. Since $\rho \tilde{u}$ is continuous, w is even C^1, and the equation $w' = \lambda \rho \tilde{u}$ holds everywhere on (x_1, y_1). This proves the assertion, because (x_1, y_1) was an arbitrary relatively compact subinterval of (x_0, y_0). ∎

Chapter 3

Markov uniqueness

Suppose we are given a densely defined, negative definite symmetric diffusion operator $(\mathcal{L}, \mathcal{A})$ on some L^2 space. Then the Friedrichs extension $L^{(0)}$ of $(\mathcal{L}, \mathcal{A})$ is the generator of a symmetric sub–Markovian C^0 contraction semigroup. We want to know under which conditions it is the only extension of $(\mathcal{L}, \mathcal{A})$ which generates such a semigroup. It turns out that in many applications, this question can be answered by proceeding in three steps :

1.) Construction of a **maximal extension** $\hat{L}$ of $(\mathcal{L}, \mathcal{A})$ which generates a symmetric sub–Markovian C^0 contraction semigroup. This construction will be carried out for arbitrary symmetric diffusion operators on general state spaces in Section c). It turns out that the domain of the Dirichlet form of $\hat{L}$ can be characterized in terms of a **weak Sobolev space** corresponding to the operator $(\mathcal{L}, \mathcal{A})$. This weak Sobolev space will be introduced in Section a) for symmetric diffusion operators on domains in $\mathbf{R}^n$ and on Banach spaces, and in Section b) for symmetric diffusion operators on general state spaces.

2.) Derivation of simple descriptions for the Dirichlet form $\hat{\mathcal{E}}$ corresponding to the extension $\hat{L}$ in concrete applications; cf. in particular Section a), and, to some extent, Section b).

3.) Proof of Markov uniqueness (or non–Markov uniqueness) by showing that the test–functions are dense (resp. not dense) in the domain of $\hat{\mathcal{E}}$, which is essentially a weak Sobolev space. Markov uniqueness holds if and only if $\hat{L} = L^{(0)}$, i.e., if and only if $\mathcal{A}$ is dense in the domain of $\hat{\mathcal{E}}$.

Step 3 is carried out in Section d) in the one–dimensional case, and in Sections e) and f) in the finite dimensional case. Markov uniqueness of infinite dimensional diffusion operators is studied in Chapter 5. In Section g), we discuss relations of Markov uniqueness to ergodicity and extremality of Gibbs states.

The results of this chapter have to a large part been announced in the Comptes rendus note [Eb 95]. However, some improvements and new applications, that have been found since the appearance of the note, are included here as well.

To describe the results presented below in detail, it is convenient to look first at a simple example. Suppose E is an open subset in $\mathbf{R}^n$, $n \in \mathbf{N}$, and $\mathcal{A} = C_0^\infty(E)$. Let $(\mathcal{L}, \mathcal{A})$ be the densely defined symmetric diffusion operator on $L^2(E; \rho\, dx)$ corresponding to the pre–Dirichlet form

$$\mathcal{E}(f, g) = \int_E \sum_{i,j=1}^n a_{ij} \frac{\partial f}{\partial x_i} \frac{\partial g}{\partial x_j}\, \rho\, dx, \qquad f, g \in \mathcal{A},$$

i.e.,

$$\mathcal{L}g = \frac{1}{\rho} \sum_{i,j=1}^n \frac{\partial}{\partial x_i} \left(\rho\, a_{ij} \frac{\partial g}{\partial x_j} \right)$$

$$= \sum_{i,j=1}^n \left(a_{ij} \frac{\partial^2 g}{\partial x_i \partial x_j} + \frac{1}{\rho} \frac{\partial}{\partial x_i} (\rho a_{ij}) \frac{\partial g}{\partial x_j} \right), \qquad g \in \mathcal{A}.$$

For the moment, let us assume that ρ and a_{ij} are sufficiently smooth (e.g. C^1) functions on E, $\rho(x) > 0$ for dx–a.e. x, the matrix $(a_{ij}(x))$ is symmetric and strictly positive definite for a.e. x, and $\mathcal{L}g$ is in $L^2(E; \rho\, dx)$ for all $g \in \mathcal{A}$. Later, we will study the same operator under partially less restrictive conditions on the coefficients.

For describing the maximal Dirichlet extension of $(\mathcal{L}, \mathcal{A})$, it turns out to be convenient to use a more geometric notation. For $x \in E$, let $(\cdot, \cdot)_{T_x' E}$ be the inner product on $(\mathbf{R}^n)^*$ given by

$$(\omega, \sigma)_{T_x' E} = \sum_{i,j=1}^n a_{ij}(x)\, \omega(e_i)\, \sigma(e_j),$$

where e_i is the i–th canonical unit vector in $\mathbf{R}^n$. Then $x \mapsto (\cdot, \cdot)_{T_x' E}$ defines a metric on the co–tangent bundle $T'E$, and we have

$$\mathcal{E}(f, g) = \int_E (d_x f, d_x g)_{T_x' E}\, \rho(x)\, dx \qquad \text{for all } f, g \in \mathcal{A}.$$

We briefly write

$$\mathcal{E}(f, g) = \int_E (df, dg)_{T'E}\, \rho\, dx.$$

The operator $\mathcal{L}$ is hence given as $\mathcal{L} = -d^*d$, where d^* is the adjoint of the densely defined linear operator $d : \mathcal{A} \subset L^2(E; \rho\, dx) \to L^2(E \to T'E; \rho\, dx)$. Here $L^2(E \to T'E; \rho\, dx)$ denotes the $\rho\, dx$–square integrable sections of the bundle $T'E$ with metric defined as above.

In the situation described, we can prove the existence of a maximal Dirichlet extension $\hat{L}$ of the operator $(\mathcal{L}, \mathcal{A})$ on $L^2(E; \rho\, dx)$, cf. Section c). The corresponding Dirichlet form $\hat{\mathcal{E}}$ can be described in the following way :

A bounded function $u \in L^2(E; \rho\,dx)$ is in the domain of $\hat{\mathcal{E}}$ if and only if there exists a section $\hat{d}u \in L^2(E \to T'E; \rho\,dx)$ such that

$$\int \left(\hat{d}u, \omega\right)_{T'E} \rho\,dx \;=\; \int u \, d^*\omega \; \rho\,dx$$

holds for all $\omega \in C_0^\infty(E \to (\mathbf{R}^n)^*)$. In this case,

$$\hat{\mathcal{E}}(u, u) \;=\; \int \left(\hat{d}u, \hat{d}u\right)_{T'E} \rho\,dx .$$

The one–form $\hat{d}u$ can be viewed as a *weak derivative* of u. In this sense, bounded functions in $L^2(E; \rho\,dx)$ are in the domain of $\hat{\mathcal{E}}$ if and only if they are in the Sobolev space $W^{1,2}(E, T'E; \rho\,dx)$ of all weakly differentiable functions in $L^2(E; \rho\,dx)$ with derivatives in $L^2(E \to T'E; \rho\,dx)$. In the sequel, we will often use the brief notation $W^{1,2}(d)$ instead of $W^{1,2}(E, T'E; \rho\,dx)$. Note that, in spite of this somehow inaccurate notation, our *definition of the weak Sobolev space depends essentially on the chosen metric on $T'E$, and on the measure $\rho\,dx$.*

To prove Markov uniqueness (or non–Markov uniqueness) of the operator $(\mathcal{L}, \mathcal{A})$, we need more explicit descriptions of the space $W^{1,2}(d)$. We first give a reformulation of the definition above in terms of directional derivatives. Consider the vector fields $X_i = \sum_{j=1}^n a_{ij} \frac{\partial}{\partial x_j}$ on E, $1 \le i \le n$. Note that for $f \in C_0^\infty(E)$, we have $X_i f = (df, dx_i)_{T'E}$. It is not difficult to verify that a function $u \in L^2(E; \rho\,dx)$ is in $W^{1,2}(d)$ if and only if :

(i) For every $1 \le i \le n$, there exists a function $\hat{X}_i u \in L^2_{\text{loc}}(E; \rho\,dx)$ such that

$$\int \hat{X}_i u \, f \, \rho\,dx \;=\; -\int u \, X_i f \; \rho\,dx \;-\; \int u \, f \, \text{div}\,(\rho X_i) \, dx$$

holds for all $f \in C_0^\infty(E)$.

(ii) There exists a section $\hat{d}u \in L^2(E \to T'E; \rho\,dx)$ such that $\hat{X}_i u = \hat{d}u\,(X_i)$ for all $1 \le i \le n$.

Now suppose for the moment in addition, that $\rho(x) > 0$ for all $x \in E$, and the matrix $(a_{ij}(x))$ is strictly positive definite for all x as well. Then, by using the above description of $W^{1,2}(d)$, we can prove that a function $u \in L^2(E; \rho\,dx)$ is in $W^{1,2}(d)$ if and only if u is in $H^{1,2}_{\text{loc}}(E; dx)$, and

$$\int_E \sum_{i,j=1}^n a_{ij} \frac{\partial u}{\partial x_i} \frac{\partial u}{\partial x_j} \, \rho\,dx < \infty ,$$

cf. Lemma 3.2. In this case, $\hat{d}u = \sum_{i=1}^n \frac{\partial u}{\partial x_i} dx_i$, where $\frac{\partial u}{\partial x_i}$ is the usual weak derivative. Using this characterization, we can now show by more or less standard techniques, how Markov uniqueness depends on the boundary behaviour of the operator coefficients, cf. Sections d) and f).

In the singular case, i.e., if $\rho(x)$ has zeros or $(a_{ij}(x))$ has points of degeneracy, the last description of $W^{1,2}(d)$ does *not* hold. In the one–dimensional case, for example, functions in $W^{1,2}(d)$ can have jumps at zeros of $\rho(x)$ and at points of degeneracy of $a(x)$, cf. Example (i) under Lemma 3.1. In Section a) we will show, how nevertheless useful characterizations of $W^{1,2}(d)$ can be derived in singular cases.

So far, we have only described how to study the Markov uniqueness problem for diffusion operators on $\mathbf{R}^n$. Our considererations relied heavily on the geometric representation $\mathcal{L} = -d^*d$. However, a similar representation holds for every symmetric diffusion operator defined on an L^2 space over an arbitrary measure space, provided we generalize the notions of a "differential" and a "cotangent bundle" in an appropriate (and in some sense rather natural) way. This allows us to carry out a large part of the considerations above for diffusion operators with very general state spaces, cf. Section b). In particular, we are able to introduce a weak Sobolev space for every symmetric diffusion operator, and to prove a maximality result as described above, cf. Section c). This can be (and has been) applied for example to particle systems, cf. [AlbKoRö 97b].

a) Weak Sobolev spaces corresponding to diffusion operators on $\mathbf{R}^n$ and on Banach spaces

In this section, we introduce weak Sobolev spaces corresponding to symmetric diffusion operators on domains in $\mathbf{R}^n$ and on Banach spaces. Moreover, we derive explicit characterizations of elements in the weak Sobolev spaces, cf. the lemmas 3.1, 3.2 and 3.3. In Subsection 1), we consider finite dimensional diffusion operators with scalar diffusion coefficient, in Subsection 2) we look at general diffusion operators in $\mathbf{R}^n$, and in Subsection 3) we study the Banach space case. In the finite dimensional cases, we allow a certain kind of degeneracy of the diffusion matrices, cf. the assumption in Subsection 2).

1) Weighted Sobolev spaces on $\mathbf{R}^n$ and diffusion operators with conformally flat geometry

We first look at a higher dimensional analogue of Sturm–Liouville type diffusion operators. The case of one–dimensional symmetric diffusion operators is completely included. Suppose E is an open subset in $\mathbf{R}^n$, $n \in \mathbf{N}$, and $\mathcal{A} = C_0^\infty(E)$. Fix $\rho \in L^1_{\mathrm{loc}}(E; dx)$, $\rho > 0$ dx–a.e., and let $m := \rho\, dx$. Let a be a function in $L^2_{\mathrm{loc}}(E; m)$, $a > 0$ dx–a.e., and let $\alpha := a \cdot \rho$. We assume that α is in $H^{1,1}_{\mathrm{loc}}(E; dx)$, and $\frac{1}{\rho}\frac{\partial \alpha}{\partial x_i}$ is in $L^2_{\mathrm{loc}}(E; m)$ for all $1 \leq i \leq n$. These conditions are for example satisfied, if $\rho = \varphi^2$ for some $\varphi \in H^{1,2}_{\mathrm{loc}}(E; dx)$, and a is a locally bounded function in $H^{1,1}_{\mathrm{loc}}(E; dx)$ with derivatives $\frac{\partial a}{\partial x_i}$ in $L^2_{\mathrm{loc}}(E; m)$, $1 \leq i \leq n$.

We consider the symmetric diffusion operator

$$\mathcal{L} \;=\; \frac{1}{\rho}\, \mathrm{div}\,(\,a\,\nabla \cdot\,)$$

$$=\; \mathrm{div}\,(\,a\,\nabla\cdot\,) + a\,\frac{\nabla\rho}{\rho}\cdot\nabla \;=\; a\,\Delta + \frac{\nabla(a\rho)}{\rho}\cdot\nabla$$

with domain $C_0^\infty(E)$ on $L^2(E\,;\,m)$. The corresponding pre–Dirichlet form is

$$\mathcal{E}\,(f,\,g) \;=\; \int_E \alpha\,\nabla f\cdot\nabla g\; dx \;=\; \int_E a\,\nabla f\cdot\nabla g\; dm,$$

$f,\,g \in C_0^\infty(E) \subset L^2(E\,;\,m)$. Weak Sobolev spaces corresponding to this type of diffusion operators are easier to characterize than weak Sobolev spaces corresponding to general diffusion operators on $\mathbf{R}^n$, because the induced metric

$$(3.1) \qquad (\omega,\,\sigma)_{T_x' E} \;=\; a(x)\sum_{i=1}^{n} \omega(e_i)\,\sigma(e_i), \qquad \omega,\,\sigma \in (\mathbf{R}^n)^*,$$

on the co–tangent bundle is *conformally flat*, i.e., it differs from the Euclidean metric only by a scalar factor.

Definition 3.1 *The weak Sobolev space $W^{1,2}(E,\,a;\,m)$ is the space consisting of all functions $u \in L^2(E\,;\,m)$ with the following property :*
There exist functions $v_i \in L^2(E;\,\alpha\,dx)$, $1 \le i \le n$, such that the integration by parts identities

$$(3.2) \qquad \int_E \alpha\,v_i\,g\; dx \;=\; -\int_E u\,\frac{\partial}{\partial x_i}(\alpha g)\; dx,$$

$1 \le i \le n$, hold for all $g \in C_0^\infty(E)$.
For $u \in W^{1,2}(E,\,a;\,m)$ and $1 \le i \le n$, the uniquely determined function $v_i \in L^2(E;\,\alpha\,dx)$ satisfying (3.2) is called the weak derivative of u in direction e_i, and denoted by $\hat{\partial}_i u$.

REMARKS.   (i) Note that the integral on the right hand side of (3.2) is defined, since u and $\frac{1}{\rho}\frac{\partial\alpha}{\partial x_i}$ are in $L^2(E;\,\rho\,dx)$.
(ii) The space $W^{1,2}(E,\,a;\,m)$ coincides with the space $W^{1,2}(E,\,T'E;\,m)$ defined in the introduction of this chapter, if the metric on $T'E$ is given by (3.1). The weak differential $\hat{d}u$ of a function $u \in W^{1,2}(E,\,a;\,m)$ as defined in the introduction is given by $\hat{d}u = \sum_{i=1}^n \hat{\partial}_i u\; dx_i$. In particular, $W^{1,2}(E,\,a;\,m)$ is a Hilbert space with inner product

$$(u,\,v)_{W^{1,2}(E,\,a;\,m)} \;=\; \int_E \alpha\sum_{i=1}^{n}\hat{\partial}_i u\,\hat{\partial}_i v\; dx + \int_E u\,v\; dm.$$

In fact : The adjoint d^* of the differential d viewed as an operator from $L^2(E\,;\,m)$ to $L^2(E \to T'E;\,m)$ is given by $d^*(g\,dx_i) = -\frac{1}{\rho}\frac{\partial}{\partial x_i}(\alpha g)$ for $1 \le i \le n$ and

$g \in C_0^\infty(E)$, because

$$\int (df, g\, dx_i)_{T'E}\, \rho\, dx = \int \alpha \frac{\partial f}{\partial x_i} g\, dx = -\int f \frac{1}{\rho} \frac{\partial}{\partial x_i} (\alpha g)\, \rho\, dx$$

for all $f \in C_0^\infty(E)$. Hence u is in $W^{1,2}(E, T'E; m)$ if and only if there exists $\hat{d}u \in L^2(E \rightarrow T'E; m)$, $\hat{d}u = \sum_{i=1}^n v_i\, dx_i$, such that

$$\int \alpha v_i g\, dx = \int \left(\hat{d}u, g\, dx_i \right)_{T'E} \rho\, dx$$

$$= \int u\, d^*(g\, dx_i)\, \rho\, dx = -\int u \frac{\partial}{\partial x_i} (\alpha g)\, dx$$

for all $1 \leq i \leq n$ and $g \in C_0^\infty(E)$. The condition $\hat{d}u \in L^2(E \rightarrow T'E; m)$ means that v_i is in $L^2(E; \alpha\, dx)$ for all $1 \leq i \leq n$. Thus $W^{1,2}(E, T'E; m) = W^{1,2}(E, a; m)$.

(iii) The definition of $W^{1,2}(E, a; m)$ seems to depend only in a marginal way on the measure m, respectively the function ρ. However, if we write down the definition in terms of a and ρ instead of α and ρ, then the dependence of the weak Sobolev space on ρ becomes much more obvious. For example, under suitable differentiability assumptions on a and ρ, the integration by parts formula (3.2) can be rewritten as

$$(3.3) \qquad \int a v_i g\, dm = -\int u \frac{\partial}{\partial x_i} (ag)\, dm - \int u g\, a\, \beta_i\, dm$$

with $\beta_i = \frac{1}{\rho} \frac{\partial \rho}{\partial x_i}$.

Fix $1 \leq i \leq n$. Let $\mathbf{R}_i^n$ denote the hyperplane in $\mathbf{R}^n$ orthogonal to the i-th canonical unit vector e_i. For $y \in \mathbf{R}_i^n$ let $J_i(y) := \{ s \in \mathbf{R};\ y + s e_i \in E \}$. The sets $J_i(y)$, $y \in \mathbf{R}_i^n$, are open subsets of $\mathbf{R}$, i.e., disjoint unions of open intervals. For a function $u : E \rightarrow \mathbf{R}$ let $u_i(y, \cdot) : J_i(y) \rightarrow \mathbf{R}$, $y \in \mathbf{R}_i^n$, be the functions defined by $u_i(y, s) = u(y + s e_i)$. Let dy denote Lebesgue measure on $\mathbf{R}_i^n$, respectively $dy := \delta_0$ if $n = 1$. It is well-known that $\alpha \in H^{1,1}_{\mathrm{loc}}(E; dx)$ implies that α has a dx-version $\tilde{\alpha}$ (possibly depending on i) such that $\tilde{\alpha}_i(y, \cdot)$ is absolutely continuous on $J_i(y)$ for every $y \in \mathbf{R}_i^n$. Let

$$S(\tilde{\alpha}_i(y, \cdot)) := \{ s \in J_i(y);\ \tilde{\alpha}_i(y, s) = 0 \}$$

denote the corresponding singularity set. Recall that we have introduced similar singularity sets in our study of L^p uniqueness of Sturm–Liouville operators on $\mathbf{R}^1$, cf. Section e) in Chapter 2 above. Note that $\{ y + s e_i;\ y \in \mathbf{R}_i^n, s \in S(\tilde{\alpha}_i(y, \cdot)) \} = \{ x \in E;\ \tilde{\alpha}(x) = 0 \}$, which is a Lebesgue measure zero set. We have the following explicit description of $W^{1,2}(E, a; m)$:

Lemma 3.1 Let u be a function in $L^2(E; m)$. Then u is in $W^{1,2}(E, a; m)$ if and only if the following condition holds for every $1 \leq i \leq n$:

For dy–a.e. $y \in \mathbf{R}_i^n$, the function $u_i(y, \cdot)$ has an absolutely continuous ds–version

$\tilde{u}_i(y,\cdot)$ on $J_i(y) \setminus S(\tilde{\alpha}_i(y,\cdot))$. *There exists a function* $\frac{\partial u}{\partial x_i} \in L^2(E; \alpha\,dx)$ *such that*

$$\frac{\partial u}{\partial x_i}(y + se_i) = \frac{\partial}{\partial s}\tilde{u}_i(y,s) \qquad \text{for } ds\text{-a.e. } s \in J_i(y) \setminus S(\tilde{\alpha}_i(y,\cdot))$$

holds for dy*-a.e.* $y \in \mathbf{R}_i^n$.

If u *is in* $W^{1,2}(E, a; m)$, *then* $\frac{\partial u}{\partial x_i}$ *is* dx*-a.e. uniquely determined, and* $\frac{\partial u}{\partial x_i} = \hat{\partial}_i u$ dx*-a.e.*

REMARK. The lemma implies in particular that the bilinear form

$$\hat{\mathcal{E}}(u, v) = \int_E \sum_{i=1}^n \hat{\partial}_i u\, \hat{\partial}_i v\, \alpha\,dx, \qquad u, v \in W^{1,2}(E, a; m),$$

is a Dirichlet form on $L^2(E; m)$, cf. the proof below.

EXAMPLES. (i) (*One-dimensional case*). Suppose E is an interval. Then α has an absolutely continuous dx–version. Without loss of generality, we may assume that α itself is absolutely continuous. Let $S(\alpha) = \{s \in E\,; \alpha(s) = 0\}$. Then a function $u \in L^2(E; \rho\,dx)$ is in the weak Sobolev space $W^{1,2}(E, a; \rho\,dx)$ corresponding to the operator $(\frac{1}{\rho}\frac{d}{dx}(\alpha\frac{d}{dx}\cdot)\,,\, C_0^\infty(E))$ on $L^2(E; \rho\,dx)$, if and only if u has a modification $\tilde{u}$ that is absolutely continuous on $E \setminus S(\alpha)$, and $\int_E \alpha\,(\tilde{u}')^2\,dx < \infty$.

(ii) (*Generalized Schrödinger operators*). Suppose again that E is an open subset in $\mathbf{R}^n$, and let φ be a function in $H_{\text{loc}}^{1,2}(E; dx)$ such that $\varphi > 0$ dx–a.e. Suppose $\rho = \varphi^2$ and $a = 1$, i.e., $\alpha = \varphi^2$ as well. Then the conditions on α and ρ imposed above are satisfied. In particular, $\frac{1}{\rho}\frac{\partial\alpha}{\partial x_i} = \frac{2}{\varphi}\frac{\partial\varphi}{\partial x_i} \in L_{\text{loc}}^2(E; \varphi^2\,dx)$ for $1 \leq i \leq n$. The corresponding diffusion operator $\mathcal{L}$ is a generalized Schrödinger operator, i.e., $\mathcal{L} = \Delta + 2\frac{\nabla\varphi}{\varphi}\cdot\nabla$ on $C_0^\infty(E) \subset L^2(E; \varphi^2 dx)$. The Markov uniqueness problem for this type of operators has been solved completely in previous work by M. Röckner and T. S. Zhang [RöZha 92, 94], and M. Takeda [Ta 92]. The corresponding weak Sobolev space $W^{1,2}(E, 1; \varphi^2 dx)$ can be viewed as a weighted Sobolev space w.r.t. the measure $\varphi^2 dx$. It has already been introduced and characterized in various ways in [AlbKusRö 90] (– although here the notion "weak Sobolev space" has not been used).

PROOF OF LEMMA 3.1. By definition, u is in $W^{1,2}(E, a; m)$ if and only if there exist $v_i \in L^2(E; \alpha\,dx)$, $1 \leq i \leq n$, such that (3.2) holds for all $g \in C_0^\infty(E)$. Since α is in $H_{\text{loc}}^{1,1}(E; dx)$, we can rewrite (3.2) as

$$(3.4) \qquad \int_E \alpha\,u\,\frac{\partial g}{\partial x_i}\,dx = -\int_E g\left(\frac{\partial\alpha}{\partial x_i}u + \alpha v_i\right)\,dx,$$

where $\frac{\partial\alpha}{\partial x_i}$ denotes the weak derivative. Note that $\frac{\partial\alpha}{\partial x_i}u$ is locally dx–integrable, because $\frac{\partial\alpha}{\partial x_i}u = \frac{1}{\rho}\frac{\partial\alpha}{\partial x_i}u\rho$, and the functions $\frac{1}{\rho}\frac{\partial\alpha}{\partial x_i}$ and u are in $L_{\text{loc}}^2(E; \rho\,dx)$ by assumption.

Now suppose first that u is in $W^{1,2}(E, a; m)$. Then there exist functions $v_i \in$

$L^2(E; \alpha \, dx)$, $1 \leq i \leq n$, such that (3.4) holds. Fix $1 \leq i \leq n$. The equation (3.4) means that the distributional derivative of the function αu in direction e_i is equal to $w := \frac{\partial \alpha}{\partial x_i} u + \alpha v_i$, which is in $L^1_{\text{loc}}(E; dx)$. It is a well–known consequence that for dy–a.e. $y \in \mathbf{R}^n_i$, the function $\alpha_i(y, \cdot) \, u_i(y, \cdot)$ has an absolutely continuous ds–version $\psi_i(y, \cdot)$ on $J_i(y)$, and

$$\frac{\partial}{\partial s} \psi_i(y, s) \; = \; w_i(y, s) \qquad \text{for } ds\text{–a.e. } s \in J_i(y),$$

cf. e.g. [Maz 85, Sect. 1.1.3]. On the other hand, the functions $\tilde{\alpha}_i(y, \cdot)$ are absolutely continuous on $J_i(y)$ for every $y \in \mathbf{R}^n$ as well. Hence for dy–a.e. $y \in \mathbf{R}^n_i$, the function $\bar{u}_i(y, \cdot) := \psi_i(y, \cdot)/\tilde{\alpha}_i(y, \cdot)$ is a ds–version of $u_i(y, \cdot)$, which is absolutely continuous on $J_i(y) \setminus S(\tilde{\alpha}_i(y, \cdot))$. Moreover, by the product rule and the definitions of w_i and ψ_i,

$$\frac{\partial}{\partial s} \bar{u}_i(y, s) \; = \; \frac{w_i(y, s)}{\tilde{\alpha}_i(y, s)} - \frac{\psi_i(y, s) \cdot (\partial \tilde{\alpha}_i / \partial s)(y, s)}{(\tilde{\alpha}_i(y, s))^2} \; = \; v_i\,(y + se_i)$$

for ds–a.e. $s \in J_i(y) \setminus S(\tilde{\alpha}_i(y, \cdot))$ and dy–a.e. $y \in \mathbf{R}^n_i$. Hence the condition in the lemma holds with $\frac{\partial u}{\partial x_i} = v_i \in L^2(E; \alpha \, dx)$.

Now let $\tilde{W}^{1,2}(E, a; m)$ denote the space of all functions $u \in L^2(E; m)$ satisfying the conditions in Lemma 3.1. We have just shown that $W^{1,2}(E, a; m) \subseteq \tilde{W}^{1,2}(E, a; m)$, and $\frac{\partial u}{\partial x_i} = \hat{\partial}_i u$ for all $u \in W^{1,2}(E, a; m)$ and $1 \leq i \leq n$.

To prove the converse inclusion , note that the bilinear form

$$\tilde{\mathcal{E}}\,(u, v) \; = \; \int_E \sum_{i=1}^n \frac{\partial u}{\partial x_i} \frac{\partial v}{\partial x_i} \; \alpha \, dx\,, \quad u, v \in \tilde{W}^{1,2}(E, a; m),$$

is a Dirichlet form. This is not difficult to show by using that the operators $\frac{\partial}{\partial x_i}$, $1 \leq i \leq n$, defined in the assertion of Lemma 3.1 satisfy a chain rule, cf. e.g. [AlbRö 90, Thm. 3.8]. In particular, the bounded functions are dense in $\tilde{W}^{1,2}(E, a; m)$ w.r.t. the norm $\|u\|_{\tilde{W}^{1,2}(E, a; m)} := \left(\tilde{\mathcal{E}}(u, u) + \int u^2 \, dm \right)^{1/2}$. Thus to complete the proof of the lemma, it suffices to show $\tilde{W}^{1,2}(E, a; m) \cap L^\infty(E; m) \subseteq W^{1,2}(E, a; m)$.

Now fix a bounded function $u \in \tilde{W}^{1,2}(E, a; m)$. We want to show that (3.2) holds for every $g \in C_0^\infty(E)$ with $v_i = \frac{\partial u}{\partial x_i}$. Fix $g \in C_0^\infty(E)$. Let $y \in \mathbf{R}^n_i$ such that $u_i(y, \cdot)$ has an absolutely continuous ds–version $\bar{u}_i(y, \cdot)$ on $J_i(y) \setminus S(\tilde{\alpha}_i(y, \cdot))$. Suppose $[x_1, y_1]$ is a compact subinterval of $J_i(y) \setminus S(\tilde{\alpha}_i(y, \cdot))$. Then $\bar{u}_i(y, \cdot)$ is absolutely continuous on $[x_1, y_1]$, whence integration by parts yields

$$\int_{t_1}^{t_2} \bar{u}_i(y, s) \frac{\partial}{\partial s}(\tilde{\alpha}_i \, g_i)(y, s) \; ds$$

$$= \; (\bar{u}_i \tilde{\alpha}_i g_i)(y, t_2) - (\bar{u}_i \tilde{\alpha}_i g_i)(y, t_1) - \int_{t_1}^{t_2} \tilde{\alpha}_i(y, s) \, g_i(y, s) \frac{\partial}{\partial s} \bar{u}_i(y, s) \; ds.$$

By letting t_1 and t_2 tend to the infimum respectively supremum of a connection component of $J_i(y) \setminus S(\tilde{\alpha}_i(y, \cdot))$, we see that

$$(3.5) \quad \int_I \tilde{u}_i(y, s) \frac{\partial}{\partial s} (\tilde{\alpha}_i g_i)(y, s) \, ds \; = \; - \int_I \tilde{\alpha}_i(y, s) g_i(y, s) \frac{\partial}{\partial s} \tilde{u}_i(y, s) \, ds$$

holds for every connection component I. Here we have used that $(\tilde{u}_i \tilde{\alpha}_i g_i)(y, s)$ converges to 0 as s tends to the boundary of a connection component of $J_i(y) \setminus S(\tilde{\alpha}_i(y, \cdot))$, because u (and thus $\tilde{u}_i(y, \cdot)$) is bounded, $\tilde{\alpha}_i(y, \cdot)$ is continuous and vanishes on $S(\tilde{\alpha}_i(y, \cdot))$, and $g_i(y, \cdot)$ is continuous and vanishes near the infimum and supremum of $J_i(y) \setminus S(\tilde{\alpha}_i(y, \cdot))$, since g is in $C_0^\infty(E)$. Summing up over all connection components, we see that (3.5) holds with I replaced by $J_i(y) \setminus S(\tilde{\alpha}_i(y, \cdot))$, i.e.,

$$\int_{J_i(y) \setminus S(\tilde{\alpha}_i(y, \cdot))} u(y + se_i) \frac{\partial}{\partial x_i} (\alpha g)(y + se_i) \, ds$$
$$= - \int_{J_i(y) \setminus S(\tilde{\alpha}_i(y, \cdot))} \alpha(y + se_i) g(y + se_i) \frac{\partial u}{\partial x_i}(y + se_i) \, ds$$

holds for dy–a.e. $y \in \mathbf{R}_i^n$ with $\frac{\partial u}{\partial x_i}$ defined as in the condition in the claim of Lemma 3.1. Integrating over y, we obtain (3.2) with $v_i = \frac{\partial u}{\partial x_i}$, because $\{ y + se_i \, ; \, y \in \mathbf{R}_i^n, \, s \in S(\tilde{\alpha}_i(y, \cdot)) \}$ has Lebesgue measure zero. Since i has been chosen arbitrarily, u is in $W^{1,2}(E, a; m)$. ∎

2) General symmetric diffusion operators on $\mathbf{R}^n$

We now characterize weak Sobolev spaces corresponding to a broad class of symmetric diffusion operators on domains in $\mathbf{R}^n$, including singular and degenerate operators with non–trivial geometry. As before, let E be an open subset in $\mathbf{R}^n$, $n \in \mathbf{N}$. We fix a function $\varphi \in H^{1,2}_{\mathrm{loc}}(E; dx)$, $\varphi > 0$ dx–a.e., and set $m := \varphi^2 dx$. The weighted Sobolev space $H_0^{1,2}(E; m)$ is defined as the completion of $C_0^\infty(E)$ w.r.t. the norm $\|f\|_{W^{1,2}(E; m)} := \left(\int_E (f^2 + |\nabla f|^2) \, dm \right)^{1/2}$, i.e., $H_0^{1,2}(E; m)$ is the closure of $C_0^\infty(E)$ in the weak Sobolev space $W^{1,2}(E; m) := W^{1,2}(E, 1; m)$ introduced in Subsection 1). The local weighted Sobolev space $H^{1,2}_{\mathrm{loc}}(E; m)$ is defined to consist of all functions $u \in L^2_{\mathrm{loc}}(E; m)$ such that for every relatively compact open subset $U \subset E$, u coincides on U m–a.e. with a function in $H_0^{1,2}(E; m)$.

REMARK.   Similarly, we could define a weak local Sobolev space $W^{1,2}_{\mathrm{loc}}(E; m)$ It is a highly non–trivial, but known fact, that under the assumptions above, $H^{1,2}_{\mathrm{loc}}(E; m) = W^{1,2}_{\mathrm{loc}}(E; m)$, cf. Theorem 3.4 below.

We fix functions $a^{ij} \in H^{1,2}_{\mathrm{loc}}(E; m) \cap L^\infty_{\mathrm{loc}}(E; m)$ such that $a^{ij} = a^{ji}$ for all $1 \leq i, j \leq n$. Instead of assuming locally strict positivity of the matrix (a^{ij}), we impose only the following weaker assumption, which allows some degeneracy provided it appears uniformly in all directions :

Assumption : For $1 \leq i, j \leq n$ and $x \in E$,

$$a^{ij}(x) \;=\; a_-(x) \cdot c^{ij}(x)$$

with functions $a_-, c^{ij} \in H^{1,2}_{\mathrm{loc}}(E; m) \cap L^{\infty}_{\mathrm{loc}}(E; m)$ satisfying $a_- > 0\ dx$–a.e., and

$$\sum_{i,j=1}^{n} c^{ij}(x)\, \xi_i \xi_j \;\geq\; |\xi|^2 \qquad \text{for all } \xi \in \mathbf{R}^n \text{ and } x \in E.$$

We consider the symmetric diffusion operator $(\mathcal{L}, C_0^{\infty}(E))$ on $L^2(E; m)$ with Dirichlet form

$$\mathcal{E}(f, g) \;=\; \int_E \sum_{i,j=1}^{n} a^{ij}\, \frac{\partial f}{\partial x_i}\, \frac{\partial g}{\partial x_j}\, dm,$$

i.e.,

$$(3.6) \quad \mathcal{L}f \;=\; \frac{1}{\varphi^2}\, \mathrm{div}\,(\varphi^2\, a\, \nabla f) \;=\; \sum_{i,j=1}^{n} \left(\frac{\partial}{\partial x_i}(a^{ij} \frac{\partial f}{\partial x_j}) + \beta_i a^{ij} \frac{\partial f}{\partial x_j} \right),$$

where $\beta_i := \frac{2}{\varphi} \frac{\partial \varphi}{\partial x_i}$. The induced (pseudo) metric on the co–tangent bundle is

$$(3.7) \quad (\omega, \sigma)_{T'_x E} \;=\; \sum_{i,j=1}^{n} a^{ij}(x)\, \omega(e_i)\, \sigma(e_j), \qquad \omega, \sigma \in (\mathbf{R}^n)^*.$$

We have

$$\mathcal{E}(f, g) \;=\; \int (d_x f, d_x g)_{T'_x E}\, m(dx) \qquad \text{for all } f, g \in C_0^{\infty}(E).$$

The following definition generalizes Definition 3.1 :

Definition 3.2 *The weak Sobolev space $W^{1,2}(E, (a^{ij}); m)$ is the space consisting of all functions $u \in L^2(E; m)$ with the following property :*
There exist functions $v_j : E \to \mathbf{R}$, $1 \leq j \leq n$, such that $\int_E \sum_{i,j=1}^{n} a^{ij}\, v_i\, v_j\, dm < \infty$, and the integration by parts identities

$$(3.8) \quad \int_E \sum_{j=1}^{n} a^{ij} v_j\, g\, dm \;=\; -\int_E u \sum_{j=1}^{n} \frac{\partial}{\partial x_j}(a^{ij} g)\, dm \;-\; \int_E u\, g \sum_{j=1}^{n} a^{ij} \beta_j\, dm,$$

$1 \leq i \leq n$, hold for all $g \in C_0^{\infty}(E)$.
For $u \in W^{1,2}(E, (a^{ij}); m)$, the functions v_j, $1 \leq j \leq n$, are uniquely determined. They are called the weak derivatives of u in $W^{1,2}(E, (a^{ij}); m)$, and they are denoted by $\hat{\partial}_j u$.

REMARK. Again, the weak Sobolev space $W^{1,2}(E, (a^{ij}); m)$ coincides with the space $W^{1,2}(d)$ corresponding to the operator $(\mathcal{L}, C_0^{\infty}(E))$ as defined in the introduction of this chapter. The weak differential $\hat{d}u$ of $u \in W^{1,2}(E, (a^{ij}); m)$

is given by $\hat{d}u = \sum_{j=1}^{n} \hat{\partial}_j u \, dx_j$. In particular,
$W^{1,2}(E, (a^{ij}); m)$ is a Hilbert space w.r.t. the inner product

$$(u, v)_{W^{1,2}(E, (a^{ij}); m)} = \int_E \left(uv + \sum_{i,j=1}^{n} a^{ij} \, \hat{\partial}_i u \, \hat{\partial}_j v \right) \, dm.$$

In fact, the left hand side of (3.8) is equal to $\int (\sum v_j dx_j, g dx_i)_{T'E} \, dm$, and the right hand side coincides with $\int u \, d^*(g dx_i) \, dm$, where d^* is the adjoint of $d : C_0^{\infty}(E) \subset L^2(E; m) \to L^2(E \to T'E; m)$.

To have a more explicit characterization of $W^{1,2}(E, (a^{ij}); m)$, we compare this space with the space $W^{1,2}(E, a_-; m)$, which has been discussed in Subsection 1) above. Let $\alpha := a_- \cdot \varphi^2$. It is not difficult to verify that α is in $H_{\mathrm{loc}}^{1,1}(E; dx)$ and $\frac{1}{\varphi^2} \frac{\partial \alpha}{\partial x_i}$ is in $L_{\mathrm{loc}}^2(E; m)$ for $1 \le i \le n$, because a_- is in $H_{\mathrm{loc}}^{1,2}(E; m) \cap L_{\mathrm{loc}}^{\infty}(E; m)$, $m = \varphi^2 dx$, and φ is in $H_{\mathrm{loc}}^{1,2}(E; dx)$. Hence the assumptions on the functions a and ρ imposed in Subsection 1) are satisfied by $a := a_-$ and $\rho := \varphi^2$.

Lemma 3.2

$$W^{1,2}(E, (a^{ij}); m) = \left\{ u \in W^{1,2}(E, a_-; m) ; \int_E \sum_{i,j=1}^{n} a^{ij} \, \hat{\partial}_i u \, \hat{\partial}_j u \, dm < \infty \right\},$$

where $\hat{\partial}_j u$, $1 \le j \le n$, denote the weak derivatives in the space $W^{1,2}(E, a_-; m)$. For a function $u \in W^{1,2}(E, (a^{ij}); m)$, the weak derivatives $\hat{\partial}_j u$ in $W^{1,2}(E, a_-; m)$ coincide with those in $W^{1,2}(E, (a^{ij}); m)$.

Together with Lemma 3.1, Lemma 3.2 yields a very explicit characterization of elements in the Sobolev space $W^{1,2}(E, (a^{ij}); m)$. This characterization will be used in Section e) and f) below to derive density theorems of smooth functions in $W^{1,2}(E, (a^{ij}); m)$, and, finally, conditions for Markov uniqueness of the corresponding operator $(\mathcal{L}, C_0^{\infty}(E))$ on $L^2(E; m)$.

PROOF OF LEMMA 3.2. Let u be a function in $W^{1,2}(E, (a^{ij}); m)$, and let v_j, $1 \le j \le n$, be the weak derivatives in $W^{1,2}(E, (a^{ij}); m)$. For brevity, we will use the Einstein summation convention, and we write ∂_i instaed of $\frac{\partial}{\partial x_i}$. Rewriting (3.8), we have

$$(3.9) \qquad \int g \, a^{ij} v_j \, dm = -\int u \, a^{ij} \partial_j g \, dm - \int u g \, (\partial_j a^{ij} + \beta_j a^{ij}) \, dm$$

for $1 \le i \le n$ and $g \in C_0^{\infty}(E)$. Since the functions a^{ij} are locally bounded, u is in $L^2(E; m)$, and $\partial_j a^{ij} + \beta_j a^{ij}$ is in $L_{\mathrm{loc}}^2(E; m)$ for all $1 \le i \le n$, (3.9) extends by continuity to all bounded compactly supported functions $g \in H_0^{1,2}(E; m)$. In fact, every such function can be approximated w.r.t. the $H_0^{1,2}(E; m)$ norm by a sequence of uniformly bounded functions in $C_0^{\infty}(E)$ with support in a joint relatively compact open subset of E.

Now recall that $a^{ij}(x) = a_-(x) \cdot c^{ij}(x)$, $1 \le i, j \le n$, $x \in E$. By assumption, the matrix $(c^{ij}(x))$ is invertible for every x. Let $(c_{ij}(x))$ denote the inverse. By Cramer's formula, the functions c_{ij}, $1 \le i, j \le n$, are in $H^{1,2}_{loc}(E; m) \cap L^\infty_{loc}(E; m)$, because the same holds for the functions c^{ij}, and the matrix-valued function (c^{ij}) is bounded from below by the identity matrix. Let $f \in C_0^\infty(E)$. By choosing $g = c_{ki} f$ in (3.9), and taking the sum over i, we obtain for $1 \le k \le n$,

$$
\begin{aligned}
(3.10) \qquad \int a_- v_k f \, dm &= \int f \, c_{ki} \, a^{ij} v_j \, dm \\
&= -\int u \, \partial_j (f c_{ki}) \, a^{ij} \, dm - \int u f \, c_{ki} \, (\partial_j a^{ij} + \beta_j a^{ij}) \, dm \\
&= -\int u \, \partial_j (f c_{ki} a^{ij}) \, dm - \int u f \, c_{ki} \, a^{ij} \beta_j \, dm \\
&= -\int u \, \partial_k (f a_-) \, dm - \int u f \, a_- \beta_k \, dm.
\end{aligned}
$$

This is precisely the formula (3.2) for $\alpha = a_- \cdot \varphi^2$, cf. also (3.3). Moreover,

$$
\int a_- v_k^2 \, dm \le \int a^{ij} v_i v_j \, dm < \infty \qquad \text{for all } 1 \le k \le n.
$$

Since (3.10) holds for every $f \in C_0^\infty(E)$, we obtain $u \in W^{1,2}(E, a_-; m)$ and $\hat\partial_k u = v_k$ for $1 \le k \le n$, where $\hat\partial_k u$ is the weak derivative in $W^{1,2}(E, a_-; m)$. This proves the inclusion "$\subseteq$" in Lemma 3.2, as well as the identification of the weak derivatives.

The inclusion "$\supseteq$" can be proven similarly : If u is a function in $W^{1,2}(E, a_-; m)$ with weak derivatives $\hat\partial_k u$, $1 \le k \le n$, then the outer integration by parts formula in (3.10) holds with $v_k = \hat\partial_k u$. Again, the equation can be extended from $f \in C_0^\infty(E)$ to every compactly supported bounded function in $H_0^{1,2}(E; m)$. Choosing $f = c^{ik} g$, $g \in C_0^\infty(E)$, and summing over k, we obtain (3.9). If

$$
\int a^{ij} \, \hat\partial_i u \, \hat\partial_j u \, dm < \infty,
$$

then this proves $u \in W^{1,2}(E, (a^{ij}); m)$. ∎

3) Weak Sobolev spaces over Banach spaces

We now describe weak Sobolev spaces corresponding to symmetric diffusion operators with trivial geoemtry on a Banach space E. In the case $E = \mathbf{R}^n$, the following considerations reduce to those from Subsection 1) with $a = 1$, i.e., $\alpha = \rho$. In infinite dimensional situations, however, it is too restrictive to assume that the underlying measure m has a density ρ w.r.t. some canonical reference measure.

We briefly introduce our infinite dimensional framework, which is described in more detail in Chapter 5, Section a). Let K be a vector space of continuous linear functionals on E, which generate the Borel σ–algebra. Let $\mathcal{A} := \mathcal{F}C_b^\infty(K)$ be the space of smooth cylinder functions based on functionals in K, cf. also Chapter 5, Section a), 1). We assume that we are given a Hilbert space H such that $E \cap H$ is dense in H, and the functionals in K are continuous on $E \cap H$ w.r.t. the norm on H. Hence

$$K \subseteq H' \xrightarrow{j} H \,,$$

where j is the Riesz isometry. The differential $DF : E \to K$ of a cylinder function $F \in \mathcal{F}C_b^\infty(K)$ is defined in the usual way, cf. Chapter 5, Section a), 1). We view H as a tangent space to E at each point, and define the H–gradient $\nabla F : E \to H$, $F \in \mathcal{F}C_b^\infty(K)$, by

$$(\nabla F)(x) \;=\; j((DF)(x)) \qquad \text{for all } x \in E.$$

Let m be a probability measure on E. We assume :

(IP) There exist functions $\beta_\ell^m \in L^2(E\,;\,m)$, $\ell \in K$, such that the integration by parts identity

$$(3.11) \qquad \int \partial_{j(\ell)} F \, dm \;=\; -\int \beta_\ell^m \, F \, dm$$

holds for all $F \in \mathcal{F}C_b^\infty(K)$ and $\ell \in K$.

Here $\partial_{j(\ell)} F$ denotes the directional derivative in direction $j(\ell)$, i.e.,

$$(\partial_{j(\ell)} F)\,(x) \;=\; \langle (DF)(x), j(\ell) \rangle \;=\; ((\nabla F)(x), j(\ell))_H \;=\; \ell\,((\nabla F)(x))$$

for all $x \in E$, where $(\,\cdot\,,\,\cdot\,)$ is the dualisation between H' and H. The integration by parts identity is a replacement for the existence of a differentiable density of m, which we have assumed in finite dimensions. Since K generates the Borel σ–algebra, $\mathcal{F}C_b^\infty(K)$ is dense in $L^2(E\,;\,m)$.

We consider the diffusion operator $(\mathcal{L}, \mathcal{F}C_b^\infty(K))$ corresponding to the pre–Dirichlet form

$$\mathcal{E}\,(F, G) \;=\; \int (DF, DG)_{H'} \; dm \;=\; \int (\nabla F, \nabla G)_H \; dm \,,$$

$F, G \in \mathcal{F}C_b^\infty(K)$, on $L^2(E\,;\,m)$. Explicitly,

$$\mathcal{L}F \;=\; \sum_{i,j=1}^{n} (\ell_i, \ell_j)_{H'} \, \frac{\partial^2 f}{\partial x_i \partial x_j} \,(\ell_1, \dots, \ell_n) \;+\; \sum_{j=1}^{n} \beta_{\ell_j}^m \, \frac{\partial f}{\partial x_j} \,(\ell_1, \dots, \ell_n)$$

whenever $F = f(\ell_1, \dots, \ell_n)$ with $n \in \mathbf{N}$, $\ell_1, \dots, \ell_n \in K$, and $f \in C_b^\infty(\mathbf{R}^n)$. Other representations of the operator, as well as examples, are given in Chapter 5. In the situation considered, S. Albeverio, S. Kusuoka and M. Röckner [AlbKusRö 90] have introduced a weak Sobolev space, derived an explicit characterization, and proven a maximality result for the operator $(\mathcal{L}, \mathcal{F}C_b^\infty(K))$. The aim of this subsection is to recall some of their considerations, and to include them into our more general framework.

Definition 3.3 *The weak Sobolev space* $W^{1,2}(E, H'; m)$ *consists of all functions* $G \in L^2(E; m)$ *with the following properties :*

(i) *For every* $\ell \in K$, *there exists a function* $\hat{\partial}_{j(\ell)}G \in L^2(E; m)$ *such that*

$$\int \hat{\partial}_{j(\ell)}G \, F \, dm \;=\; -\int G \, \partial_{j(\ell)}F \, dm \;-\; \int \beta_\ell^m \, G \, F \, dm$$

for all $F \in \mathcal{F}C_b^\infty(K)$.

(ii) *There exists a section* $\hat{D}G \in L^2(E \to H'; m)$ *(or, equivalently, a section* $\hat{\nabla}G \in L^2(E \to H; m)$ *) such that*

$$\left(\hat{D}G, \ell \right)_{H'} \;=\; \hat{\partial}_{j(\ell)}G$$

(respectively $\left(\hat{\nabla}G, j(\ell) \right)_H = \hat{\partial}_{j(\ell)}G$ *)* m-*a.e. for all* $\ell \in K$.

REMARK. The definition coincides with the general definition of a weak Sobolev space corresponding to a symmetric diffusion operator, which will be given in the next section, cf. the example below Lemma 3.6. In the notation used there, D is a generalized differential corresponding to the operator $(\mathcal{L}, \mathcal{F}C_b^\infty(K))$ w.r.t. the (trivial) measurable co-tangent bundle given by $T_z'E = H'$ for all $z \in E$, cf. also Appendix D. The operator $\hat{D}$ is the correpoding weak differential. $W^{1,2}(E, H'; m)$ is a Hilbert space with inner product

$$\left(G, \bar{G} \right)_{W^{1,2}(E, H'; m)} \;=\; \int \{ G\bar{G} + \left(\hat{D}G, \hat{D}\bar{G} \right)_{H'} \} \, dm.$$

We now want to sketch how to derive a characterization of elements in $W^{1,2}(E, H'; m)$ similar to Lemma 3.1. We closely follow the arguments of [AlbKusRö 90]. Fix $h \in E \cap H$, $h \neq 0$. Let E_h be a closed subspace of E such that $E = E_h \oplus \mathrm{span}\,\{h\}$. The space E_h replaces the space $\mathbf{R}_i^n$ from Subsection 1). Let $\Pi_h : E \to E_h$ be the canonical projection, and let $m_h := m \circ \Pi_h^{-1}$. Then there exists a kernel $\rho_h : E_h \times \mathcal{B}(\mathbf{R}) \to [0, 1]$ such that

$$\int_E F(z) \, m\,(dz) \;=\; \int_{E_h} \int_{\mathbf{R}} F(y + sh) \, \rho_h(y, ds) \, m_h(dy)$$

for every bounded function F on E. Now suppose $h = j(\ell)$ for some $\ell \in K$. The crucial point for the further analysis is, that in this case, the integration by parts identity (IP) *implies* that for m_h-a.e. $y \in E_h$, $\rho_h(y, ds) = \bar{\rho}_h(y, s)\, ds$ for some absolutely continuous function $\bar{\rho}_h(y, \cdot)$ on $\mathbf{R}$ such that

$$\frac{\partial}{\partial s}\, \bar{\rho}_h(y, s)/\bar{\rho}_h(y, s) \;=\; \beta_\ell^m\,(y + sh) \qquad \text{for } \rho_h(y, ds)\text{-a.e. } s \in \mathbf{R}.$$

We refer to [AlbKusRö 90, Thm. 2.5] for the proof. In analogy to Subsection 1), we introduce the singularity sets

$$S(\bar{\rho}_h(y, \cdot)) \;:=\; \{\, s \in \mathbf{R};\; \bar{\rho}_h(y, s) = 0 \,\}, \qquad y \in E_h.$$

For a function G on E, $y \in E_h$ and $s \in \mathbf{R}$, let $G_h(y, s) := G(y + sh)$. The following infinite dimensional counterpart to Lemma 3.1 is implicitly contained in the proofs of [AlbKusRö 90], although the result itself has not been stated explicitly there. Actually, in [AlbKusRö 90], the space $W^{1,2}(E, H'; m)$ is defined through the description given now.

Lemma 3.3 *Suppose that* $j(K) \subset E$. *Let* G *be a function in* $L^2(E; m)$. *Then* G *is in* $W^{1,2}(E, H'; m)$ *if and only if the following conditions hold :*

(i) *Let* $h = j(\ell)$ *for some* $\ell \in K$. *Then for* m_h-*a.e.* $y \in E_h$, *the function* $G_h(y, \cdot) : \mathbf{R} \to \mathbf{R}$ *has an absolutely continuous* ds-*version* $\bar{G}_h(y, \cdot)$ *on* $\mathbf{R} \setminus S(\bar{\rho}_h(y, \cdot))$. *There exists a function* $\partial_h G \in L^2(E; m)$ *such that*

$$(\partial_h G)(y + sh) = \frac{\partial}{\partial s} \bar{G}_h(y, s)$$

for ds-*a.e.* $s \in \mathbf{R} \setminus S(\bar{\rho}_h(y, \cdot))$ *and* m_h-*a.e.* $y \in E_h$.

(ii) *There exists* $\bar{D}G \in L^2(E \to H'; m)$ *(or, equivalently,* $\tilde{\nabla}G \in L^2(E \to H; m)$ *), such that*

$$\left(\bar{D}G \right)(h) = \partial_h G$$

(respectively $\left(\tilde{\nabla}G, h \right)_H = \partial_h G$ *)* m-*a.e. for all* $h \in j(K)$.

If G *is in* $W^{1,2}(E, H'; m)$, *then* $\bar{D}G$ *is* m-*a.e. uniquely determined, and* $\bar{D}G = \dot{D}G$ m-*a.e.*

The proof of Lemma 3.3 can be carried out in a similar way as the proof of Lemma 3.1 above. Essentially, the only additional consideration needed is to show that for G, $V \in L^2(E; m)$, $\ell \in K$, and $h = j(\ell)$, an integration by parts identity of type

$$\int V F \, dm = -\int G \, \partial_h F \, dm - \int \beta_\ell^m \, G \, F \, dm \quad \text{for all } F \in \mathcal{F}C_b^\infty(K)$$

imples that the corresponding one–dimensional integration by parts formula

$$\int_{\mathbf{R}} V_h(y, s) \, f(s) \, \bar{\rho}_h(y, s) \, ds$$

$$= -\int_{\mathbf{R}} G_h(y, s) \, f'(s) \, \bar{\rho}_h(y, s) \, ds - \int_{\mathbf{R}} \frac{\partial}{\partial s} \tilde{\rho}_h(y, s) \, G_h(y, s) \, f(s) \, ds$$

$$\text{for all } f \in C_0^\infty(\mathbf{R})$$

holds for m_h–a.e. y. The absolute continuity of $G_h(y, \cdot)$ for $G \in W^{1,2}(E, H'; m)$ then follows easily by a one–dimensional argument. We refer to [RöZha 92, Remark 1.5 (i)] for details.

b) Weak and strong Sobolev spaces corresponding to diffusion operators on general state spaces

In this section, we demonstrate how the concept of a weak Sobolev space can be carried over to an arbitrary symmetric diffusion operator over some measure space $(E, \mathcal{B}, m)$. More precisely, we introduce a strong and a weak Sobolev space corresponding to a differential taking values in some measurable co-tangent bundle. As demonstrated in Appendix D, we can associate with every symmetric diffusion operator $\mathcal{L}$ on an L^2 space a measurable co-tangent bundle and a differential d such that $\mathcal{L}$ takes the form $-d^*d$, cf. Theorem 3.11. Hence for each symmetric diffusion operator, we obtain a corresponding weak and strong Sobolev space. We first give the general definitions, and then characterize elements in weak Sobolev spaces in terms of integration by parts identities for directional derivatives. At the end of this section, we describe weak Sobolev spaces corresponding to differentials defined on cylinder functions.

1) Basic definitions

We fix a σ-finite measure space $(E, \mathcal{B}, m)$ such that $L^2(E; m)$ is separable. Let $\mathcal{A}$ be a set of bounded m-square integrable functions on E, which generates $\mathcal{B}$ and is dense in $L^2(E; m)$. We assume that $\phi(u_1, \ldots, u_k)$ is in $\mathcal{A}$ for all $k \in \mathbf{N}$, $u_1, \ldots, u_k \in \mathcal{A}$, and $\phi \in C^\infty(\mathbf{R}^k)$ such that $\phi(0) = 0$. In particular, $\mathcal{A}$ is an algebra.

Let $T'E = (T'_z E)_{z \in E}$ be a measurable field of Hilbert spaces over E, see Appendix D for the definition. $T'E$ should be viewed as a *measurable co-tangent bundle* to E. The *direct integral* of $T'E$ w.r.t. m is denoted by $L^2(E \to T'E; m)$, cf. Appendix D for this and other notations used here and below. Elements in $L^2(E \to T'E; m)$ will be called *1-forms*. For two 1-forms $\omega, \sigma \in L^2(E \to T'E; m)$, the function $z \mapsto (\omega(z), \sigma(z))_{T'_z E}$ in $L^1(E; m)$ will in the sequel be denoted by (ω, σ).

Let $d : \mathcal{A} \to L^2(E \to T'E; m)$ be a map with the following properties :

I) d is an L^2 differential w.r.t. the co-tangent bundle $T'E$, i.e.,

    (i) The span of $\{ f\, dg \,; \, f, g \in \mathcal{A} \}$ is dense in $L^2(E \to T'E; m)$.

    (ii) d is linear.

    (iii) $d(f \cdot g) = f \cdot dg + g \cdot df$ for all $f, g \in \mathcal{A}$.

II) For every $f \in \mathcal{A}$, the m-a.e. defined function $z \mapsto (df(z), df(z))_{T'_z E}$ is in $L^2(E; m)$.

III) For every $f \in \mathcal{A}$, df is contained in the domain of the adjoint d^* of the densely defined linear operator $d : \mathcal{A} \subset L^2(E; m) \to L^2(E \to T'E; m)$.

Examples of measurable co-tangent bundles and differentials are given in Appendix D.

REMARK. The properties I)–III) imply that the operator $-d^*d$ is a symmetric diffusion operator with domain $\mathcal{A}$ on $L^2(E\,;\,m)$, cf. Remark (i) under Lemma 3.4 below. The carré du champ of $-d^*d$ is given by

$$\Gamma(f,\,g)\;=\;(df,\,dg)\quad m\text{--a.e. for all }f,\,g\in\mathcal{A}.$$

Conversely, by Theorem 3.11 in Appendix D, we can construct for any given symmetric diffusion operator $(\mathcal{L},\,\mathcal{A})$ with invariant measure m a measurable co-tangent bundle and a corresponding L^2 differential d such that $\mathcal{L}=-d^*d$, and I)–III) hold.

Let

$$\Omega(\mathcal{A})\;:=\;\operatorname{span}\{\,f\,dg\,;\,f,\,g\in\mathcal{A}\,\}.$$

$\Omega(\mathcal{A})$ is a dense subspace of $L^2(E\to T'E;\,m)$. Thinking of the functions in $\mathcal{A}$ as test functions, we may view the elements in $\Omega(\mathcal{A})$ as "test 1–forms".

EXAMPLE. Suppose E is an open subset in $\mathbf{R}^n$, $\mathcal{A}=C_0^\infty(E)$, and d is the ordinary differential. Then $\Omega(\mathcal{A})=C_0^\infty(E\to(\mathbf{R}^n)^*)$. The inclusion "$\subseteq$" is obvious. The inclusion "$\supseteq$" holds, because for every $f\in C_0^\infty(E)$ and $1\le i\le n$, $f\,dx_i=f\,dg$, where g is a function in $C_0^\infty(E)$ such that $g(x)=x_i$ for all x in the support of f. In Subsection 3) below, we describe the space $\Omega(\mathcal{A})$ in typical infinite dimensional situations.

Using that df is in the domain of d^*, and $(df,\,dg)$ is in $L^2(E\,;\,m)$ for all $f,\,g\in\mathcal{A}$, one immediately verifies that $\Omega(\mathcal{A})$ is a subspace of the domain of d^*, and

(3.12) $d^*(f\,dg)\;=\;f\,d^*dg\;-\;(df,\,dg)$ for all $f,\,g\in\mathcal{A}.$

There are two ways to extend the differential d to a broader class of functions in $\mathcal{A}$:

Definition 3.4   (i) *The adjoint $\hat{d}$ of the densely defined linear operator d^* :* $\Omega(\mathcal{A})\subset L^2(E\to T'E;\,m)\to L^2(E\,;\,m)$ *is called the* **weak differential** *w.r.t.* $(d,\,\mathcal{A})$. *Its domain $W^{1,2}(d)$ is called the corresponding* **weak Sobolev space.** *(ii) The closure $\bar{d}$ of the operator $d:\mathcal{A}\subset L^2(E\,;\,m)\to L^2(E\to T'E;\,m)$ is called the* **strong differential** *w.r.t.* $(d,\,\mathcal{A})$. *Its domain $H_0^{1,2}(d)$ is called the corresponding* **strong Sobolev space.**

Hence the weak Sobolev space $W^{1,2}(d)$ consists of all functions $f\in L^2(E\,;\,m)$ for which there exists a section $\hat{d}f$ in $L^2(E\to T'E;\,m)$ such that

$$\int f\,d^*\omega\,dm\;=\;\int(\hat{d}f,\,\omega)\,dm\qquad\text{for all }\omega=g\,dh,\;g,\,h\in\mathcal{A}.$$

The strong Sobolev space $H_0^{1,2}(d)$ consists of all functions $f\in L^2(E\,;\,m)$ for which there exists a sequence $(f_n)_{n\in\mathbf{N}}$ in $\mathcal{A}$ such that $f_n\to f$ in $L^2(E\,;\,m)$, and $(df_n)_{n\in\mathbf{N}}$ is a Cauchy sequence in $L^2(E\to T'E;\,m)$.

REMARKS.   (i) Since $(d^*, \Omega(\mathcal{A}))$ is a restriction of $(d^*, \mathrm{Dom}\,(d^*))$, the weak differential $(\hat{d}, W^{1,2}(d))$ is a closed extension of the operator $(d, \mathcal{A})$. In particular, $(d, \mathcal{A})$ is closable, and the weak differential $(\hat{d}, W^{1,2}(d))$ extends the closure $(\bar{d}, H_0^{1,2}(d))$.

(ii) The notations $H_0^{1,2}(d)$ and $W^{1,2}(d)$ are used, because if E is a domain in $\mathbf{R}^n$, $\mathcal{A} = C_0^\infty(E)$, d is the ordinary differential, and the metric on the co–tangent space $T_z'E = (\mathbf{R}^n)^*$ is Euclidean, then $H_0^{1,2}(d)$ and $W^{1,2}(d)$ are the usual strong and weak Sobolev space $H_0^{1,2}(E; dx)$, $W^{1,2}(E; dx)$ respectively. In this case, the functions in $H_0^{1,2}(d)$ vanish at the boundary of E, but in more general, in particular infinite dimensional cases, the definition of $H_0^{1,2}(d)$ is not necessarily related to vanishing at the boundary.

The following chain and product rules hold for the operators $\bar{d}$, d^* and $\hat{d}$:

Lemma 3.4 *(i) Let $k \in \mathbf{N}$, $f_1, \ldots, f_k \in H_0^{1,2}(d) \cap L^\infty(E; m)$, and $\phi \in C^\infty(\mathbf{R}^k)$ such that $\phi(0) = 0$. Then $\phi(u_1, \ldots, u_k)$ is again in $H_0^{1,2}(d)$, and*

$$\bar{d}\,(\phi(f_1, \ldots, f_k)) \;=\; \sum_{i=1}^k \frac{\partial \phi}{\partial x_i}\,(f_1, \ldots, f_k)\;\bar{d}f_i \quad m\text{--}a.e.$$

(ii) Let $f \in H_0^{1,2}(d) \cap L^\infty(E; m)$, and let ω be a bounded 1–form in the domain of d^*. Then $f\omega$ is in the domain of d^*, and

$$d^*\,(f\omega) \;=\; f\,d^*\omega \;-\; (\bar{d}f, \omega).$$

(iii) Let $f \in H_0^{1,2}(d) \cap L^\infty(E; m)$ and $g \in W^{1,2}(d) \cap L^\infty(E; m)$. Then fg is in $W^{1,2}(d)$, and

$$\hat{d}(fg) \;=\; f\,\hat{d}g \;+\; g\,\bar{d}f.$$

REMARKS.   (i) The chain rule in $H_0^{1,2}(d)$ implies in particular that the closed quadratic form $(\tilde{\mathcal{E}}_{(0)}, H_0^{1,2}(d))$ on $L^2(E; m)$ defined by $\tilde{\mathcal{E}}_{(0)}(f, g) = \int (\bar{d}f, \bar{d}g)\, dm$ is a local Dirichlet form. As a consequence, the operator $\bar{d}$ has the following energy **image density property**, cf. [BouHi 91, Ch. I, Thm. 7.1.1] : For every function $f \in H_0^{1,2}(d)$, the law of f under the measure $(\bar{d}f, \bar{d}f) \cdot m$ is absolutely continuous w.r.t. Lebesgue measure. In particular, $\bar{d}f$ vanishes m–a.e. on $\{z \in E; f(z) = 0\}$. The operator $(-d^*d, \mathcal{A})$ is a diffusion operator on $L^2(E; m)$, and its Friedrichs extension is the generator $-d^*\bar{d}$ of $(\tilde{\mathcal{E}}_{(0)}, H_0^{1,2}(d))$.

(ii) In general, it is not clear if a chain rule holds in $W^{1,2}(d)$, and if $W^{1,2}(d)$ is a Dirichlet space. In esentially all applications we are interested in, we can, however, extend the chain rule to bounded functions in $W^{1,2}(d)$, cf. for example the finite dimensional cases considered in Section a) above, and see Remark (vi) in Section c), 2), in Chapter 5 for the infinite–dimensional case.

PROOF OF LEMMA 3.4.   (i) We first prove the chain rule for $f_1, \ldots, f_k \in \mathcal{A}$. By the product rule for d, we have

$$(3.13) \qquad d\,(\,P(f_1, \ldots, f_k)) \;=\; \sum_{i=1}^k \frac{\partial P}{\partial x_i}\,(f_1, \ldots, f_k)\; df_i \quad m\text{--a.e.}$$

for every polynomial $P : \mathbf{R}^k \to \mathbf{R}$ such that $P(0) = 0$. Let $R \subset \mathbf{R}^k$ be the range of $(f_1, \ldots, f_k)$. Since the functions in $\mathcal{A}$ are bounded, R is relatively compact. Hence there exist polynomials $P_n : \mathbf{R}^k \to \mathbf{R}$, $n \in \mathbf{N}$, such that $P_n(0) = 0$ for all n, $P_n \to \phi$ uniformly on R, and $\frac{\partial P_n}{\partial x_i} \to \frac{\partial \phi}{\partial x_i}$ uniformly on R as $n \to \infty$ for all $1 \leq i \leq k$. In particular, there exists a finite constant C such that $|P_n(x)| \leq C \cdot |x|$ holds for all $x \in R$ and $n \in \mathbf{N}$. Thus

$$|P_n(f_1, \ldots, f_k)| \leq C \cdot |(f_1, \ldots, f_k)| \in L^2(E; m),$$

whence by dominated convergence,

$$P_n(f_1, \ldots, f_k) \to \phi(f_1, \ldots, f_k) \quad \text{in } L^2(E; m)$$

as $n \to \infty$. Moreover, by (3.13), $d(P_n(f_1, \ldots, f_k))$, $n \in \mathbf{N}$, converges to $\sum_{i=1}^k \frac{\partial \phi}{\partial x_i}(f_1, \ldots, f_k) \, df_i$ in $L^2(E \to T'E; m)$ as $n \to \infty$. Since $d : \mathcal{A} \subset L^2(E; m) \to L^2(E \to T'E; m)$ is a closable operator, we obtain

$$d(\phi(f_1, \ldots, f_k)) = \sum_{i=1}^k \frac{\partial \phi}{\partial x_i}(f_1, \ldots, f_k) \, df_i \quad m\text{-a.e.}$$

The chain rule on $\mathcal{A}$ implies in particular that for every bounded function f in $H_0^{1,2}(d)$ there exist a sequence $(f_n)_{n \in \mathbf{N}}$ of *uniformly bounded* functions in $\mathcal{A}$ such that $f_n \to f$ and $df_n \to \bar{d}f$ in $L^2(E; m)$. In fact, let $\left(\tilde{f}_n\right)_{n \in \mathbf{N}}$ be a not necessarily bounded sequence in $\mathcal{A}$ with these properties, and let ϕ be a smooth function on $\mathbf{R}$ such that $|\phi'(x)| \leq 1$ for all x, and $\phi(x) = x$ for x in the range of f. Let $f_n := \phi(\tilde{f}_n)$, $n \in \mathbf{N}$. Then by dominated convergence, $df_n \to \bar{d}f$ in $L^2(E; m)$.

Now let f, g be bounded functions in $H_0^{1,2}(d)$. By approximating f and g by uniformly bounded functions f_n, g_n in $\mathcal{A}$ as just described, one verifies that $f \cdot g$ is again in $\mathcal{A}$, and the product rule

$$\bar{d}(f \cdot g) = f \, \bar{d}g + g \, \bar{d}f$$

holds. In fact, $(f_n g_n)_{n \in \mathbf{N}}$ converges to fg in $L^2(E; m)$, and

$$d(f_n g_n) = f_n \, dg_n + g_n \, df_n \to f \, \bar{d}g + g \, \bar{d}f$$

in $L^2(E; m)$. Hence the product rule holds for bounded functions in $H_0^{1,2}(d)$. The chain rule on $H_0^{1,2}(d) \cap L^\infty(E; m)$ can now be proven similarly as for functions in $\mathcal{A}$.

(ii) The operator d^* is also the adjoint of the operator $(\bar{d}, H_0^{1,2}(d))$. Hence by the product rule for $\bar{d}$, we obtain

$$\int (dg, f\omega) \, dm = \int (\bar{d}(gf), \omega) \, dm - \int g \, (\bar{d}f, \omega) \, dm$$

$$= \int g \, (f \, d^*\omega - (\bar{d}f, \omega)) \, dm$$

for all $g \in \mathcal{A}$.

Assertion (iii) can be verified similarly using the product rule for $\bar{d}$. ∎

2) Directional derivative characterizations of weak Sobolev spaces

To make the notation more transparent, it is convenient to introduce the tangent bundle $TE = (T_z E)_{z \in E}$ formally as the dual bundle of $T'E = (T'_z E)_{z \in E}$, i.e., $T_z E$ is the dual space of $T'_z E$ for every $z \in E$. Clearly, TE is again a measurable field of Hilbert spaces with measurable structure induced by the isometry $j = (j_z)_{z \in E}$, where $j_z : T'_z E \to T_z E$ is the Riesz isometry for each z. A vector field X in $L^2(E \to TE; m)$ can be viewed as an L^2 derivation on $\mathcal{A}$ by defining

$$(Xf)(z) := \langle df(z), X(z) \rangle$$

for m–a.e. $z \in E$ and all $f \in \mathcal{A}$, where $\langle \cdot, \cdot \rangle$ denotes the dualisation between $T'_z E$ and $T_z E$. In concrete applications, tangent vectors in $T_z E$, $z \in E$, can usually be identified with derivatives of curves in E passing through z.

For $f \in \mathcal{A}$, the section $z \mapsto j_z(df(z))$ in $L^2(E \to T'E; m)$ will be denoted by ∇f. The **gradient operator** $\nabla : \mathcal{A} \subset L^2(E; m) \to L^2(E \to T'E; m)$ thus defined is a densely defined linear operator. Let ∇^* denote the adjoint.

REMARKS. (i) Since j is an isometry between $T'E$ and TE, the gradient ∇ is closable, and the strong Sobolev space $H_0^{1,2}(d)$ is the domain of the closure $\bar{\nabla}$.

(ii) For the same reason, the vector fields ∇f, $f \in \mathcal{A}$, are in the domain of ∇^*, and

$$\nabla^* \nabla f = d^* d f \qquad \text{for all } f \in \mathcal{A}.$$

If X is a vector field in the domain of ∇^*, then the integration by parts identities

$$(3.14) \qquad \int Xf \, dm = \int f \, \nabla^* X \, dm , \qquad \text{and, hence,}$$

$$(3.15) \qquad \int f \, Xg \, dm = -\int g \, Xf \, dm + \int fg \, \nabla^* X \, dm$$

hold for all $f, g \in \mathcal{A}$. The second equation means that $\mathcal{A}$ is contained in the domain of the adjoint X^* of the operator $X : \mathcal{A} \subset L^2(E; m) \to L^2(E; m)$, and

$$(3.16) \qquad X^* f = -Xf + f \nabla^* X \qquad \text{for all } f \in \mathcal{A}.$$

In particular, this is the case for all vector fields X that are gradients of functions in $\mathcal{A}$.

Definition 3.5 *Let X be a vector field in the domain of ∇^*. Then a function u in $L^2(E; m)$ is called **weakly differentiable in direction X** if and only if there exists a function $\hat{X}u$ in $L^2(E; m)$ such that*

$$(3.17) \qquad \int \hat{X}u \, f \, dm = -\int u \, Xf \, dm + \int u \, f \, \nabla^* X \, dm$$

for all $f \in A$. *The up to m–equivalence uniquely determined function* $\hat{X}u$ *is called the* **weak derivative** *of* u *in direction* X. *The vector space consisting of all functions* u *in* $L^2(E\,;\,m)$ *that are weakly differentiable in direction* X *is denoted by* $W^{1,2}(X)$.

Remarks. (i) Note that by (3.16), the right hand side of (3.17) is equal to $\int u\, X^* f\, dm$.

(ii) Let X be a vector field in the domain of ∇^*. Then, by the product rule, it is easy to verify that $W^{1,2}(g\,X) = W^{1,2}(X)$ for all $g \in A$. Moreover, for $u \in W^{1,2}(X)$ and $g \in A$, the weak derivative of u in direction $g\,X$ is $g\,\hat{X}u$.

Lemma 3.5 *Let* u *be a function in* $L^2(E\,;\,m)$. *Then* u *is contained in* $W^{1,2}(d)$ *if and only if*

(i) u *is in* $W^{1,2}(X)$ *for every vector field* $X = \nabla g$, $g \in A$.

(ii) *There exists a section* $\hat{d}u$ *in* $L^2(E \to T'E;\, m)$ *(or, equivalently, a vector field* $\hat{\nabla}u$ *in* $L^2(E \to T'E;\, m)$ *), such that*

$$\hat{X}u \;=\; \langle\, \hat{d}u\,,\, X\,\rangle \qquad m\text{--}a.e.$$

(respectively $\hat{X}u = \left(\hat{\nabla}u,\, X\right)_{TE} \quad m\text{--}a.e.$ *) for all* $X = \nabla g,\ g \in A$.

It can be verified immediately that the lemma is just a reformulation of the definition of $W^{1,2}(d)$ in terms of directional derivatives. The operator $\hat{d}$ appearing in (ii) is the weak differential.

3) Weak Sobolev spaces w.r.t. differentials defined on cylinder functions

In Lemma 3.5, we have given a directional derivative characterization of elements in $W^{1,2}(d)$. However, usually we want to characterize elements in $W^{1,2}(d)$ by integration by parts identities w.r.t. a *smaller* or *more natural* class of vector fields than the class $\{\nabla g;\ g \in A\}$ used in Lemma 3.5. For example, we have already seen in Section a), 2), how the weak Sobolev space corresponding to a diffusion operator on $\mathbf{R}^n$ with diffusion matrix (a^{ij}) can be described by integration by parts identities w.r.t. the n vector fields $\sum_{i=1}^{n} a^{ij} \frac{\partial}{\partial x_j}$, $1 \le i \le n$. We will now give a similar description of $W^{1,2}(d)$ in the case where d is a differential defined on some space of cylinder functions over a typically infinite dimensional state space.

We consider the same situation as above, but in addition we assume that m is a finite measure, and

$$A \;=\; \mathcal{F}C_b^\infty(K) \;=\; \{z \mapsto f(g_1(z),\dots,g_n(z));\ n \in \mathbf{N},\ f \in C_b^\infty(\mathbf{R}^n),$$
$$g_1,\dots,g_n \in K\,\},$$

where K is a set of functions on E. For example, $K \subseteq E'$ if E is a Banach space, $K = \{\omega \mapsto \omega_i(s); 0 \leq s \leq 1, 1 \leq i \leq d\}$ if $E \subseteq C([0,1], M)$ for some submanifold $M \subseteq \mathbf{R}^d$, or $K = \{\mu \mapsto \int \varphi \, d\mu; \ \varphi \in C_0^\infty(M)\}$ if K is a space of measures over a manifold M.

Suppose the differential d on $\mathcal{A}$ is given by

$$(3.18) \qquad d\big(f(g_1, \ldots, g_n)\big) \ = \ \sum_{i=1}^{n} \frac{\partial f}{\partial x_i}(g_1, \ldots, g_n) \, dg_i,$$

for all $n \in \mathbf{N}$, $g_1, \ldots, g_n \in K$, and $f \in C_b^\infty(\mathbf{R}^n)$, where

$$d : \ K \ \rightarrow \ L^\infty(E \rightarrow T'E; m)$$

is some naturally defined differential on K. Let $(f_k)_{k \in \mathbf{N}}$ be a sequence of functions in $C_b^\infty(\mathbf{R})$ such that $f_k(s) = s$ if $|s| \leq k$, $0 \leq f_k'(s) \leq 1$ for all s, and the functions f_k'', $k \in \mathbf{N}$, are uniformly bounded. We need the following assumption :

Assumption : For every $g \in K$, $(\mathcal{L}(f_k \circ g))_{k \in \mathbf{N}}$ is a Cauchy sequence in $L^2(E; m)$.

For an m–square integrable function $g \in K$, the assumption means that g is in the domain of the closure of the operator $(\mathcal{L}, \mathcal{A})$. We do not assume that functions in K are square integrable, although in most applications this will be the case. In any case, the assumption implies that for every $g \in K$, the differential dg is in the domain of the adjoint d^* of $d : \mathcal{A} \subset L^2(E; m) \rightarrow L^2(E \rightarrow T'E; m)$, and $-d^* dg = \lim_{k \to \infty} \mathcal{L}(f_k \circ g)$.

As above, we define the gradient ∇F of a function $F \in \mathcal{F}C_b^\infty(K)$ or $F \in K$ by setting $(\nabla F)(z) = j_z(df(z))$, where $j_z : T_z'E \rightarrow T_zE$ is the Riesz isometry. Let

$$\mathcal{V}_0 \ := \ \{\nabla g; \ g \in K\} \ \subseteq \ L^2(E \rightarrow TE; m).$$

Lemma 3.6 *Under the assumptions above, a function* $u \in L^2(E; m)$ *is in* $W^{1,2}(d)$ *if and only if*

(i) *u is in* $W^{1,2}(X)$ *for all* $X \in \mathcal{V}_0$.

(ii) *There exists* $\hat{d}u \in L^2(E \rightarrow T'E; m)$ *such that* $\hat{X}u = \langle \hat{d}u, X \rangle$ *m–a.e. for all* $X \in \mathcal{V}_0$.

REMARK. Consider the following space of finitely based 1–forms :

$$\mathcal{F}C_b^\infty(K, T'E) \ := \ \mathrm{span}\,\{F\,dg; \ F \in \mathcal{F}C_b^\infty(K), \ g \in K\}$$

Note that by Condition (i) in the definition of an L^2 differential, and by (3.18), $\{dg(z); \ g \in K\}$ is dense in $T_z'E$ for m–a.e. z. Therefore, $\mathcal{F}C_b^\infty(K, T'E)$ seems to be in some sense a reasonable infinite dimensional replacement for corresponding spaces of smooth 1–forms over finite–dimensional manifolds. Obviously,

$$(3.19) \qquad\qquad \Omega(\mathcal{A}) \ \subseteq \ \mathcal{F}C_b^\infty(K, T'E).$$

The assumption above implies that $\mathcal{F}C_b^\infty(K, T'E)$ is a subset of the domain of d^*. Therefore, we could replace the space $\Omega(\mathcal{A})$ in the definition of the weak Sobolev space by the space $\mathcal{F}C_b^\infty(K, T'E)$. Because of (3.19), the modified weak Sobolev space

$$\bar{W}^{1,2}(d) := \mathrm{Dom}\left(\left(d^*|_{\mathcal{F}C_b^\infty(K,T'E)}\right)^*\right)$$

obtained in this way is a priori *smaller* than $W^{1,2}(d)$. The assertion of Lemma 3.6 means that, actually, $\bar{W}^{1,2}(d) = W^{1,2}(d)$.

EXAMPLE. (*Flat case*) Consider the Banach space framework described in Section a), 3). Let $T'E$ be the trivial bundle given by $T'_zE = H'$ for all $z \in E$, and let $d := D$ be the ordinary differential on E. The set K is now a vector space of continuous linear functionals on E, cf. Section a), 3). Hence $V_0 = \{\partial_{j(\ell)}; \ell \in K\}$, where $\partial_{j(\ell)}$ denotes the constant vector field $z \mapsto j(\ell)$ on E. It is not difficult to verify the assumption. In fact,

$$\mathcal{L}(f_k \circ \ell) = (\ell, \ell)_{H'} \, f_k'' \circ \ell + \beta_\ell^m \, f_k' \circ \ell \;\to\; \beta_\ell^m \quad \text{in } L^2(E; m) \text{ as } k \to \infty.$$

Thus Lemma 3.6 applies. The assertion of Lemma 3.6 in this case means that $W^{1,2}(d)$ coincides with the space $W^{1,2}(E, H'; m)$ introduced in Definition 3.3.

PROOF OF LEMMA 3.6. Let $u \in W^{1,2}(d)$ and $g \in K$. Let $X := \nabla g$, and $X_k := \nabla(f_k \circ g)$, $k \in \mathbf{N}$. By Lemma 3.5, the integration by parts formula

$$(3.20) \qquad \int \langle \hat{d}u, X_k \rangle F \, dm \;=\; -\int u \, \langle dF, X_k \rangle \, dm \,+\, \int u \, F \, \nabla^* X_k \, dm$$

holds for all $F \in \mathcal{F}C_b^\infty(K)$ and $k \in \mathbf{N}$. The assumption implies that X is in the domain of ∇^*, $X_k \to X$ in $L^2(E \to T'E; m)$, and $\nabla^* X_k \to \nabla^* X$ in $L^2(E; m)$ as $k \to \infty$. Moreover, we have assumed that dF is bounded for all $F \in K$, and, hence, for all $F \in \mathcal{F}C_b^\infty(K)$. Thus letting k tend to infinity in (3.20), we obtain

$$\int \langle \hat{d}u, X \rangle F \, dm \;=\; -\int u \, \langle dF, X \rangle \, dm \,+\, \int u \, F \, \nabla^* X \, dm$$

for all $F \in \mathcal{F}C_b^\infty(K)$. This proves that u is in $W^{1,2}(X)$, and $\hat{X}u = \langle \hat{d}u, X \rangle$.

Conversely, fix a function $u \in L^2(E; m)$ satisfying the conditions (i) and (ii) in Lemma 3.6. Let G be a function in $\mathcal{F}C_b^\infty(K)$, and let $X := \nabla G$. If $G = f(g_1, \ldots, g_n)$ with $n \in \mathbf{N}$, $f \in C_b^\infty(\mathbf{R}^n)$, and $g_1, \ldots, g_n \in K$, then $X = \sum_{i=1}^n G_i X_i$, where $G_i := \frac{\partial f}{\partial x_i}(g_1, \ldots, g_n) \in \mathcal{F}C_b^\infty(K)$, and $X_i := \nabla g_i \in V_0$. Since u is in $W^{1,2}(X_i)$ for all $1 \le i \le n$, we have

$$\int G_i \, \hat{X}_i u \, F \, dm \;=\; -\int u \, X_i(G_i F) \, dm \,+\, \int u \, F G_i \, \nabla^* X_i \, dm$$

$$= \; -\int u \, G_i \, X_i F \, dm \,+\, \int u \, F \, \nabla^*(G_i X_i) \, dm$$

for all $F \in \mathcal{F}C_b^\infty(K)$ and $1 \leq i \leq n$. Summing over i, we see that u is in $W^{1,2}(X)$, and

$$\hat{X}u = \sum_{i=1}^{n} G_i \hat{X}_i u = \langle \hat{d}u, \sum_{i=1}^{n} G_i X_i \rangle,$$

where $\hat{d}u \in L^2(E \to T'E; m)$ is the 1-form appearing in Condition (ii) in Lemma 3.6. Since $G \in \mathcal{F}C_b^\infty(K)$ ha been chosen arbitrarily, u is in $W^{1,2}(d)$ by Lemma 3.5, and $\hat{d}u$ is the weak differential. ∎

c)   Maximal Dirichlet extensions and the basic criterion for Markov uniqueness

We fix a σ-finite measure space $(E, \mathcal{B}, m)$ such that $L^2(E; m)$ is separable. Let $(\mathcal{L}, \mathcal{A})$ be a densely defined symmetric diffusion operator on $L^2(E; m)$ such that $\mathcal{A}$ consists of bounded, m-square integrable functions on E, which generate $\mathcal{B}$. We assume that m is an invariant measure for $(\mathcal{L}, \mathcal{A})$. By Theorem 3.11 in Appendix D, there exist a measurable field of Hilbert spaces $T'E = (T'_z E)_{z \in E}$ over E, and an L^2 differential $d : \mathcal{A} \subset L^2(E; m) \to L^2(E \to T'E; m)$ such that

$$\mathcal{L} = -d^* d,$$

and properties I) — III) from Section b), 1), hold. The measurable co-tangent bundle and the differential are uniquely determined up to equivalence, cf. Theorem 3.11 (ii). We have

$$(3.21) \qquad (df, dg) = \Gamma(f, g) = (\mathcal{L}(fg) - f\mathcal{L}g - g\mathcal{L}f)/2$$

m-a.e. for all $f, g \in \mathcal{A}$.

We now fix a measurable co-tangent bundle $T'E$ and an L^2 differential d as above. Since d is essentially uniquely determined by $\mathcal{L}$, it should be possible to characterize the Sobolev spaces $H_0^{1,2}(d)$ and $W^{1,2}(d)$ in terms of $\mathcal{L}$. In fact, we have :

Lemma 3.7 *(i) The space $H_0^{1,2}(d)$ is the domain of the closure $\bar{\mathcal{E}}^{(0)}$ of the quadratic form $(\mathcal{E}^{(0)}, \mathcal{A})$ given by $\mathcal{E}^{(0)}(f, g) = -\int \mathcal{L}f \, g \, dm$. For $f, g \in H_0^{1,2}(d)$, $(\bar{d}f, \bar{d}g) = \Gamma(f, g)$ m-a.e. Here Γ denotes the unique extension of the carré du champ operator to a bilinear map from $H_0^{1,2}(d) \times H_0^{1,2}(d)$ to $L^1(E; m)$, which is continuous w.r.t. the inner product $\bar{\mathcal{E}}^{(0)}(\cdot, \cdot) + (\cdot, \cdot)_{L^2(E; m)}$.*

(ii) A function $u \in L^2(E; m)$ is in $W^{1,2}(d)$ if and only if there exists a finite constant C such that

$$\left| \sum_{i=1}^{n} \left(\int u \, \Gamma(f_i, g_i) \, dm + \int u \, f_i \, \mathcal{L}g_i \, dm \right) \right| \leq C \cdot \left(\sum_{i,j=1}^{n} \int f_i \, f_j \, \Gamma(g_i, g_j) \, dm \right)^{1/2}$$

(3.22)

holds for all $n \in \mathbf{N}$, $f_1, \ldots, f_n \in A$, and $g_1, \ldots, g_n \in A$. In this case,
$\left(\int (\hat{d}u, \hat{d}u)\, dm \right)^{1/2}$ *is the infimum of all $C \geq 0$ for which (3.22) holds.*

In particular, the spaces $H_0^{1,2}(d)$ and $W^{1,2}(d)$, and the quadratic forms $(f,g) \mapsto \int (\bar{d}f, \bar{d}g)\, dm$ on $H_0^{1,2}(d)$ and $(u,v) \mapsto \int (\hat{d}u, \hat{d}v)\, dm$ on $W^{1,2}(d)$ are independent of the choice of a measurable co–tangent bundle $T'E$ and an L^2 differential d corresponding to the operator $(\mathcal{L}, A)$.

PROOF. (i) is obvious by (3.21), the symmetry of the operator $(\mathcal{L}, A)$ on $L^2(E; m)$, and the invariance of the measure m. To prove (ii), we fix $u \in L^2(E; m)$. By definition of $W^{1,2}(d)$, and by the Riesz representation theorem, u is in $W^{1,2}(d)$ if and only if there exists a finite constant C such that

$$(3.23) \qquad \left| \int u\, d^*(\sum_{i=1}^{n} f_i\, dg_i)\, dm \right| \leq C \cdot \left(\int (\sum_{i=1}^{n} f_i\, dg_i, \sum_{i=1}^{n} f_i\, dg_i)\, dm \right)^{1/2}$$

holds for all $n \in \mathbf{N}$, $f_1, \ldots, f_n \in A$, and $g_1, \ldots, g_n \in A$. But

$$d^*\left(\sum_{i=1}^{n} f_i\, dg_i\right) = \sum_{i=1}^{n} (f_i\, d^* dg_i - (df_i, dg_i)) = - \sum_{i=1}^{n} (f_i\, \mathcal{L}g_i + \Gamma(f_i, g_i))$$

by (3.12), and

$$\left(\sum_{i=1}^{n} f_i\, dg_i, \sum_{i=1}^{n} f_i\, dg_i\right) = \sum_{i,j=1}^{n} f_i f_j\, (dg_i, dg_j) = \sum_{i,j=1}^{n} f_i f_j\, \Gamma(g_i, g_j).$$

Hence (3.23) and (3.22) are the same equations, i.e., u is in $W^{1,2}(d)$ if and only if (3.22) holds. Moreover, for $u \in W^{1,2}(d)$, the $L^2(E \to T'E; m)$ norm of $\hat{d}u$ is the infimum of all $C \geq 0$ satisfying (3.23) respectively (3.22). ∎

Let $W_\infty^{1,2}(d)$ be the closure of the space $W^{1,2}(d) \cap L^\infty(E; m)$ in $W^{1,2}(d)$ w.r.t. the norm $\|u\|_{1,2,d} := \left(\int (u^2 + (\hat{d}u, \hat{d}u))\, dm \right)^{1/2}$. The reason for introducing the space $W_\infty^{1,2}(d)$ is, that *in general* it is not clear if $W^{1,2}(d)$ is a Dirichlet space, respectively if the bounded functions are dense in $W^{1,2}(d)$. In many applications, however, we can show that $W_\infty^{1,2}(d)$ is a Dirichlet space.

REMARK. If $W^{1,2}(d)$ is a Dirichlet space (which can be verified in several but not in all applications we are interested in), then $W^{1,2}(d)$ and $W_\infty^{1,2}(d)$ coincide, cf. e.g. [MaRö 92, Ch. I, Prop. 4.17].

Recall that there is a one–to–one correspondence between negative definite self–adjoint operators $(L, \text{Dom}(L))$ and closed quadratic forms $(\mathcal{E}, \text{Dom}(\mathcal{E}))$ on $L^2(E; m)$, cf. Chapter 1. The correspondence is determined by the relation

$$\mathcal{E}(f, g) = - \int \mathcal{L}f\, g\, dm \qquad \text{for all } f \in \text{Dom}(L) \text{ and } g \in \text{Dom}(\mathcal{E}).$$

We now state the main result of this section.

Theorem 3.1 *Let $(L, \text{Dom}(L))$ be a negative definite self–adjoint operator extending $(\mathcal{L}, \mathcal{A})$, and let $(\mathcal{E}, \text{Dom}(\mathcal{E}))$ be the associated closed quadratic form. Suppose that the C^0 semigroup $(e^{tL})_{t\geq 0}$ generated by L is sub–Markov. Then*

$$H_0^{1,2}(d) \ \subseteq \ \text{Dom}(\mathcal{E}) \ \subseteq \ W_\infty^{1,2}(d),$$

$$\mathcal{E}(u, u) \ = \ \int (\bar{d}u, \bar{d}u)\, dm \qquad \textit{for all } u \in H_0^{1,2}(d), \textit{ and}$$

$$\mathcal{E}(u, u) \ \geq \ \int (\hat{d}u, \hat{d}u)\, dm \qquad \textit{for all } u \in \text{Dom}(\mathcal{E}).$$

Before proving Theorem 3.1, we give a different formulation, and we comment on consequences for the Markov uniqueness problem. Let $L^{(0)} := -d^*\bar{d}$. Clearly, $L^{(0)}$ is the *Friedrichs extension* of the operator $(\mathcal{L}, \mathcal{A})$, i.e., it is the negative definite self–adjoint operator associated with the quadratic form $(\bar{\mathcal{E}}^{(0)}, H_0^{1,2}(d))$ defined in Lemma 3.7. Since $(\bar{\mathcal{E}}^{(0)}, H_0^{1,2}(d))$ is a Dirichlet form, the symmetric C^0 contraction semigroup $(e^{tL^{(0)}})_{t\geq 0}$ generated by $L^{(0)}$ is *sub–Markov*, cf. Lemma 1.10 in Appendix B.

Similarly, let $\hat{L}$ be the generator of the closed quadratic form $(\hat{\mathcal{E}}, W_\infty^{1,2}(d))$ defined by

$$\hat{\mathcal{E}}(f, g) \ = \ \int (\hat{d}f, \hat{d}g)\, dm,$$

i.e., $\hat{L} = -\hat{d}^*\hat{d}$, where $\hat{d}$ denotes the restriction of $\hat{d}$ to $W_\infty^{1,2}(d)$. The operator $\hat{L}$ is a negative definite self–adjoint extension of $(\mathcal{L}, \mathcal{A})$. In fact, for $f \in \mathcal{A}$ and $g \in W_\infty^{1,2}(d)$, we have

$$\hat{\mathcal{E}}(f, g) \ = \ \int (df, \hat{d}g)\, dm \ = \ \int d^*df \, g\, dm.$$

In general, it is not clear if $(\hat{\mathcal{E}}, W_\infty^{1,2}(d))$ is a Dirichlet form, but in many applications we are interested in, this is the case, cf. for example the cases considered in Section a) above, and see Remark (vi) in Section c), 2) in Chapter 5 below.

Let $\text{Ext}(\mathcal{L}, \mathcal{A})$ be the set of all negative definite self–adjoint extensions of the operator $(\mathcal{L}, \mathcal{A})$ on $L^2(E; m)$. Let $L^{(1)}$ and $L^{(2)}$ be operators in $\text{Ext}(\mathcal{L}, \mathcal{A})$, and let $\mathcal{E}^{(1)}$ and $\mathcal{E}^{(2)}$ be the corresponding quadratic forms. We define the following order on $\text{Ext}(\mathcal{L}, \mathcal{A})$:

$$L^{(1)} \ \leq \ L^{(2)} \ :\Leftrightarrow \ \left\{ \begin{array}{l} \text{Dom}(\mathcal{E}^{(1)}) \subseteq \text{Dom}(\mathcal{E}^{(2)}), \quad \text{and} \\ \mathcal{E}^{(1)}(u, u) \geq \mathcal{E}^{(2)}(u, u) \quad \text{for all } u \in \text{Dom}(\mathcal{E}^{(1)}). \end{array} \right.$$

We call an operator L in $\text{Ext}(\mathcal{L}, \mathcal{A})$ **Dirichlet**, iff the semigroup $(e^{tL})_{t\geq 0}$ is sub–Markov. An operator is Dirichlet if and only if its quadratic form is a Dirichlet form, cf. e.g. [MaRö 92]. We can now rephrase the assertion of Theorem 3.1 :

Corollary 3.1 (Maximality result) *For every Dirichlet operator L in $\text{Ext}(\mathcal{L}, \mathcal{A})$, we have*

$$L^{(0)} \ \leq \ L \ \leq \ \hat{L}.$$

In particular, $L^{(0)}$ is the **minimal Dirichlet extension** *of $(\mathcal{L}, \mathcal{A})$. If $\hat{L}$ is a Dirichlet operator, then it is the* **maximal Dirichlet extension** *of $(\mathcal{L}, \mathcal{A})$.*

REMARKS. (i) If $\hat{L}$ is a Dirichlet operator, or, equivalently, $(\hat{\mathcal{E}}, W_\infty^{1,2}(d))$ is a Dirichlet form, then Theorem 3.1 respectively Corollary 3.1 solves the **maximality problem** for $(\mathcal{L}, \mathcal{A})$. This problem was first posed and answered in the case where $(\mathcal{L}, \mathcal{A})$ is a symmetric diffusion operator with *"flat geometry"* on a topological *vector space* (i.e., the generator of a so–called "classical Dirichlet form") by S. Albeverio, S. Kusuoka and M. Röckner [AlbKusRö 90], cf. Section a), 3) above. Our extension to the general non–flat case relies heavily on the geometric representation $\mathcal{L} = -d^*d$ of the diffusion operator. In our proof, we use techniques that have been developed by S. Song to give a simplified proof of the maximality result in the flat case, cf. [So 94].

(ii) Clearly, $L^{(0)}$ is also the minimal element in Ext $(\mathcal{L}, \mathcal{A})$. The maximal negative definite self–adjoint extension of a symmetric linear operator has been identified long ago by M. G. Krein [Kr 47], cf. also [FuOshTa 94, Sect. 3.3]. As the examples of Markov unique operators that are not essentially self–adjoint already indicate, the Krein extension is often much larger than $\hat{L}$, cf. [FuOshTa 94].

As a consequence of the maximality result, we obtain the following criterion for Markov uniqueness :

Corollary 3.2 (Basic criterion for Markov uniqueness)
If $H_0^{1,2}(d) = W_\infty^{1,2}(d)$, then the operator $(\mathcal{L}, \mathcal{A})$ is Markov unique. Conversely, if $(\mathcal{L}, \mathcal{A})$ is Markov unique, and $(\hat{\mathcal{E}}, W_\infty^{1,2}(d))$ is a Dirichlet form, then $H_0^{1,2}(d) = W_\infty^{1,2}(d)$.

PROOF. If $H_0^{1,2}(d) = W_\infty^{1,2}(d)$, then all quadratic forms corresponding to Dirichlet operators in Ext $(\mathcal{L}, \mathcal{A})$ coincide. Since the correspondence between negative definite self–adjoint operators and closed quadratic forms on $L^2(E\,;\,m)$ is one–to–one, there is only one Dirichlet operator in Ext $(\mathcal{L}, \mathcal{A})$.
Conversely, if $(\mathcal{L}, \mathcal{A})$ is Markov unique and $(\hat{\mathcal{E}}, W_\infty^{1,2}(d))$ is a Dirichlet form, then $(\hat{\mathcal{E}}, W_\infty^{1,2}(d)) = (\bar{\mathcal{E}}^{(0)}, H_0^{1,2}(d))$, because the generators $\hat{L}$ and $L^{(0)}$ of both forms are Dirichlet operators extending $(\mathcal{L}, \mathcal{A})$. ∎

REMARK. In many applications, it can even be shown that $(\hat{\mathcal{E}}, W^{1,2}(d))$ is a Dirichlet form. In this case, $W^{1,2}(d) = W_\infty^{1,2}(d)$, whence $(\mathcal{L}, \mathcal{A})$ is Markov unique if and only if $H_0^{1,2}(d) = W^{1,2}(d)$.

We will now prove Theorem 3.1 in several steps.

PROOF OF THEOREM 3.1. Since the operator $(L, \text{Dom}\,(L))$ extends $(\mathcal{L}, \mathcal{A})$, the associated closed quadratic form $(\mathcal{E}, \text{Dom}\,(\mathcal{E}))$ extends $(\mathcal{E}^{(0)}, \mathcal{A})$, whence it also extends the closure $(\bar{\mathcal{E}}^{(0)}, H_0^{1,2}(d))$. This proves the first half of the assertion. The crucial and much harder part of the proof, however, is to show the maximality of $(\hat{\mathcal{E}}, W_\infty^{1,2}(d))$. This will now be done in several steps.

Step 1 : *It suffices to show that every bounded function $u \in \text{Dom}\,(\mathcal{E})$ is contained* in $W^{1,2}(d)$, and $\mathcal{E}(u,u) \geq \int (\hat{d}u, \hat{d}u)\, dm$.

In fact, suppose we can show this. Then the bounded functions in Dom $(\mathcal{E})$ are in $W^{1,2}_\infty(d)$. Since the semigroup $(e^{tL})_{t\geq 0}$ is sub–Markovian, the corresponding quadratic form $(\mathcal{E}, \mathrm{Dom}\,(\mathcal{E}))$ is a Dirichlet form. Hence for an arbitrary function $u \in \mathrm{Dom}\,(\mathcal{E})$, the functions $(u \wedge n) \vee (-n)$, $n \in \mathbf{N}$, are again in Dom $(\mathcal{E})$ (and thus in $W^{1,2}_\infty(d)$), and $(u \wedge n) \vee (-n) \to u$ w.r.t. the norm $(\mathcal{E}(\cdot,\cdot)+(\cdot,\cdot)_{L^2(E;m)})^{1/2}$, cf. e.g. [MaRö 92, Ch. I, Prop. 4.17]. Since this norm restricted to $\mathrm{Dom}\,(\mathcal{E}) \cap L^\infty(E; m)$ dominates the norm in $W^{1,2}(d)$ from above, $(u \wedge n) \vee (-n)$ is also a Cauchy sequence w.r.t. the $W^{1,2}(d)$ norm, whence the L^2 limit u is in $W^{1,2}_\infty(d)$. Thus the domain of $\mathcal{E}$ is contained in $W^{1,2}_\infty(d)$. By continuity, the inequality $\mathcal{E}(u,u) \geq \int (\hat{d}u, \hat{d}u)\, dm$ extends to all u in the domain of $\mathcal{E}$.

From now on, we fix a bounded function u in Dom $(\mathcal{E})$. We have to show :

Claim : u is in $W^{1,2}(d)$, and $\mathcal{E}(u,u) \geq \int (\hat{d}u, \hat{d}u)\, dm$.

Step 2 : *Formulation of the claim in terms of $\mathcal{E}$.*

By Lemma 3.7, the claim holds, if and only if (3.22) is satisfied with $C := \mathcal{E}(u,u)^{1/2}$ for all $n \in \mathbf{N}$, $f_1, \ldots, f_n \in \mathcal{A}$, and $g_1, \ldots, g_n \in \mathcal{A}$. Since $\Gamma(f,g) = (\mathcal{L}(fg) - f\mathcal{L}g - g\mathcal{L}f)/2$ for all $f, g \in \mathcal{A}$, and the generator of $(\mathcal{E}, \mathrm{Dom}\,(\mathcal{E}))$ extends $(\mathcal{L}, \mathcal{A})$, we can rewrite (3.22) in the following way :

$$(3.24)\quad \frac{1}{2}\left| \sum_{i=1}^{n} \left(\mathcal{E}(u, f_i g_i) + \mathcal{E}(u f_i, g_i) - \mathcal{E}(u g_i, f_i) \right) \right|$$

$$\leq\ \mathcal{E}(u,u)^{1/2} \cdot \left\{ \sum_{i,j=1}^{n} \left(\mathcal{E}(f_i f_j g_i, g_j) - \frac{1}{2}\mathcal{E}(f_i f_j, g_i g_j) \right) \right\}^{1/2}$$

Here we have used that the product of two bounded functions in Dom $(\mathcal{E})$ is again in Dom $(\mathcal{E})$, because $\mathcal{E}$ is a Dirichlet form, cf. e.g. [MaRö 92, Ch. I, Cor. 4.15].

REMARK.   Under suitable regularity conditions, one can show that there exist signed measures $\mu_{\langle v,w \rangle}$, $v, w \in \mathrm{Dom}\,(\mathcal{E}) \cap L^\infty(E; m)$, such that

$$\int f\, d\mu_{\langle v,w \rangle} = (\mathcal{E}(vf, w) + \mathcal{E}(wf, v) - \mathcal{E}(vw, f))/2$$

holds for all $f \in \mathcal{A}$. For example, this is the case with $\mu_{\langle v,w \rangle} = \Gamma(v,w) \cdot m$ provided the Dirichlet form $(\mathcal{E}, \mathrm{Dom}\,(\mathcal{E}))$ admits a carré du champ Γ. Then, the inequality (3.22) can be deduced from the following Kunita–Watanabe type inequality :

$$\left| \sum_{i=1}^{n} \int f_i\, d\mu_{\langle u,g_i \rangle} \right| \leq \left(\int d\mu_{\langle u,u \rangle} \right)^{1/2} \cdot \left(\sum_{i,j=1}^{n} \int f_i f_j\, d\mu_{\langle g_i,g_j \rangle} \right)^{1/2}$$

To avoid regularity assumptions, we give a direct proof of (3.24), which does not use the existence of the "energy measures" $\mu_{\langle v,w \rangle}$. It is based on a discretized version of the Kunita–Watanabe inequality.

Step 3 : *Semigroup approximation of $\mathcal{E}$.*

The inequality (3.24) is the infinitesimal version of an inequality for the C^0 semigroup $(e^{tL})_{t \geq 0}$ generated by L. In fact, for $t > 0$ and $v, w \in \text{Dom}\,(\mathcal{E})$, we have

$$\mathcal{E}\,(v, w) \;=\; \lim_{t \downarrow 0} \frac{1}{t}\,\mathcal{E}^t\,(v, w),$$

where

$$\mathcal{E}^t\,(v, w) \;:=\; \int v \cdot (w - e^{tL} w)\ dm,$$

cf. e.g. [FuOshTa 94, Lemma 1.3.4]. Thus it is sufficient to prove (3.24) with $\mathcal{E}$ replaced by $\mathcal{E}^t$ for every $t > 0$. From now on, we fix $t > 0$.

Step 4 : *Discretization.*

Since the operator e^{tL} is a contraction both on $L^2(E\,;\,m)$ and on $L^\infty(E\,;\,m)$, it suffices to show that (3.24) with $\mathcal{E}$ replaced by $\mathcal{E}^t$ holds for *elementary functions* $u, f_1, \ldots, f_n$, and $g_1, \ldots, g_n$, such that

$$u = \sum_{k=1}^{p} u^k \chi_{A_k}, \quad g_i = \sum_{k=1}^{p} g_i^k \chi_{A_k}, \quad f_i = \sum_{k=1}^{p} f_i^k \chi_{A_k}, \quad 1 \leq i \leq n,$$

for some $p \in \mathbb{N}$, disjoint sets $A_1, \ldots, A_p \in \mathcal{B}$ with $m(A_k) < \infty$ for all k, and constants $u^k, g_i^k, f_i^k \in \mathbb{R}$, $1 \leq i \leq n$, $1 \leq k \leq p$. For $1 \leq k, l \leq p$ let $m_k := m(A_k)$ and $P_{kl} := \int \chi_{A_k}\, e^{tL} \chi_{A_l}\, dm$. Since the bilinear form $\mathcal{E}^t$ is symmetric, we have for $1 \leq i \leq n$:

$$(3.25) \qquad \mathcal{E}^t(u, f_i g_i) + \mathcal{E}^t(u f_i, g_i) - \mathcal{E}^t(u g_i, f_i)$$

$$= \sum_{k,l=1}^{p} \left(u^k f_i^l g_i^l + u^l f_i^l g_i^k - u^k g_i^k f_i^l \right) \left(m_k \delta_{kl} - P_{kl} \right)$$

$$= \sum_{k,l=1}^{p} \left(u^k - u^l \right) \left(g_i^k - g_i^l \right) f_i^l P_{kl} + \sum_{l=1}^{p} u^l g_i^l f_i^l \left(m_l - \sum_{k=1}^{p} P_{kl} \right)$$

For $1 \leq l \leq p$ let $\bar{m}_l := m_l - \sum_{k=1}^{p} P_{kl}$. Since the operator e^{tL} is symmetric and positivity preserving, (P_{kl}) is a symmetric matrix with positive entries. Moreover, since e^{tL} is sub–Markov, we have

$$(3.26) \qquad \sum_{k=1}^{p} P_{kl} = \int e^{tL} \chi_{\bigcup_k A_k}\, \chi_{A_l}\, dm \;\leq\; \int \chi_{A_l}\, dm \;=\; m_l$$

for all l, whence $\bar{m}_l \geq 0$. Therefore, the bilinear form

$$((A, a), (B, b)) \;\longmapsto\; \sum_{k,l=1}^{p} A_{kl}\, B_{kl}\, P_{kl} + \sum_{k=1}^{p} a_l\, b_l\, \bar{m}_l$$

is a (pseudo) inner product on $\mathbf{R}^{p^2} \oplus \mathbf{R}^p$. By applying the Cauchy–Schwarz inequality w.r.t. this inner product, we can estimate the right hand side of (3.25). If we first take the sum over i, and then apply the Cauchy–Schwarz estimate, we obtain

$$\left| \sum_{i=1}^{n} \left(\mathcal{E}^t(u, f_i g_i) + \mathcal{E}^t(u f_i, g_i) - \mathcal{E}^t(u g_i, f_i) \right) \right|$$

$$= \left| \sum_{k,l=1}^{p} (u^k - u^l) \left(\sum_{i=1}^{n} (g_i^k - g_i^l) f_i^l \right) P_{kl} + \sum_{l=1}^{p} u^l \left(\sum_{i=1}^{n} g_i^l f_i^l \right) \bar{m}_l \right|$$

$$\leq \left\{ \sum_{k,l=1}^{p} (u^k - u^l)^2 P_{kl} + \sum_{l=1}^{p} (u^l)^2 \bar{m}_l \right\}^{1/2}$$

$$\times \left\{ \sum_{k,l=1}^{p} \sum_{i,j=1}^{n} (g_i^k - g_i^l)(g_j^k - g_j^l) f_i^l f_j^l P_{kl} + \sum_{l=1}^{p} \sum_{i,j=1}^{n} g_i^l g_j^l f_i^l f_j^l \bar{m}_l \right\}^{1/2}$$

$$\leq \left\{ 2 \sum_{k,l=1}^{p} u_k u_l \left(m_k \delta_{kl} - P_{kl} \right) \right\}^{1/2}$$

$$\times \left\{ \sum_{i,j=1}^{n} \left[\sum_{k,l=1}^{p} (g_i^k g_j^l f_i^l f_j^l + g_i^l g_j^k f_i^l f_j^l - g_i^k g_j^k f_i^l f_j^l) (m_k \delta_{kl} - P_{kl}) \right] \right\}^{1/2}$$

$$= \left(2 \, \mathcal{E}^t(u, u) \right)^{1/2}$$

$$\times \left\{ \sum_{i,j=1}^{n} \left(\mathcal{E}^t(f_i f_j g_j, g_i) + \mathcal{E}^t(f_i f_j g_i, g_j) - \mathcal{E}^t(f_i f_j, g_i g_j) \right) \right\}^{1/2}.$$

Here we have once more used the estimate (3.26), as well as the symmetry of the matrix (P_{kl}). Dividing by 2, we obtain (3.24) with $\mathcal{E}$ replaced by $\mathcal{E}^t$. Since $t > 0$, and the elementary functions u, f_i and g_i have been chosen arbitrarily, this completes the proof of Theorem 3.1. ■

d) $H_0^{1,2} = W^{1,2}$ and Markov uniqueness : The one–dimensional case

In this section, we prove a necessary and sufficient condition for Markov uniqueness of one–dimensional symmetric diffusion operators by applying the general results from the last section.

Let (x_0, y_0), $-\infty \leq x_0 < y_0 \leq \infty$, be an interval, and let $\mathcal{A} := C_0^\infty(x_0, y_0)$.

We consider a symmetric diffusion operator $(\mathcal{L}, \mathcal{A})$ of type

$$\mathcal{L} = \frac{1}{\rho} \frac{d}{dx} \left(\alpha \frac{d}{dx} \cdot \right)$$

on $L^2(x_0, y_0\,;\,\rho\,dx)$, where ρ is a continuous function on (x_0, y_0), α is absolutely continuous, ρ and α are strictly positive dx–a.e., and the non–divergence form coefficients α/ρ and α'/ρ are in $L^2_{\mathrm{loc}}(x_0, y_0\,;\,\rho\,dx)$. For simplicity, we also assume that the singularity set

$$S(\alpha) := \{x \in (x_0, y_0)\,;\ \alpha(x) = 0\}$$

is finite, i.e., the zeros of α do not accumulate.

Let $a := \alpha/\rho$. The weak Sobolev space $W^{1,2}(d)$ corresponding to the operator $(\mathcal{L}, \mathcal{A})$ is the space $W^{1,2}((x_0, y_0), a;\,\rho\,dx)$ introduced in Section a), 1), above. Consequently, we will denote the space $H_0^{1,2}(d)$ (i.e., the closure of the test–functions in $W^{1,2}((x_0, y_0), a;\,\rho\,dx)$) by $H_0^{1,2}((x_0, y_0), a;\,\rho\,dx)$. By Lemma 3.1, a function $u \in L^2(x_0, y_0\,;\,\rho\,dx)$ is in $W^{1,2}((x_0, y_0), a;\,\rho\,dx)$ if and only if u has a modification $\tilde{u}$ that is absolutely continuous on $(x_0, y_0) \setminus S(\alpha)$, and $\int_{x_0}^{y_0} \alpha\,(\tilde{u}')^2\,dx < \infty$. On such functions u, the quadratic form $\hat{\mathcal{E}}$ introduced in Section c) is given by

$$\hat{\mathcal{E}}(u, u) = \int_{x_0}^{y_0} \alpha\,(\tilde{u}')^2\,dx.$$

The explicit representation shows that $(\hat{\mathcal{E}}, W^{1,2}((x_0, y_0), a;\,\rho\,dx))$ is a Dirichlet form, cf. the remark below Lemma 3.1. In particular, $W_\infty^{1,2}((x_0, y_0), a;\,\rho\,dx) = W^{1,2}((x_0, y_0), a;\,\rho\,dx)$. By Corollary 3.1, the generator $\hat{L}$ of the form $\hat{\mathcal{E}}$ is the maximal Dirichlet extension of $(\mathcal{L}, \mathcal{A})$. In particular, $(\mathcal{L}, \mathcal{A})$ is Markov unique on $L^2(x_0, y_0\,;\,\rho\,dx)$ if and only if

$$H_0^{1,2}((x_0, y_0), a;\,\rho\,dx) = W^{1,2}((x_0, y_0), a;\,\rho\,dx),$$

i.e., if and only if every function (or, equivalently, every bounded function) in $W^{1,2}((x_0, y_0), a;\,\rho\,dx)$ can be approximated w.r.t. the $W^{1,2}$ norm by functions in $C_0^\infty(x_0, y_0)$. We will use this fact to prove the following result :

Theorem 3.2   *The operator* $(\mathcal{L}, C_0^\infty(x_0, y_0))$ *is Markov unique on* $L^2(x_0, y_0\,;\,\rho\,dx)$ *if and only if the following conditions hold :*

(i) *The functions* ρ *and* $1/\alpha$ *are not both* dx–*integrable at* x_0.

(ii) *The functions* ρ *and* $1/\alpha$ *are not both* dx–*integrable at* y_0.

(iii) *For every* $s \in S(\alpha)$, *the function* $1/\alpha$ *is not* dx–*integrable at* s.

REMARK. Conditions (i) and (ii) mean that x_0 and y_0 are not regular boundaries in Feller's sense, cf. Section a) in Chapter 4 below. Condition (iii) means that for every $s \in S(\alpha)$, s is not both a regular boundary for (x_0, s) and (s, y_0). However, it may be a regular boundary for one of the intervals. In Chapter 4, Section a), 3), we give a probabilistic "explanation" for this result.

Before proving Theorem 3.2, we show some properties of functions in the strong Sobolev space $H_0^{1,2}((x_0, y_0), a; \rho\, dx)$:

Lemma 3.8 *Let u be a bounded function in $H_0^{1,2}((x_0, y_0), a; \rho\, dx)$, and let $\tilde{u}$ denote its continuous dx–version on $(x_0, y_0) \setminus S(\alpha)$.*

(i) Suppose $1/\alpha$ is dx–integrable at x_0. Then $\lim_{x \downarrow x_0} \tilde{u}(x) = 0$. Similarly, if $1/\alpha$ is dx–integrable at y_0, then $\lim_{x \uparrow y_0} \tilde{u}(x) = 0$.

(ii) Let $s \in S(\alpha)$, and suppose that $1/\alpha$ is dx–integrable in a neighbourhood of s. Then $\lim_{x \uparrow s} \tilde{u}(x) = \lim_{x \downarrow s} \tilde{u}(x)$.

PROOF.   (i) Let $(f_n)_{n \in \mathbf{N}}$ be a sequence of functions in $C_0^\infty(x_0, y_0)$ such that $f_n \to u$ in $H_0^{1,2}((x_0, y_0), a; \rho\, dx)$. Then for dx–a.e. $x \in (x_0, y_0)$, we have

$$
\begin{aligned}
|\tilde{u}(x)| &= \left| \lim_{n \to \infty} f_n(x) \right| = \lim_{n \to \infty} \left| \int_{x_0}^x f_n'(x)\, dx \right| \\
&\le \left(\int_{x_0}^x (1/\alpha)\, dx \right)^{1/2} \cdot \sup_{n \in \mathbf{N}} \left(\int_{x_0}^{y_0} \alpha\, (f_n')^2\, dx \right)^{1/2}.
\end{aligned}
$$

The supremum is finite, because $(f_n)_{n \in \mathbf{N}}$ is a bounded sequence w.r.t. the $H_0^{1,2}((x_0, y_0), a; \rho\, dx)$ norm. Since $1/\alpha$ is dx–integrable and u is continuous near x_0, we see that $\tilde{u}(x) \to 0$ as $x \downarrow x_0$. The assertion for y_0 can be proven in the same way.

(ii) For $(f_n)_{n \in \mathbf{N}}$ as in (i), we have

$$
\begin{aligned}
|\tilde{u}(y) - \tilde{u}(x)| &= \lim_{n \to \infty} |f_n(y) - f_n(x)| \\
&= \lim_{n \to \infty} \left| \int_x^y f_n'(x)\, dx \right| \le \left(\int_x^y (1/\alpha)\, dx \right)^{1/2} \cdot \sup_{n \in \mathbf{N}} \left(\int_{x_0}^{y_0} \alpha\, (f_n')^2\, dx \right)^{1/2}
\end{aligned}
$$

for dx–a.e. $x, y \in (x_0, y_0)$. This implies the assertion, because $1/\alpha$ is dx–integrable near s, and $\tilde{u}$ is continuous both to the left and to the right of s. ∎

After these preparations, we prove Theorem 3.2 :

PROOF OF THEOREM 3.2.   For brevity, we use the notations $W^{1,2}(d)$ and $H_0^{1,2}(d)$ instead of $W^{1,2}((x_0, y_0), a; \rho\, dx)$ and $H_0^{1,2}((x_0, y_0), a; \rho\, dx)$. Suppose first that one of the conditions in Theorem 3.2 does not hold. There are three possibilities :

Case I : *There exists $s \in S(\alpha)$ such that $1/\alpha$ is dx–integrable in a neighbourhood of s.*

Let f be an arbitrary function in $C_0^\infty(x_0, y_0)$ such that $f(s) \neq 0$. Then by Lemma 3.1, the function $f \cdot \chi_{(x_0, s)}$ is in $W^{1,2}(d)$. On the other hand, by Lemma 3.8, every bounded function in $H_0^{1,2}(d)$ has a modification that is continuous at s, whence $f \cdot \chi_{(x_0, s)}$ is not in $H_0^{1,2}(d)$. Thus $H_0^{1,2}(d) \neq W^{1,2}(d)$, i.e., Markov uniqueness does not hold.

Case II : $1/\alpha$ *and* ρ *are* dx*–integrable at* x_0.

Let f be a smooth function on (x_0, y_0) such that $f(x) = 1$ for x in a neighbourhood of x_0, and $f(x) = 0$ for x in a neighbourhood of y_0. Since ρ is dx-integrable at x_0, f is in $L^2(x_0, y_0 \,;\, \rho\, dx)$. Moreover, f' is in $C_0^\infty(x_0, y_0)$. Thus by Lemma 3.1, f is in $W^{1,2}(d)$. On the other hand, by Lemma 3.8, f is not in $H_0^{1,2}(d)$. Thus $H_0^{1,2}(d) \neq W^{1,2}(d)$.

Case III : $1/\alpha$ *and* ρ *are* dx*–integrable at* y_0.

Here $H_0^{1,2}(d) \neq W^{1,2}(d)$ can be shown as in Case II.

Thus in all cases, the operator $(\mathcal{L}, C_0^\infty(x_0, y_0))$ is not Markov unique.

Now suppose conversely, that the conditions (i)–(iii) in Theorem 3.2 are satisfied. We want to show that every bounded function in $W^{1,2}(d)$ is in $H_0^{1,2}(d)$. Since $W^{1,2}(d)$ is a Dirichlet space, this implies $H_0^{1,2}(d) = W^{1,2}(d)$. We first note that every compactly supported absolutely continuous function $g : (x_0, y_0) \to \mathbf{R}$ is in $H_0^{1,2}(d)$, provided $\int_{x_0}^{y_0} (g')^2 \, dx < \infty$. In fact, let $(\varphi_\varepsilon)_{\varepsilon > 0}$ be a Dirac sequence of mollifiers. Then the functions $g * \varphi_\varepsilon$ are in $C_0^\infty(x_0, y_0)$ for small ε, $g * \varphi_\varepsilon \to g$ in $L^2(x_0, y_0; \, dx)$, and $(g * \varphi_\varepsilon)' \to g'$ in $L^2(x_0, y_0; \, dx)$ as $\varepsilon \downarrow 0$. Since ρ and α are locally bounded, we obtain that $g * \varphi_\varepsilon \to g$ in $L^2(x_0, y_0; \, \rho\, dx)$, and $\int_{x_0}^{y_0} \alpha \left((g * \varphi_\varepsilon)' - g' \right)^2 dx \to 0$ as $\varepsilon \downarrow 0$, which shows that g is in $H_0^{1,2}(d)$. Thus to prove $H_0^{1,2}(d) = W^{1,2}(d)$ it is enough to show the following claim :

Claim : *Every bounded function* u *in* $W^{1,2}(d)$ *can be approximated w.r.t. the* $W^{1,2}(d)$ *norm by compactly supported absolutely continuous functions with square integrable derivatives.*

To prove the claim, we proceed in two steps :

Step 1 : *Approximation by compactly supported functions.*

We fix a bounded function u in $W^{1,2}(d)$. By Lemma 3.1, we may assume that u is absolutely continuous on $(x_0, y_0) \setminus S(\alpha)$, and $\int_{x_0}^{y_0} \alpha\, (u')^2 \, dx < \infty$. We first show how to approximate u by bounded functions in $W^{1,2}(d)$ that vanish in neighbourhoods of x_0. By Condition (i), ρ and $1/\alpha$ are not both dx–integrable at x_0. We distinguish two cases :

Case I : $1/\alpha$ *is not* dx*–integrable at* x_0.

We fix a point $x_1 \in (x_0, y_0)$ such that $\alpha(x) > 0$ for all $x \in (x_0, x_1]$. Let $\psi := \int_\bullet^{x_1} (1/\alpha) \, dx$. For $k \in \mathbf{N}$, let $e_k = (k - \psi)^+ \wedge 1$, i.e.,

$$
e_k(x) = \begin{cases} 1 & \text{if } \psi(x) \leq k - 1 \\ 1 - \psi(x) & \text{if } k - 1 \leq \psi(x) \leq k \\ 0 & \text{if } \psi(x) \geq k \end{cases}
$$

Obviously, the functions e_k, $k \in \mathbf{N}$, are absolutely continuous, and

$$e'_k = -(1/\alpha) \, \chi_{\{k-1 \leq \psi \leq k\}} .$$

Since $\psi(x_0) = \infty$, e_k vanishes in a neighbourhood of x_0 for every k. Moreover, $e_k(x) \uparrow 1$ as $k \to \infty$ for every x. Let $u_k := u \cdot e_k$. Then $(u_k)_{k \in \mathbf{N}}$ is a sequence of functions on (x_0, y_0) that converges pointwise to u as $k \to \infty$. Since $|u_k(x)| \leq |u(x)|$ for all x and k, $u_k \to u$ in $L^2(x_0, y_0 ; \rho \, dx)$. For every k, u_k is absolutely continuous on $(x_0, y_0) \setminus S(\alpha)$, and

$$u'_k \; = \; e_k \, u' + e'_k \, u \; = \; e_k \, u' - \chi_{\{k-1 \leq \psi \leq k\}} \cdot (u/\alpha) .$$

In particular,

$$\frac{1}{2} \int_{x_0}^{y_0} \alpha \, (u'_k)^2 \, dx \; \leq \; \int_{x_0}^{y_0} \alpha \, (u')^2 \, dx + \int_{x_0}^{y_0} (\chi_{\{k-1 \leq \psi \leq k\}}/\alpha) \, dx \cdot \sup |u|^2$$

$$= \; \int_{x_0}^{y_0} \alpha \, (u')^2 \, dx + \sup |u|^2 \; < \; \infty .$$

Hence, by Lemma 3.1, $(u_k)_{k \in \mathbf{N}}$ is a uniformly bounded sequence in the Hilbert space $W^{1,2}(d)$. Since $u_k \to u$ in $L^2(x_0, y_0 ; \rho \, dx)$, the theorems of Banach/Alaoglu and Banach/Saks now imply that $u_k \to u$ weakly in $W^{1,2}(d)$, and the Césaro means $w_n := \frac{1}{n} \sum_{i=1}^{n} u_{k_i}$ of a subsequence $(u_{k_i})_{i \in \mathbf{N}}$ converge to u strongly in $W^{1,2}(d)$ as $n \to \infty$, cf. e.g. [MaRö 92, Appendix 2]. Obviously, the functions w_n, $n \in \mathbf{N}$, are bounded, and vanish in a neighbourhood of x_0.

Case II : $1/\alpha$ *is* dx*–integrable at* x_0, *but* ρ *is not.*

For $x_0 < x \leq y < y_0$ such that $[x, y] \cap S(\alpha) = \emptyset$, we have

$$|u(y) - u(x)| \; = \; \left| \int_x^y u' \, dx \right| \; \leq \; \left(\int_x^y (1/\alpha) \, dx \right)^{1/2} \cdot \left(\int_{x_0}^{y_0} \alpha \, (u')^2 \, dx \right)^{1/2} .$$

In particular, the limit of $u(x)$ as $x \downarrow x_0$ exists. Since ρ is not dx–integrable at x_0, and u is in $L^2(x_0, y_0 ; \rho \, dx)$, we obtain $\lim_{x \downarrow x_0} u(x) = 0$. Let $(x_k)_{k \in \mathbf{N}}$ be a decreasing sequence such that $x_k \downarrow x_0$, and $(x_0, x_1] \cap S(\alpha) = \emptyset$. For $k \in \mathbf{N}$, let u_k be the bounded absolutely continuous function on $(x_0, y_0) \setminus S(\alpha)$ defined by $u_k(x) = u(x)$ for $x \geq x_k$, and $u_k(x) = \left(|u(x_k)| - 2 \int_x^{x_k} |u'| \, dx \right)^+ \cdot \mathrm{sgn}\,(u(x_k))$ for $x \leq x_k$. Then u_k vanishes in a neighbourhood of x_0, $|u_k(x)| \leq |u(x)|$, and $|u'_k(x)| \leq 2 |u'(x)|$ for all $x \in (x_0, y_0) \setminus S(\alpha)$. In particular, u_k is in $W^{1,2}(d)$. Moreover, dominated convergence shows that $u_k \to u$ in $L^2(x_0, y_0 ; \rho \, dx)$, and $u'_k \to u'$ in $L^2((x_0, y_0) \setminus S(\alpha); \alpha \, dx)$. Hence $(u_k)_{k \in \mathbf{N}}$ converges to u in $W^{1,2}(d)$.

We have shown that every bounded function in $W^{1,2}(d)$ can be approximated by bounded functions in $W^{1,2}(d)$ that vanish in neighbourhoods of x_0. By applying a similar argument at the boundary y_0, we can show that every bounded function in $W^{1,2}(d)$ which vanishes in a neighbourhood of x_0 can be approximated by bounded functions in $W^{1,2}(d)$ that vanish both in neighbourhoods of x_0 and y_0.

Hence the compactly supported bounded functions are dense in $W^{1,2}(d)$. We now show that these functions can be approximated in $W^{1,2}(d)$ by compactly supported absolutely continuous functions that are constant in a neighbourhood of s for every $s \in S(\alpha)$.

Step 2 : *Approximation by compactly supported absolutely continuous functions that are constant in neighbourhoods of the singularities.*

Let u be a compactly supported bounded function in $W^{1,2}(d)$. We may assume that u is absolutely continuous on $(x_0, y_0) \setminus S(\alpha)$. Fix a point $s \in S(\alpha)$. Fix $x_1, y_1 \in (x_0, y_0)$ such that $x_1 < s < y_1$, and $[x_1, y_1] \cap S(\alpha) = \emptyset$. We show that u can be approximated by compactly supported, bounded functions in $W^{1,2}(d)$ which are constant in neighbourhoods of s, and which coincide with u on $(x_0, x_1] \cup [y_1, y_0)$. Then, by applying the argument succesively to all singularities $s \in S(\alpha)$, we see that u can be approximated by compactly supported, bounded functions u_k, $k \in \mathbf{N}$, in $W^{1,2}(d)$, each of which is constant in a neighbourhood of every singularity $s \in S(\alpha)$. By Lemma 3.1, these approximating functions have absolutely continuous modifications $\tilde{u}_k$, $k \in \mathbf{N}$, on (x_0, y_0). Moreover, their derivatives $\tilde{u}'_k$ vanish in neighbourhoods of singularities and boundary points, and $\int_{x_0}^{y_0} \alpha \, (\tilde{u}'_k)^2 \, dx < \infty$ for all k. Hence the derivatives are square–integrable. Together with Step 1, this completes the proof of the claim.

It remains to show how to approximate u by functions that are constant in a neighbourhood of the fixed singularity s. By assumption, $1/\alpha$ is not dx–integrable at s. Without loss of generality, we may assume that $\int_s^{s+\varepsilon} (1/\alpha) \, dx = \infty$ for all $\varepsilon > 0$. As in Step 1, Case I, we define cut–off functions e_k, $k \in \mathbf{N}$, by $e_k := (k - \psi)^+ \wedge 1$ where $\psi := \int_s^{y_1} (1/\alpha) \, dx$. We define approximating functions u_k, $k \in \mathbf{N}$, in the following way :

$$u_k(x) := \begin{cases} u(x) & \text{if } x \leq s - k^{-1} \\ u(x) \cdot e_k(x) + u(s - k^{-1}) \cdot (1 - e_k(x)) & \text{if } x \geq s - k^{-1} \end{cases}$$

The functions u_k coincide with u on $(x_0, x_1] \cup [y_1, y_0)$ if k is large enough. In particular, they have compact support. In a neighbourhood of s, u_k is constant. As $k \to \infty$, $u_k(x) \to u(x)$ for dx–a.e. x. Moreover,

$$|u_k(x)| \leq |u(x)| + \chi_{[x_1, y_1]}(x) \cdot \sup |u| \qquad \text{for all } x,$$

if k is sufficiently large, whence $u_k \to u$ in $L^2(x_0, y_0; \rho \, dx)$. Finally, the functions u_k are absolutely continuous on $(x_0, y_0) \setminus S(\alpha)$, and

$$u'_k(x) = \begin{cases} u'(x) & \text{if } x \leq s - k^{-1} \text{ or } x \geq \psi^{-1}(k - 1) \\ 0 & \text{if } s - k^{-1} < x < \psi^{-1}(k) \\ u'(x) \cdot e_k(x) + (u(s - k^{-1}) - u(x)) \cdot (\alpha(x))^{-1} \\ \qquad \text{if } \psi^{-1}(k) \leq x < \psi^{-1}(k - 1) \end{cases}$$

holds for dx–a.e. x. In particular,

$$|u'_k| \leq |u'| + \chi_{\{k-1 \leq \psi \leq k\}} \cdot (2/\alpha) \cdot \sup |u|,$$

whence

$$\frac{1}{2} \int_{x_0}^{y_0} \alpha \left(u_k'\right)^2 \, dx \;\leq\; \int_{x_0}^{y_0} \alpha \left(u'\right)^2 \, dx + 4 \sup |u|^2 \cdot \int_{x_0}^{y_0} (\chi_{\{k-1 \leq \psi \leq k\}}/\alpha) \, dx$$

$$= \int_{x_0}^{y_0} \alpha \left(u'\right)^2 \, dx + 4 \sup |u|^2 \;<\; \infty.$$

Hence, by Lemma 3.1, $(u_k)_{k \in \mathbf{N}}$ is a uniformly bounded sequence in the Hilbert space $W^{1,2}(d)$. The theorems of Banach/Saks and Banach/Alaoglu again imply that $u_k \to u$ weakly in $W^{1,2}(d)$, and the Césaro means $w_n = \frac{1}{n} \sum_{i=1}^{n} u_{k_i}$ of a subsequence $(u_{k_i})_{i \in \mathbf{N}}$ converge to u strongly in $W^{1,2}(d)$ as $n \to \infty$. Obviously, the functions w_n have all the desired properties, which completes the proof of Step 2, and thus, that of the claim. ∎

e) Density of smooth functions in weak Sobolev spaces over $\mathbf{R}^n$

A major advantage of the "Weak–equals–strong–criterion" for Markov uniqueness of diffusion operators in $\mathbf{R}^n$ is that weak Sobolev spaces corresponding to different operators can be compared with each other rather easily. Therefore, Markov uniqueness for one operator, i.e, coincidence of the corresponding weak and strong Sobolev spaces, can be used to prove density theorems in weak Sobolev spaces corresponding to "comparable" operators. In this section, we first state an $H = W$ result for a certain class of weighted Sobolev spaces. This result is originally due to M. Röckner and T. S. Zhang [RöZha 92, 94]. We then demonstrate how $H = W$ results for a more genral class of weighted Sobolev spaces, and, finally, for Sobolev spaces corresponding to diffusion operators with non–trivial diffusion matrices, can be derived from the first result by comparison techniques.

1) The spaces $W^{1,2}(\mathbf{R}^n; \varphi^2 \, dx)$ for $\varphi \in H^{1,2}(\mathbf{R}^n; dx)$

Let $n \in \mathbf{N}$, and let φ be a function in $H^{1,2}(\mathbf{R}^n; dx)$ such that $\varphi > 0$ dx–a.e. We denote the space $W^{1,2}(\mathbf{R}^n, 1; \varphi^2 \, dx)$ as introduced in Definition 3.1 briefly by $W^{1,2}(\mathbf{R}^n; \varphi^2 \, dx)$. Note that all the assumptions imposed in Section a), 1), are satisfied for $E = \mathbf{R}^n$, $\rho = \varphi^2$, $a = 1$, and, hence, $\alpha = \rho = \varphi^2$. A more explicit description of elements in $W^{1,2}(\mathbf{R}^n; \varphi^2 \, dx)$ has been given in Lemma 3.1. Let $H_0^{1,2}(\mathbf{R}^n; \varphi^2 \, dx)$ be the closure of $C_0^\infty(\mathbf{R}^n)$ in $W^{1,2}(\mathbf{R}^n; \varphi^2 \, dx)$ w.r.t. the canonical inner product

$$(u, v)_{W^{1,2}(\mathbf{R}^n; \varphi^2 \, dx)} \;=\; \int_{\mathbf{R}^n} u \, v \, \varphi^2 \, dx + \sum_{i=1}^{n} \int_{\mathbf{R}^n} \hat{\partial}_i u \, \hat{\partial}_i v \, \varphi^2 \, dx.$$

The following result has first been proven by M. Röckner and T. S. Zhang :

Theorem 3.3 *Let* $\varphi \in H^{1,2}(\mathbf{R}^n; dx)$, $\varphi > 0$ *dx-a.e. Then*

$$H_0^{1,2}(\mathbf{R}^n; \varphi^2\, dx) \;=\; W^{1,2}(\mathbf{R}^n; \varphi^2\, dx).$$

The coincidence of the Sobolev spaces is equivalent to Markov uniqueness of the corresponding diffusion operator $(\mathcal{L}, C_0^\infty(\mathbf{R}^n))$, $\mathcal{L} = \Delta + 2\frac{\nabla\varphi}{\varphi} \cdot \nabla$, on $L^2(\mathbf{R}^n; \varphi^2\, dx)$, cf. Corollary 3.2; note that $W^{1,2}(\mathbf{R}^n; \varphi^2\, dx)$ is a Dirichlet space by the remark below Lemma 3.1, whence $W_\infty^{1,2}(\mathbf{R}^n; \varphi^2\, dx) = W^{1,2}(\mathbf{R}^n; \varphi^2\, dx)$. The operator $\mathcal{L}$ is a *generalized Schrödinger operator*, i.e., operators of this type appear as ground state transformations of Schrödinger operators, cf. [RöZha 92].

In the last years, various proofs of Theorem 3.3 have been obtained. We give a short overview :

* The idea of the original proof in [RöZha 92, 94] is to localize on sets where φ is bounded from above and below by strictly positive constants. On these "level sets", the weighted Sobolev spaces can be compared with Sobolev spaces w.r.t. Lebesgue measure. However, to carry out the localization procedure, an assumption on the logarithmic derivative of φ is required, which could only be verified by a probabilistic argument based on the Girsanov theorem.

* P. Cattiaux and M. Fradon [CatFra 97] obtained a proof based on smoothing functions in $W^{1,2}(\mathbf{R}^n; \varphi^2\, dx)$ by convolution. It should be pointed out that the estimates needed to show convergence of the convoluted functions are much more involved than in the proof of the classical theorem of Meyers and Serrin [MeySer 64], i.e., Theorem 3.3 with $\varphi(x) = 1$ for all x.

* V. Liskevich and Y. Semenov [LiSem 96] gave a direct proof for Markov uniqueness of the operator $(\mathcal{L}, C_0^\infty(\mathbf{R}^n))$ on $L^2(\mathbf{R}^n; \varphi^2\, dx)$. As remarked, Markov uniqueness of this operator is equivalent to the coincidence of the spaces $H_0^{1,2}(\mathbf{R}^n; \varphi^2\, dx)$ and $W^{1,2}(\mathbf{R}^n; \varphi^2\, dx)$.

For completeness, we demonstrate now, how Theorem 3.3 can be easily derived from our results on strong uniqueness in Chapter 2. The proof of Theorem 3.3 thus obtained is very similar to that in [LiSem 96].

PROOF OF THEOREM 3.3. By Theorem 2.5, the closure of the operator $(\Delta + 2\frac{\nabla\varphi}{\varphi} \cdot \nabla, C_0^\infty(\mathbf{R}^n))$ on $L^1(\mathbf{R}^n, \varphi^2\, dx)$ generates a C^0 semigroup. Note that the proof of Theorem 2.5 simplifies considerably in the special case considered, because $p = 1$, and the drift $\nabla\varphi/\varphi$ is *globally* in $L^2(\mathbf{R}^n; \varphi^2\, dx)$. Hence the operator above is $L^1(\mathbf{R}^n, \varphi^2\, dx)$ unique, and thus Markov unique on $L^2(\mathbf{R}^n; \varphi^2\, dx)$, cf. Section e) in Chapter 1. By the easy half of Corollary 3.2, we obtain $H_0^{1,2}(\mathbf{R}^n; \varphi^2\, dx) = W^{1,2}(\mathbf{R}^n; \varphi^2\, dx)$. ∎

2) The spaces $W^{1,2}(E, \sigma^2; \varphi^2\, dx)$, $E \subseteq \mathbf{R}^n$

Let E be an open subset in $\mathbf{R}^n$, $n \in \mathbf{N}$. Supppose we are given a function $\varphi \in H^{1,2}_{\mathrm{loc}}(E; dx)$ such that $\varphi > 0$ dx-a.e., and a locally bounded function $\sigma \in H^{1,2}_{\mathrm{loc}}(E; \varphi^2\, dx)$ such that $\sigma > 0$ dx-a.e. as well. The local Sobolev space w.r.t. the measure $\varphi^2\, dx$ has been defined in Section a), 2). Let $a := \sigma^2$, $\rho := \varphi^2$, and $\alpha := a \cdot \rho$. The assumption on σ implies that a is in $H^{1,2}_{\mathrm{loc}}(E; \varphi^2\, dx) \cap L^\infty_{\mathrm{loc}}(E; dx)$. Hence, by the assumption on φ, α is in $H^{1,1}_{\mathrm{loc}}(E; dx)$, and $\frac{1}{\rho}\frac{\partial a}{\partial x_i}$ is in $L^2_{\mathrm{loc}}(E; \varphi^2\, dx)$, i.e., the conditions imposed in Section a), 1), are satisfied. The corresponding weak Sobolev space $W^{1,2}(E, \sigma^2; \varphi^2\, dx)$ is a Hilbert space with inner product

$$ (u, v)_{W^{1,2}(E, \sigma^2; \varphi^2\, dx)} = \int_E uv\, \varphi^2\, dx + \int_E \sigma^2 \sum_{i=1}^n \hat{\partial}_i u\, \hat{\partial}_i v\ \varphi^2\, dx. $$

Let $H^{1,2}_0(E, \sigma^2; \varphi^2\, dx)$ and $H^{1,2}(E, \sigma^2; \varphi^2\, dx)$ denote the closures of $C^\infty_0(E)$, $C^\infty(E) \cap W^{1,2}(E, \sigma^2; \varphi^2\, dx)$ respectively in $W^{1,2}(E, \sigma^2; \varphi^2\, dx)$. The following density theorem of type Meyers–Serrin holds :

Theorem 3.4   Let $\varphi \in H^{1,2}_{\mathrm{loc}}(E; dx)$, $\varphi > 0$ dx-a.e., and let $\sigma \in H^{1,2}_{\mathrm{loc}}(E; \varphi^2\, dx) \cap L^\infty_{\mathrm{loc}}(E; dx)$, $\sigma > 0$ dx-a.e. Then :

(i) Every compactly supported function $u \in W^{1,2}(E, \sigma^2; \varphi^2\, dx)$ is in $H^{1,2}_0(E, \sigma^2; \varphi^2\, dx)$.

(ii) $H^{1,2}(E, \sigma^2; \varphi^2\, dx) = W^{1,2}(E, \sigma^2; \varphi^2\, dx)$.

REMARK.   The theorem enables us to derive a necessary and sufficient condition for the coincidence of $H^{1,2}_0(E, \sigma^2; \varphi^2\, dx)$ and $W^{1,2}(E, \sigma^2; \varphi^2\, dx)$, cf. Theorem 3.6 below. We would expect coincidence, if the diffusion process generated by the Friedrichs extension of the corresponding diffusion operator does not "notice" the boundary of E. This is in fact true, cf. Corollary 3.4.

Before proving the theorem, we briefly comment on the assumptions on φ and σ. If the condition $\varphi \in H^{1,2}_{\mathrm{loc}}(E; dx)$ is not satisfied, then even for $\sigma = 1$, the corresponding diffusion operator $(\Delta + 2\frac{\nabla \varphi}{\varphi} \cdot \nabla, C^\infty_0(E))$ does not take values in $L^2(E; \varphi^2\, dx)$. Hence the assumption $\varphi \in H^{1,2}_{\mathrm{loc}}(E; dx)$ is already needed to pose the Markov uniqueness problem we are interested in. In contrast to this, the Markov uniqueness problem would also be well–posed if we would only assume $\sigma^2 \in H^{1,2}_{\mathrm{loc}}(E; \varphi^2\, dx) \cap L^\infty_{\mathrm{loc}}(E; dx)$ instead of $\sigma \in H^{1,2}_{\mathrm{loc}}(E; \varphi^2\, dx) \cap L^\infty_{\mathrm{loc}}(E; dx)$. However, the following example shows that under this assumption, Theorem 3.4 does not hold in genral. In fact, the condition $\sigma \in H^{1,2}_{\mathrm{loc}}(E; \varphi^2\, dx)$ turns out to be rather sharp :

EXAMPLE.   Let $\gamma > 0$. Suppose $E = \mathbf{R}^1$, $\varphi(x) = 1$ and $\sigma(x) = |x|^\gamma$ for all x. The assumption $\sigma^2 \in H^{1,2}_{\mathrm{loc}}(\mathbf{R}; dx)$, which guarantees that the corresponding diffusion operator $(\mathcal{L}, C^\infty_0(\mathbf{R}))$, $\mathcal{L}f = (\sigma^2 f')'$, takes values in $L^2(\mathbf{R}; dx)$, is satisfied whenever $\gamma > 1/4$. However, the two Sobolev spaces $H^{1,2}(\mathbf{R}, \sigma^2; dx)$ and $W^{1,2}(\mathbf{R}, \sigma^2; dx)$ coincide (if and) only if $1/\sigma^2$ is not dx–integrable at 0, i.e., if and only if $\gamma \geq 1/2$. This can be shown similarly to the proof of Theorem 3.2 above. Hence for $\frac{1}{4} < \gamma < \frac{1}{2}$, the assertion of Theorem 3.4 does not hold. On the other hand, the assumption $\sigma \in H_{\text{loc}}^{1,2}(\mathbf{R}; dx)$ imposed in Theorem 3.4 is satisfied if and only if $\gamma > 1/2$. Thus the theorem yields the coincidence of the Sobolev spaces in every case where it holds, except $\gamma = 1/2$.

PROOF OF THEOREM 3.4. (i) The assumptions on φ and σ imply that $\sigma \cdot \varphi$ is in $H_{\text{loc}}^{1,2}(E; dx)$, and $\frac{\partial(\sigma \cdot \varphi)}{\partial x_i} = \sigma \frac{\partial \varphi}{\partial x_i} + \varphi \frac{\partial \sigma}{\partial x_i}$ dx–a.e. for $1 \leq i \leq n$, where $\frac{\partial \sigma}{\partial x_i}$ denotes the (dx–a.e. defined) derivative in $H_{\text{loc}}^{1,2}(E; \varphi^2 \, dx)$. Let U be a relatively compact open subset of E. Then we can find a function $\psi \in H^{1,2}(\mathbf{R}^n; dx)$ such that $\psi = \sigma \cdot \varphi$ dx–a.e. on U. In particular, $\alpha = \psi^2$ dx–a.e. on U.
Now let u be a bounded function on $\mathbf{R}^n$ such that the support of u is contained in U. Then by Lemma 3.1, the restriction of u to E is in $W^{1,2}(E, \sigma^2; \varphi^2 \, dx)$ if and only if u is in $W^{1,2}(\mathbf{R}^n; \psi^2 \, dx)$. In this case, the weak derivatives $\hat{\partial}_i u$, $1 \leq i \leq n$, in $W^{1,2}(E, \sigma^2; \varphi^2 \, dx)$ and in $W^{1,2}(\mathbf{R}^n; \psi^2 \, dx)$ coincide on U, and vanish outside U. In particular,

$$\int_E \sigma^2 \sum_{i=1}^{n} \left(\hat{\partial}_i u \right)^2 \varphi^2 \, dx \; = \; \int_{\mathbf{R}^n} \sum_{i=1}^{n} \left(\hat{\partial}_i u \right)^2 \psi^2 \, dx \, .$$

After this application of the crucial Lemma 3.1, the claim of (i) can be deduced easily from Theorem 3.3. By the remark below Lemma 3.1, the weak Sobolev space $W^{1,2}(E, \sigma^2; \varphi^2 \, dx)$ is a Dirichlet space. Hence every compactly supported function in $W^{1,2}(E, \sigma^2; \varphi^2 \, dx)$ can be approximated by bounded compactly supported functions w.r.t. the Sobolev norm. It remains to show that every bounded compactly supported function in $W^{1,2}(E, \sigma^2; \varphi^2 \, dx)$ can be approximated by functions in $C_0^{\infty}(\mathbf{R}^n)$. Fix such a function u, and let U be a relatively compact open subset in E such that $\text{supp}\, u \subset U$. Choose a function $\psi \in H^{1,2}(\mathbf{R}^n; dx)$ as above. As remarked above, the trivial extension of u to $\mathbf{R}^n$ is in $W^{1,2}(\mathbf{R}^n; \psi^2 \, dx)$. We denote it by u as well. By Theorem 3.3, $W^{1,2}(\mathbf{R}^n; \psi^2 \, dx) = H_0^{1,2}(\mathbf{R}^n; \psi^2 \, dx)$. Hence there exist functions $f_k \in C_0^{\infty}(\mathbf{R}^n)$, $k \in \mathbf{N}$, such that $f_k \to u$ in $W^{1,2}(\mathbf{R}^n; \psi^2 \, dx)$ as $k \to \infty$.
Let η be a function in $C_0^{\infty}(\mathbf{R}^n)$ such that $0 \leq \eta \leq 1$, $\eta(x) = 1$ for $x \in \text{supp}\, u$, and $\eta(x) = 0$ for $x \in \mathbf{R}^n \setminus U$. Moreover, let $h : \mathbf{R} \to \mathbf{R}$ be a smooth increasing function such that $h(s) = s$ if $|s| \leq \|u\|_{L^{\infty}(E; dx)}$, and $|h(s)| \leq 1 + \|u\|_{L^{\infty}(E; dx)}$ for all s. For $k \in \mathbf{N}$ let $g_k(x) := \eta(x) \cdot h(f_k(x))$, $x \in \mathbf{R}^n$. Obviously, $(g_k)_{k \in \mathbf{N}}$ is a sequence of uniformly bounded functions in $C_0^{\infty}(\mathbf{R}^n)$. The support of g_k is contained in U for all k, and $g_k \to u$ in $L^2(\mathbf{R}^n; \psi^2 \, dx)$ as $k \to \infty$. The uniform boundedness implies that $g_k \to u$ in $L^2(E; \varphi^2 \, dx)$ as well. Moreover, $\nabla g_k = h \circ f_k \nabla \eta + \eta \, h' \circ f_k \nabla f_k$, whence

$$|\nabla g_k| \; \leq \; (1 + \|u\|_{L^{\infty}(E; dx)}) \cdot |\nabla \eta| + \eta \, |\nabla f_k| \qquad \text{for all } k.$$

Therefore,

$$\sup_{k\in\mathbf{N}} \int_E \sigma^2 \, |\nabla g_k|^2 \, \varphi^2 \, dx \;=\; \sup_{k\in\mathbf{N}} \int_U |\nabla g_k|^2 \, \psi^2 \, dx \;<\; \infty \,.$$

Hence the sequence $(g_k)_{k\in\mathbf{N}}$ is bounded w.r.t. the $W^{1,2}(E, \sigma^2; \varphi^2\,dx)$ norm. Now the usual application of the theorems of Banach/Alaoglu and Banach/Saks (cf. e.g. [MaRö 92, Appendix]) implies that the $L^2(E; \varphi^2\,dx)$ limit u is in $H_0^{1,2}(E, \sigma^2; \varphi^2\,dx)$, and the Césaro means of a subsequence of $(g_k)_{k\in\mathbf{N}}$ converge to u in $H_0^{1,2}(E, \sigma^2; \varphi^2\,dx)$.

(ii) Let $u \in W^{1,2}(E, \sigma^2; \varphi^2\,dx)$. By the definition of the weak Sobolev space, it is easy to see that for every $f \in C_0^\infty(E)$, $u\cdot f$ is in $W^{1,2}(E, \sigma^2; \varphi^2\,dx)$. By (i), $u\cdot f$ is even in $H_0^{1,2}(E, \sigma^2; \varphi^2\,dx)$, and for every relatively compact open subset $U \subset E$ such that $\operatorname{supp} f \subset U$, there exists a sequence $(g_k)_{k\in\mathbf{N}}$ in $C_0^\infty(E)$ such that $\operatorname{supp} g_k \subset U$ for all k, and $g_k \to u\cdot f$ in $H_0^{1,2}(E, \sigma^2; \varphi^2\,dx)$. Now, by using a partition of unity $(f_i)_{i\in\mathbf{N}}$, one easily shows that there exist functions $u_k \in C^\infty(E)\cap W^{1,2}(E, \sigma^2; \varphi^2\,dx)$ that converge to u in $W^{1,2}(E, \sigma^2; \varphi^2\,dx)$, cf. e.g. [FuOshTa 94, Lemma 3.3.3] for a similar argument. ∎

3) The spaces $W^{1,2}(E, (a^{ij}); \varphi^2\,dx)$, $E \subseteq \mathbf{R}^n$

We finally consider weak Sobolev spaces corresponding to general symmetric diffusion operators with controlled degeneracy on a domain in $\mathbf{R}^n$. By using Lemma 3.2, a corresponding density theorem can be easily deduced from Theorem 3.4.

Let E be an open subset in $\mathbf{R}^n$. We fix functions $\varphi \in H_{\mathrm{loc}}^{1,2}(E; dx)$, $\varphi > 0$ dx–a.e., and $a^{ij} \in H_{\mathrm{loc}}^{1,2}(E; \varphi^2\,dx)\cap L_{\mathrm{loc}}^\infty(E; dx)$ such that $a^{ij} = a^{ji}$ for all $1 \le i, j \le n$. Similarly to the assumption imposed in Section a), 2), we assume that

$$a^{ij}(x) \;=\; (\sigma(x))^2 \cdot c^{ij}(x) \qquad \text{for all } x \in E \text{ and } 1 \le i, j \le n$$

with functions σ, $c^{ij} \in H_{\mathrm{loc}}^{1,2}(E; \varphi^2\,dx)\cap L_{\mathrm{loc}}^\infty(E; dx)$, $\sigma > 0$ dx–a.e., and $(c^{ij}) \ge (\delta_{ij})$ dx–a.e. in the form sense. Note that in comparison to Section a), 2), we again make the stronger assumption $\sigma \in H_{\mathrm{loc}}^{1,2}(E; \varphi^2\,dx)\cap L_{\mathrm{loc}}^\infty(E; dx)$ instead of only assuming $\sigma^2 \in H_{\mathrm{loc}}^{1,2}(E; \varphi^2\,dx)\cap L_{\mathrm{loc}}^\infty(E; dx)$. Without this stronger assumption, the following theorem is false, cf. the example in Subsection 2) above.

Let $H_0^{1,2}(E, (a^{ij}); \varphi^2\,dx)$ and $H^{1,2}(E, (a^{ij}); \varphi^2\,dx)$ denote the closures of $C_0^\infty(E)$, $C^\infty(E)\cap W^{1,2}(E, (a^{ij}); \varphi^2\,dx)$ respectively, in the Hilbert space $W^{1,2}(E, (a^{ij}); \varphi^2\,dx)$. The following theorem extends Theorem 3.4 :

Theorem 3.5 *Suppose the assumptions above hold. Then :*

(i) *Every compactly supported function* $u \in W^{1,2}(E, (a^{ij}); \varphi^2\,dx)$ *is in* $H_0^{1,2}(E, (a^{ij}); \varphi^2\,dx)$.

(ii) $H^{1,2}(E, (a^{ij}); \varphi^2 \, dx) = W^{1,2}(E, (a^{ij}); \varphi^2 \, dx)$.

REMARK. For smooth, locally strictly positive functions φ, (a^{ij}), Theorem 3.5 has been proven in [FuOshTa 94, Sect. 3.3]. In this case, the proof is much simpler, and can be carried out by the ordinary smoothing by convolution technique.

PROOF. By Lemma 3.2 and the local boundedness of the functions C^{ij}, $1 \leq i, j \leq n$, a compactly supported function $u \in L^2(E; \varphi^2 \, dx)$ is in $W^{1,2}(E, (a^{ij}); \varphi^2 \, dx)$ if and only if it is in $W^{1,2}(E, \sigma^2; \varphi^2 \, dx)$. Moreover, for every relatively compact open subset $U \subset E$, there exists a finite constant $C > 0$ such that

$$\| u \|_{W^{1,2}(E, \sigma^2; \varphi^2 \, dx)} \leq \| u \|_{W^{1,2}(E, (a^{ij}); \varphi^2 \, dx)} \leq C \cdot \| u \|_{W^{1,2}(E, \sigma^2; \varphi^2 \, dx)}$$

for all $u \in W^{1,2}(E, (a^{ij}); \varphi^2 \, dx)$ with $\operatorname{supp} u \subset U$. Therefore, Assertion (i) can be easily deduced from Theorem 3.4, cf. the proof of Theorem 3.4 (i) for a similar argument. Assertion (ii) follows from (i) by the same type of argument as used in the proof of Assertion (ii) in Theorem 3.4. ∎

f) $H_0^{1,2} = W^{1,2}$ and Markov uniqueness : The finite dimensional case

Let E be an open subset in $\mathbf{R}^n$. We now apply our results from Sections c) and e) to derive conditions for Markov uniqueness of symmetric diffusion operators defined on $C_0^\infty(E)$. By Section c), Markov uniqueness of such operators is equivalent to the coincidence of the corresponding Sobolev spaces $H_0^{1,2}(d)$ and $W^{1,2}(d)$. In Section e), we have already derived a weak sufficient condition, which guarantees that every compactly supported function in $W^{1,2}(d)$ is in $H_0^{1,2}(d)$, cf. Theorem 3.5. Suppose this condition is satisfied. Then Markov uniqueness holds if and only if the compactly supported functions are dense in $W^{1,2}(d)$. Density of compactly supported functions, however, can be proved or disproved by more or less standard localization techniques.

We fix functions $\varphi \in H_{\text{loc}}^{1,2}(E; dx)$, $\varphi > 0$ dx-a.e., and $a^{ij} \in H_{\text{loc}}^{1,2}(E; \varphi^2 \, dx) \cap L_{\text{loc}}^\infty(E; dx)$, $1 \leq i, j \leq n$, such that $a^{ij} = a^{ji}$. We assume that the functions a^{ij} satisfy the conditions assumed in Theorem 3.5, i.e., $a^{ij} = \sigma^2 \cdot c^{ij}$ with functions $\sigma, c^{ij} \in H_{\text{loc}}^{1,2}(E; \varphi^2 \, dx) \cap L_{\text{loc}}^\infty(E; dx)$, $\sigma > 0$ dx-a.e., and $(c^{ij}) \geq (\delta^{ij})$ dx-a.e. in the form sense. These conditions guarantee that the singularities do not destroy Markov uniqueness of the corresponding diffusion operator $(\mathcal{L}, C_0^\infty(E))$ on $L^2(E; \varphi^2 \, dx)$,

$$\mathcal{L}f = \sum_{i,j=1}^{n} \left(\frac{\partial}{\partial x_i} \left(a^{ij} \frac{\partial f}{\partial x_j} \right) + \beta_i \, a^{ij} \frac{\partial f}{\partial x_j} \right),$$

$\beta_i := \frac{2}{\varphi} \frac{\partial \varphi}{\partial x_i}$. In other words : It only depends on the boundary behaviour of the coefficients whether $(\mathcal{L}, C_0^\infty(E))$ is Markov unique or not.

In Subsection 1), we show that under the conditions imposed Markov uniqueness of $(\mathcal{L}, C_0^\infty(E))$ is equivalent to *Silverstein uniqueness* of the corresponding pre–Dirichlet form $(\mathcal{E}, C_0^\infty(E))$. Moreover, we show that Markov uniqueness holds if and only if a condition due to V. Maz'ja is satisfied. In Subsection 2), we prove that completeness of E w.r.t. the *intrinsic metric* corresponding to the operator $(\mathcal{L}, C_0^\infty(E))$ implies Markov uniqueness. This fact can be viewed as a non–smooth counterpart to Gaffney's classical result showing the essential self–adjointness of the Laplacian on a complete Riemannian manifold [Ga 51]. Finally, in Subsection 3), we show that conservativity of the semigroup generated by the Friedrichs extension of $(\mathcal{L}, C_0^\infty(E))$ implies Markov uniqueness. If the measure $\varphi^2 \, dx$ is finite, Markov uniqueness and conservativity are even equivalent under the assumptions imposed above.

1) Some conditions equivalent to Markov uniqueness

Let $(\mathcal{E}, C_0^\infty(E))$ be the pre–Dirichlet form corresponding to the operator $(\mathcal{L}, C_0^\infty(E))$ on $L^2(E; \varphi^2 \, dx)$, i.e.,

$$\mathcal{E}(f, g) = - \int_E f \, \mathcal{L}g \, \varphi^2 \, dx = \int_E \sum_{i,j=1}^n a^{ij} \frac{\partial f}{\partial x_i} \frac{\partial g}{\partial x_j} \varphi^2 \, dx \, .$$

Recall the definition of Silverstein uniqueness of $(\mathcal{E}, C_0^\infty(E))$ from Chapter 1, Section d). We say that a domain $G \subseteq E$ has finite $H^{1,2}(E, (a^{ij}); \varphi^2 \, dx)$ **capacity** if and only if there exists a function $u \in H^{1,2}(E, (a^{ij}); \varphi^2 \, dx)$ such that $u \geq 1$ dx–a.e. on G.

Theorem 3.6 *Under the assumptions imposed above, the following assertions are equivalent :*

(i) *The operator $(\mathcal{L}, C_0^\infty(E))$ is Markov unique on $L^2(E; \varphi^2 \, dx)$.*

(ii) $H_0^{1,2}(E, (a^{ij}); \varphi^2 \, dx) = H^{1,2}(E, (a^{ij}); \varphi^2 \, dx)$.

(iii) *The form $(\mathcal{E}, C_0^\infty(E))$ is Silverstein unique on $L^2(E; \varphi^2 \, dx)$.*

(iv) *For every domain $G \subseteq E$ of finite $H^{1,2}(E, (a^{ij}); \varphi^2 \, dx)$ capacity, there exists a sequence $f_k \in C_0^\infty(E)$, $k \in \mathbf{N}$, such that $f_k \to 1$ in $\varphi^2 \, dx$–measure on G, and $\int_G \sum_{i,j=1}^n a^{ij} \frac{\partial f_k}{\partial x_i} \frac{\partial f_k}{\partial x_j} \varphi^2 \, dx \to 0$ as $k \to \infty$.*

REMARKS. (i) Recall that under the assumptions imposed

$$W^{1,2}(E, (a^{ij}); \varphi^2 \, dx) = H^{1,2}(E, (a^{ij}); \varphi^2 \, dx)$$

by Theorem 3.5. Therefore, the weak Sobolev space does no longer appear explicitly in Theorem 3.6.

(ii) In the case where the functions φ and (a^{ij}) are smooth and non–degenerate, Theorem 3.6 has been proven in [FuOshTa 94, Sect. 3.3]. By applying our results above, it becomes possible to carry out the proof similarly in the singular non–smooth case.

(iii) In the case $E = \mathbf{R}^n$, $\varphi = 1$, and for smooth coefficients a^{ij}, V. Maz'ja has shown that Condition (iv) is equivalent to the coincidence of the Sobolev spaces $H_0^{1,2}(\mathbf{R}^n; (a^{ij}); dx)$ and $H^{1,2}(\mathbf{R}^n; (a^{ij}); dx)$, cf. [Maz 85, Par. 2.7].

(iv) In the case $a^{ij} = \delta^{ij}$, the assertions in the theorem are always satisfied, cf. e.g. Corollary 3.3 below. In this case, the coincidence of the Sobolev spaces $W^{1,2}$ and $H_0^{1,2}$ (and hence Markov uniqueness) has been proven already in [RöZha 94].

PROOF OF THEOREM 3.6. Let $(\hat{\mathcal{E}}, H^{1,2}(E, (a^{ij}); \varphi^2 dx))$ be the closure of the quadratic form

$$\hat{\mathcal{E}}(f, f) = \int_E \sum_{i,j=1}^n a^{ij} \frac{\partial f}{\partial x_i} \frac{\partial f}{\partial x_j} \varphi^2 dx,$$

$$f \in C^\infty(E) \cap L^2(E; \varphi^2 dx) \text{ such that } \hat{\mathcal{E}}(f, f) < \infty.$$

One easily verifies that the closure is a Dirichlet form. By Theorem 3.5 and Corollary 3.1, the generator $\hat{L}$ of this form is the maximal Dirichlet extension of the operator $(\mathcal{L}, C_0^\infty(E))$. This proves the equivalence of (i) and (ii).

For proving the implication (iii) $\Rightarrow$ (ii) it suffices to show that the Dirichlet form $(\hat{\mathcal{E}}, H^{1,2}(E, (a^{ij}); \varphi^2 dx))$ is a Silverstein extension of $(\mathcal{E}, C_0^\infty(E))$. But this is the case, since for every bounded function $u \in H^{1,2}(E, (a^{ij}); \varphi^2 dx)$ and $f \in C_0^\infty(E)$, $u \cdot f$ is a bounded compactly supported function in $H^{1,2}(E, (a^{ij}); \varphi^2 dx)$, and, thus, contained in $H_0^{1,2}(E, (a^{ij}); \varphi^2 dx)$.

The implication (i) $\Rightarrow$ (iii) is always true for symmetric diffusion operators $\mathcal{L}$ with domain $C_0^\infty(E)$, cf. [Ta 96, Proof of Lemma 3.2].

Finally, the equivalence between (ii) and (iv) can be shown similarly to the proof of Theorem 1 in [Maz 85, Par. 2.7], cf. also [FuOshTa 94, Sect. 3.3]. ∎

Examples how to verify or disprove Condition (iv) in Theorem 3.6 are given in Maz'ja's book [Maz 85]. We just point out the following facts :

- If the measure $\varphi^2 dx$ is finite, then every domain $G \subseteq E$ has finite capacity. Hence Condition (iv) holds if and only if there exist cut–off functions f_k, $k \in \mathbf{N}$, satisfying the condition for $G = E$.

- Condition (iv) holds, if $E = \mathbf{R}^n$, and every domain with finite $H^{1,2}(\mathbf{R}^n, (a^{ij}); \varphi^2 dx)$ capacity is bounded. For $n \leq 2$, this is always the case if φ and (a^{ij}) are bounded from below by strictly positive constants, cf. [Maz 85, Par. 2.7, Thm. 2], but for $n \geq 3$ there are counterexamples even in this case.

Other sufficient conditions for the assertions in Theorem 3.6 are given e.g. in [Maz 85] and [KawTa 95]. In the next subsection, we show that the assertions hold if E is complete w.r.t. the intrinsic metric corresponding to the operator $\mathcal{L}$. It has first been proven in [KawTa 95] that this kind of completeness implies Silverstein uniqueness.

2) Completeness implies Markov uniqueness

We introduce a distance function d associated with the operator $\mathcal{L}$. Suppose first that (a^{ij}) is locally strictly positive definite. Then we define $d : E \times E \to [0, \infty]$ by

$$d(x, y) := \sup \left\{ f(x) - f(y) \, ; \, f \in C^\infty(E), \, \sum_{i,j=1}^n a^{ij} \frac{\partial f}{\partial x_i} \frac{\partial f}{\partial x_j} \leq 1 \right\}.$$

The local boundedness and locally strict positivity of (a^{ij}) guarantee that d is a separable metric on E, that generates the original topology. It is finite on every component of E. Under additional smoothness assumptions on the coefficients a^{ij}, d is the ordinary distance function on the Riemannian manifold (E, g), where $(g_{ij}) = (a^{ij})^{-1}$.

If (a^{ij}) is not locally strictly positive definite, then we consider the distance function $\bar{d}_\varepsilon$ corresponding to $\bar{a}_\varepsilon^{ij} := a^{ij} + \varepsilon \cdot \delta^{ij}$ instead, where ε is some strictly positive, bounded function on E. The following corollary is a consequence of Theorem 3.6.

Corollary 3.3 *Under the assumptions imposed in Subsection 1), the operator* $(\mathcal{L}, C_0^\infty(E))$ *is Markov unique on* $L^2(E; \varphi^2 \, dx)$*, whenever one of the following conditions holds :*

(i) *(a^{ij}) is locally strictly positive definite, and the metric space (E, d) is complete.*

(ii) *There exists a locally strictly positive, bounded function ε on E, such that the metric space $(E, \bar{d}_\varepsilon)$ is complete.*

The proof is given below.

REMARKS.   (i) It has been shown in [KawTa 95] that Condition (i) implies Silverstein uniqueness of the form $(\mathcal{E}, C_0^\infty(E))$. The key ingredient in the proof is the construction of cut–off functions in Dirichlet spaces with the help of intrinsic metrics. This method has been worked out before in detail in articles of M. Biroli and U. Mosco [BirMos 95], and K. T. Sturm [Stu 95]. Intrinsic metrics similar to d are nowadays a standard tool in the analysis of Dirichlet forms.

   (ii) In a classical article, Gaffney [Ga 51] has proven in particular the essential self–adjointness of the Laplacian on a complete Riemannian manifold. Under the minimal smoothness assumptions we impose, the operator $(\mathcal{L}, C_0^\infty(E))$ is not necessarily essentially self–adjoint even if (E, d) is complete, cf. the counterexamples given in Chapter 2. Theorem 3.3 shows that nevertheless, a weakened

Gaffney type result with essential self–adjointness replaced by Markov uniqueness holds under the assumptions from Subsection 1).

EXAMPLES. 1) Suppose $E = \mathbf{R}^n$. Then (E, d) is complete if

$$\left(a^{ij}(x) \right) \leq C \cdot (|x| + 2)^2 \, (\log (|x| + 2))^2 \cdot (\delta^{ij})$$

in the form sense for dx–a.e. x, cf. [KawTa 95].

2) Suppose that $E = (x_0, y_0)$, $-\infty \leq x_0 < y_0 \leq \infty$, and a and ρ are locally
strictly positive functions on E. Then (E, d) is complete if and only if $a^{-1/2}$ is
neither dx–integrable at x_0 nor at y_0. Note that for $c \in (x_0, y_0)$,

$$(3.27) \qquad \int_{x_0}^c a^{-1/2} \, dx \;\leq\; \left(\int_{x_0}^c (a\rho)^{-1} \, dx \right)^{1/2} \cdot \left(\int_{x_0}^c \rho \, dx \right)^{1/2}, \;\; \text{and}$$

$$(3.28) \qquad \int_c^{y_0} a^{-1/2} \, dx \;\leq\; \left(\int_c^{y_0} (a\rho)^{-1} \, dx \right)^{1/2} \cdot \left(\int_c^{y_0} \rho \, dx \right)^{1/2}.$$

It has been shown in Theorem 3.2, that the operator $(\mathcal{L}, C_0^\infty(x_0, y_0))$ is Markov
unique *if and only if* both one of the integrals on the right hand side of (3.27),
and one of the integrals on the right hand side of (3.28) is infinite.

Corollary 3.3 could be deduced immediately from Theorem 3.6 and the Silverstein uniqueness result in [KawTa 95], cf. Remark (i) above. For the convenience
of the reader, we nevertheless sketch the proof in more detail :

PROOF OF COROLLARY 3.3. We first assume that E is connected. By The-
orem 3.6, it suffices to show $H_0^{1,2}(E, (a^{ij}); \varphi^2 \, dx) = H^{1,2}(E, (a^{ij}); \varphi^2 \, dx)$ for
proving Markov uniqueness of $(\mathcal{L}, C_0^\infty(E))$. Fix $u \in H^{1,2}(E, (a^{ij}); \varphi^2 \, dx)$. Let
$(g_k)_{k \in \mathbf{N}}$ be an increasing sequence of smooth decreasing functions $g_k : [0, \infty] \to$
$[0, 1]$ such that $g_k(s) = 1$ for $0 \leq s \leq k$, $g_k(s) = 0$ for $s \geq k+2$, and $|g_k'(s)| \leq 1$
for all $0 < s < \infty$. Suppose first that Condition (i) holds. Fix a point $x_0 \in E$.
It can be shown that the functions $u_k : E \to \mathbf{R}$, $u_k(x) := u(x) \cdot g_k(d(x_0, x))$, are
again in $H^{1,2}(E, (a^{ij}); \varphi^2 \, dx)$, and

$$(3.29) \qquad \sum_{i,j=1}^n a^{ij} \frac{\partial u_k}{\partial x_i} \frac{\partial u_k}{\partial x_j} \;\leq\; 2 \cdot \left(u^2 + \sum_{i,j=1}^n a^{ij} \frac{\partial u}{\partial x_i} \frac{\partial u}{\partial x_j} \right) \qquad dx\text{–a.e.}$$

for all $k \in \mathbf{N}$. Here $\frac{\partial u_k}{\partial x_i}$ denotes the dx–a.e. defined directional derivative in
$H^{1,2}(E, (a^{ij}); \varphi^2 \, dx)$. A detailed proof is, for example, given in the appendix
of [Eb 96], cf. also [BirMos 95] and [Stu 95]. The key ingredient in the proof
is the stability of Dirichlet forms and carré du champ operators w.r.t. taking
suprema.

Since (E, d) is connected, $d(x, y) < \infty$ for all $x, y \in E$. In particular, $u_k \to u$
pointwise, and, by dominated convergence, in $L^2(E; \varphi^2 \, dx)$. Moreover, since

(E, d) is complete, the d-balls are relatively compact. This follows by a general version of the Hopf–Rinow theorem, cf. [Stu 95, Thm. 2]. In particular, the functions u_k, $k \in \mathbb{N}$, have compact support, whence they are contained in $H_0^{1,2}(E, (a^{ij}); \varphi^2\, dx)$. By the estimate (3.29), they form a uniformly bounded sequence in this Hilbert space. Hence the $L^2(E; \varphi^2\, dx)$ limit u is also contained in $H_0^{1,2}(E, (a^{ij}); \varphi^2\, dx)$. Since u has been chosen arbitrarily in $H^{1,2}(E, (a^{ij}); \varphi^2\, dx)$, we obtain

$$H^{1,2}(E, (a^{ij}); \varphi^2\, dx) \;=\; H_0^{1,2}(E, (a^{ij}); \varphi^2\, dx),$$

and thus the assertion.

If Condition (ii) holds, then we can argue in the same way with d replaced by $\bar{d}_\varepsilon$. Note that, since $(a^{ij}(x)) \leq (\bar{a}_\varepsilon^{ij}(x))$ for all x in the form sense, the supremum in the definition of $\bar{d}_\varepsilon$ is taken over *less* functions than that in the definition of d. Therefore, the functions u_k are still in $H^{1,2}(E, (a^{ij}); \varphi^2\, dx)$ and satisfy (3.29), if they are defined w.r.t. $\bar{d}_\varepsilon$ instead of d. See [Eb 96, Appendix] for details.

Finally, if E is not connected, then similar arguments as above show, that for every component Λ of E and $u \in H^{1,2}(E, (a^{ij}); \varphi^2\, dx)$, the function $u \cdot \chi_\Lambda$ can be approximated by functions in $C_0^\infty(\Lambda)$ w.r.t. the Sobolev norm. Now, it is not difficult to approximate u by functions in $C_0^\infty(E)$, which again shows that $H^{1,2}(E, (a^{ij}); \varphi^2\, dx)$ and $H_0^{1,2}(E, (a^{ij}); \varphi^2\, dx)$ coincide. ∎

3) Conservativity implies Markov uniqueness

Let $L^{(0)}$ be the Friedrichs extension of $(\mathcal{L}, C_0^\infty(E))$ on $L^2(E; \varphi^2\, dx)$, i.e., $L^{(0)}$ is the generator of the closure $(\mathcal{E}, H_0^{1,2}(E, (a^{ij}); \varphi^2\, dx))$ of the pre–Dirichlet form $(\mathcal{E}, C_0^\infty(E))$. Let $T_t^{(0)} := e^{tL^{(0)}}$, $t \geq 0$, denote the symmetric sub–Markovian C^0 contraction semigroup on $L^2(E; \varphi^2\, dx)$ generated by $L^{(0)}$. It is well–known from the work of M. Silverstein [Sil 74] that conservativity of $T_t^{(0)}$ implies Silverstein uniqueness of the form $(\mathcal{E}, C_0^\infty(E))$. Therefore, by Theorem 3.6 above, it also implies Markov uniqueness of the operator $(\mathcal{L}, C_0^\infty(E))$.

To state the result precisely, note that by the sub–Markov property and symmetry, $T_t^{(0)}$ induces a linear contraction on $L^p(E; \varphi^2\, dx)$ for every $1 \leq p \leq \infty$ and $t \geq 0$. We say that the semigroup $\left(T_t^{(0)}\right)_{t \geq 0}$ is **conservative** if and only if $T_t^{(0)} 1 = 1$ dx–a.e. for every $t \geq 0$. Obviously, because of the $\varphi^2\, dx$–symmetry of $\left(T_t^{(0)}\right)_{t \geq 0}$, conservativity holds if and only if $\varphi^2\, dx$ is a **stationary distribution** for $\left(T_t^{(0)}\right)_{t \geq 0}$, i.e.,

$$\int_E T_t^{(0)} f \; \varphi^2\, dx \;=\; \int_E f \, \varphi^2\, dx \qquad \text{for all } f \in L^1(E; \varphi^2\, dx).$$

Corollary 3.4 *(i) Suppose the semigroup $\left(T_t^{(0)}\right)_{t \geq 0}$ generated by the Friedrichs extension of $(\mathcal{L}, C_0^\infty(E))$ on $L^2(E; \varphi^2 \, dx)$ is conservative. Then the operator $(\mathcal{L}, C_0^\infty(E))$ is Markov unique.*

(ii) Conversely, if $\int_E \varphi^2 \, dx < \infty$, and $(\mathcal{L}, C_0^\infty(E))$ is Markov unique, then $\left(T_t^{(0)}\right)_{t \geq 0}$ is conservative.

REMARK. If the measure $\varphi^2 \, dx$ is infinite, then the converse implication is not necessarily true. In fact, if E is an interval, then, by the results in Section d), Markov uniqueness holds if and only if both boundaries are not regular in Feller's sense, cf. also the discussion in Section a) of Chapter 4. Hence if one boundary is exit/no entrance, and the other is an arbitrary non–regular boundary, then Markov uniqueness holds although $\left(T_t^{(0)}\right)_{t \geq 0}$ is not conservative.

For the reader's convenience, we sketch the proof of the corollary :

PROOF OF COROLLARY 3.4. (i) By Theorem 3.6, it suffices to show Silverstein uniqueness of $(\mathcal{E}, C_0^\infty(E))$. Let $(\mathcal{E}^\sharp, \mathcal{F}^\sharp)$ be a Silverstein extension of $(\mathcal{E}, C_0^\infty(E))$. Then, by Theorem A.4.4 in the appendix of [FuOshTa 94], there exists a *regular representation* $(\tilde{E}, \tilde{m}, \tilde{\mathcal{E}}^\sharp, \tilde{\mathcal{F}}^\sharp)$ of the Dirichlet space $(E, m, \mathcal{E}^\sharp, \mathcal{F}^\sharp)$, $m := \varphi^2 \, dx$, and an open subset $\tilde{G} \subseteq \tilde{E}$ such that the *part of $(E, m, \mathcal{E}^\sharp, \mathcal{F}^\sharp)$ on $\tilde{G}$ is equivalent to the Dirichlet space* $(E, m, \mathcal{E}, H_0^{1,2}(E, (a^{ij}); \varphi^2 \, dx))$. We refer to [FuOshTa 94] for the vocabulary used here. Roughly speaking, the result stated means, that after some minor modifications of the state spaces, the Dirichlet form $(\mathcal{E}, H_0^{1,2}(E, (a^{ij}); \varphi^2 \, dx))$ can be obtained from the form $(\mathcal{E}^\sharp, \mathcal{F}^\sharp)$ in a similar way, as the Dirichlet form of absorbed Brownian motion on a smooth domain Ω in $\mathbf{R}^n$ is obtained from the Dirichlet form of reflected Brownian motion on $\bar{\Omega}$. The space $\tilde{E}$ replaces $\bar{\Omega}$, and the open subset $\tilde{G}$ replaces Ω. The important consequence for us is that the result described above immediately implies, that the sub–Markovian C^0 semigroup $\left(T_t^\sharp\right)_{t \geq 0}$ corresponding to the Dirichlet form $(\mathcal{E}^\sharp, \mathcal{F}^\sharp)$ dominates the sub–Markovian C^0 semigroup $\left(T_t^{(0)}\right)_{t \geq 0}$ from above, i.e.,

$$(3.30) \qquad T_t^{(0)} f \leq T_t^\sharp f \qquad dx\text{--a.e.}$$

for all positive functions f on E, and $t \geq 0$, cf. [FuOshTa 94, Thm. 4.4.2].

Now suppose the semigroups $\left(T_t^{(0)}\right)_{t \geq 0}$ and $\left(T_t^\sharp\right)_{t \geq 0}$ on $L^2(E; \varphi^2 \, dx)$ do not coincide. Then there exist $t > 0$, and a function $f : E \to \mathbf{R}$, $0 \leq f \leq 1$, such that $T_t^{(0)} f < T_t^\sharp f$ on a set B with strictly positive Lebesgue measure. Hence, by (3.30) and the sub–Markov property of $T_t^\sharp$,

$$T_t^{(0)} 1 = T_t^{(0)} f + T_t^{(0)}(1 - f) < T_t^\sharp f + T_t^\sharp(1 - f) = T_t^\sharp 1 \leq 1$$

dx–a.e. on B. This contradicts the conservativity of $\left(T_t^{(0)}\right)_{t\geq 0}$, whence $\left(T_t^{(0)}\right)_{t\geq 0}$ and $\left(T_t^{\sharp}\right)_{t\geq 0}$ coincide. Since the correspondence between symmetric Dirichlet forms and symmetric sub–Markovian C^0 semigroups on $L^2(E; \varphi^2\, dx)$ is one–to–one, $(\mathcal{E}^{\sharp}, \mathcal{F}^{\sharp})$ and $(\mathcal{E}, H_0^{1,2}(E, (a^{ij}); \varphi^2\, dx))$ coincide as well. Thus $(\mathcal{E}, C_0^{\infty}(E))$ is Silverstein unique.

(ii) If $\int_E \varphi^2\, dx < \infty$, then the constant function 1 is in $H^{1,2}(E, (a^{ij}); \varphi^2\, dx)$, and $\hat{\mathcal{E}}(1, f) = 0$ for all $f \in C_0^{\infty}(E)$. If, moreover, $(\mathcal{L}, C_0^{\infty}(E))$ is Markov unique, then $(\hat{\mathcal{E}}, H^{1,2}(E, (a^{ij}); \varphi^2\, dx)) = (\mathcal{E}, H_0^{1,2}(E, (a^{ij}); \varphi^2\, dx))$. In this case, 1 is in the domain of $L^{(0)}$, and $L^{(0)}1 = 0$ m–a.e. Thus $T_t^{(0)}1 = 1$ m–a.e. for all $t \geq 0$. ∎

g) Ergodicity, Markov uniqueness, and extremality of symmetrizing measures

We have shown in the last section that, if the semigroup generated by the Friedrichs extension of a symmetric diffusion operator on a domain in $\mathbf{R}^n$ is conservative, then boundary effects do not destroy Markov uniqueness, i.e., if the coefficients of the operator satisfy some local regularity condition then Markov uniqueness holds. Similarly to this relation between *conservativity* and the *influence of boundaries* on Markov uniqueness, there is a relation between *ergodicity* and the *influence of singularities* on Markov uniqueness. This relation is, however, not as direct as the implication "conservativity ⇒ Markov uniqueness". In Subsection 1) below, we use our techniques developed above, to derive a *necessary and sufficient condition for L^2 ergodicity of the semigroup generated by the maximal Dirichlet extension of a symmetric diffusion operator* $(\mathcal{L}, \mathcal{A})$ on $L^2(E; m)$, where E is an open subset in $\mathbf{R}^n$, and m is a probability measure. Moreover, we prove a *sufficient condition for ergodicity of the semigroup generated by the Friedrichs extension* of the operator, which is sharp in the one–dimensional case. In many non–Markov unique situations, the maximal semigroup is not ergodic, whereas the Friedrichs semigroup is ergodic.

Ergodicity of the maximal semigroup on $L^2(E; m)$ is, in fact, equivalent to extremality of the measure m in the convex set of all symmetrizing probability measures for the diffusion operator defined on test–functions. This relation, which has first been pointed out for a special class of infinite dimensional diffusion operators in [AlbKoRö 97c], is proven for general symmetric diffusion operators in Subsection 2).

As a consequence, non–uniqueness of the symmetrizing measure of a diffusion operator implies non–ergodicity of the semigroup generated by the Friedrichs extension of the operator on $L^2(E; m)$, or non–Markov uniqueness on $L^2(E; m)$ (i.e., "non–uniqueness of the Friedrichs semigroup"), provided m is a non–extremal element in the set of all symmetrizing probability measures. In Subsection 3), we discuss in the finite–dimensional case, and, briefly, for lattice systems from classical statistical mechanics, which of the consequences occurs. It turns out that *for lattice systems from statistical mechanics that have a phase transition, usually the Friedrichs semigroup is unique but not ergodic, provided the interactions are bounded. If, however, the interactions are unbounded, then, conversely, the Friedrichs semigroup can be ergodic and non–unique.*

1) Ergodicity and Markov–uniqueness : The finite dimensional case

Let E be an open subset in $\mathbf{R}^n$, $n \in \mathbf{N}$, and let $\mathcal{A} := C_0^\infty(E)$. We consider the situation described in Section a), 2), above. In particular, we fix a measure m on E such that $m = \varphi^2 \, dx$ for some $\varphi \in H^{1,2}_{\text{loc}}(E; dx)$, $\varphi > 0$ dx–a.e., and we consider a symmetric diffusion operator $(\mathcal{L}, \mathcal{A})$ on $L^2(E; m)$ of type (3.6), such that the coefficients a^{ij} satisfy the assumptions from Section a), 2). For simplicity, we assume that m is finite, i.e., $\int_E \varphi^2 \, dx < \infty$. We point out, however, that by using localization techniques, similar considerations as below can be carried out for non–finite measures. In this case, the Dirichlet spaces appearing below would have to be replaced by Dirichlet spaces in an extended sense.

Since our focus in this section is only on the influence of *singularities* on ergodicity and Markov uniqueness, we assume that the C^0 semigroup $\left(T_t^{(0)}\right)_{t \geq 0}$ generated by the Friedrichs extension $L^{(0)}$ of the operator $(\mathcal{L}, \mathcal{A})$ on $L^2(E; m)$ is *conservative*, i.e.,

$$(3.31) \qquad T_t^{(0)} 1 \; = \; 1 \qquad m\text{–a.e. for all } t \geq 0.$$

This assumption rules out non–Markov uniqueness caused by boundary effects, cf. Corollary 3.4, where we have shown that (3.31) and appropriate local conditions on the operator coefficients guarantee Markov uniqueness. Note that, since m is finite, (3.31) holds if and only if the constant function 1 is in the kernel of $L^{(0)}$, or, equivalently, 1 is in $H_0^{1,2}(E, (a^{ij}); m)$ and $\mathcal{E}^{(0)}(1,1) = 0$, where $(\mathcal{E}^{(0)}, H_0^{1,2}(E, (a^{ij}); m))$ is the closure of the pre–Dirichlet form

$$\mathcal{E}^{(0)}(f, g) \; = \; \int_E \sum_{i,j=1}^n a^{ij} \frac{\partial f}{\partial x_i} \frac{\partial g}{\partial x_j} \, dm$$

with domain $C_0^\infty(E)$. As a consequence, in this case every function $f \in C^\infty(E)$ such that $\mathcal{E}^{(0)}(f, f) + \int f^2 \, dm < \infty$ is in $H_0^{1,2}(E, (a^{ij}); m)$, i.e.,

$$(3.32) \qquad H_0^{1,2}(E, (a^{ij}); m) \; = \; H^{1,2}(E, (a^{ij}); m).$$

Before proceeding, we recall some basic definitions. Let $(\mathcal{E}, \mathcal{F})$ be a symmetric Dirichlet form on $L^2(E; m)$ such that 1 is in $\mathcal{F}$ and $\mathcal{E}(1,1) = 0$. Let

$(T_t)_{t \geq 0}$ be the C^0 semigroup generated by the form generator. $(T_t)_{t \geq 0}$ is called L^2 ergodic, iff every function $u \in L^2(E; m)$ such that $T_t u = u$ for all t is (m–a.e.) constant, or, equivalently, iff

$$T_t u \; \rightarrow \; \int u \, dm \qquad \text{in } L^2(E; m) \text{ as } t \rightarrow \infty$$

for all $u \in L^2(E; m)$. The semigroup $(T_t)_{t \geq 0}$ is L^2 ergodic if and only if the Dirichlet form $(\mathcal{E}, \mathcal{F})$ is **irreducible**, i.e., every function $u \in \mathcal{F}$ such that $\mathcal{E}(u, u) = 0$ is (m–a.e.) constant. This connection follows easily from the spectral theorem. We refer to [AlbKoRö 97c] for a discussion of relations between L^2 ergodicity of transition semigroups, and different notions of ergodicity for the associated Markov processes.

We now return to our concrete finite–dimensional situation. By (3.32), $(\mathcal{E}^{(0)}, H^{1,2}(E, (a^{ij}); m))$ is the Dirichlet form of the minimal Dirichlet extension $L^{(0)}$ of the operator $(\mathcal{L}, A)$ on $L^2(E; m)$. In Section a), 2), we have introduced the weak Sobolev space $W^{1,2}(E, (a^{ij}); m)$ correspondint to $(\mathcal{L}, A)$. In particular, Lemma 3.2 and 3.1 show that the bilinear form $(\hat{\mathcal{E}}, W^{1,2}(E, (a^{ij}); m))$ on $L^2(E; m)$,

$$(3.33) \qquad \hat{\mathcal{E}}(u, v) \; = \; \int_E \sum_{i,j=1}^{n} a^{ij} \, \hat{\partial}_i u \, \hat{\partial}_j v \, dm,$$

which extends $(\mathcal{E}^{(0)}, H^{1,2}(E, (a^{ij}); m))$ is a *Dirichlet form*, cf. also the remark below Lemma 3.1. Thus

$$W_\infty^{1,2}(E, (a^{ij}); m) \; = \; W^{1,2}(E, (a^{ij}); m),$$

and, by Corollary 3.1, the generator $\hat{L}$ of $(\hat{\mathcal{E}}, W^{1,2}(E, (a^{ij}); m))$ is the maximal Dirichlet extension of $(\mathcal{L}, A)$.

Let $\left(\hat{T}_t\right)_{t \geq 0}$ be the C^0 semigroup on $L^2(E; m)$ generated by $\hat{L}$. We are interested in ergodicity of the semigroups $\left(T_t^{(0)}\right)_{t \geq 0}$ and $\left(\hat{T}_t\right)_{t \geq 0}$, and consequences for Markov uniqueness of the operator $(\mathcal{L}, A)$. If, for example, one of the semigroups is ergodic, and the other is not, then $(\mathcal{L}, A)$ is obviously not Markov unique. Clearly, L^2 ergodicity of $\left(\hat{T}_t\right)_{t \geq 0}$ implies L^2 ergodicity of $\left(T_t^{(0)}\right)_{t \geq 0}$, because irreducibility of $(\hat{\mathcal{E}}, W^{1,2}(E, (a^{ij}); m))$ implies irreducibility of $(\mathcal{E}^{(0)}, H^{1,2}(E, (a^{ij}); m))$. On the other hand, $\left(T_t^{(0)}\right)_{t \geq 0}$ can be ergodic even if $\left(\hat{T}_t\right)_{t \geq 0}$ is not ergodic. In fact, we have :

Theorem 3.7 *Let a_- be the function controlling the degeneracy of (a^{ij}), cf. Section a), 2). Let $\alpha := a_- \cdot \varphi^2$. Suppose that α is continuous. Then the following assertions hold :*

(i) The semigroup $\left(\hat{T}_t\right)_{t \geq 0}$ is $L^2(E; m)$ ergodic if and only if the open set $\{x \in E;\ \alpha(x) > 0\}$ is connected.

(ii) Let B denote the unit ball around 0 in $\mathbf{R}^{n-1}$. Suppose that for any two components U, V of $\{x \in E;\ \alpha(x) > 0\}$, there exist a finite sequence $U_0 = U, U_1, U_2, \dots, U_{k-1}, U_k = V$, $k \in \mathbf{N}$, of components U_i of $\{x \in E;\ \alpha(x) > 0\}$, and C^1 diffeomorphisms $\phi_i : (-1,1) \times B \to E$, $1 \leq i \leq k$, such that $\phi_i\left((-1,0) \times B\right) \subset U_{i-1}$, $\phi_i\left((0,1) \times B\right) \subset U_i$, and

$$\int_{-1}^{1} \int_{B} \left(\alpha(\phi_i(s,y))\right)^{-1} dy\, ds \ < \ \infty \qquad for\ all\ 1 \leq i \leq k.$$

Then the semigroup $\left(T_t^{(0)}\right)_{t \geq 0}$ is $L^2(E; m)$ ergodic.

REMARKS. (i) In the case $a^{ij} = \delta^{ij}$, the connection between zeros of α respectively φ and the ergodic decomposition of the corresponding diffusion process has been discussed in [Fu 87].

(ii) The condition in (ii) is satisfied in particular, if the set $\{x \in E;\ \alpha(x) = 0\}$ is sufficiently regular, and $1/\alpha$ is locally dx–integrable on E. However, for $n \geq 2$, the condition is of course much weaker than local integrability of $1/\alpha$.

EXAMPLE. (One–dimensional case) Suppose $E = (x_0, y_0)$, $-\infty \leq x_0 < y_0 \leq \infty$. Then we can choose $a_- = a$. By the assumptions on φ and a, α is in $H_{\text{loc}}^{1,1}(x_0, y_0;\ dx)$, and has thus an absolutely continuous modification. Therefore, we can assume without loss of generality that α itself is continuous. The first assertion in Theorem 3.7 now means that $\left(\hat{T}_t\right)_{t \geq 0}$ is L^2 ergodic if and only if $\alpha(x) > 0$ for all x. By the second assertion, $\left(T_t^{(0)}\right)_{t \geq 0}$ is L^2 ergodic if $1/\alpha$ is locally dx–integrable. In the one–dimensional case, this condition is also sharp. In fact, if $1/\alpha$ is not dx–integrable at a point $s_0 \in (x_0, y_0)$, then it can be shown that $\chi_{(x_0, s_0)}$ is in $H^{1,2}(E, a;\ \varphi^2\, dx)$, and $\mathcal{E}^{(0)}(\chi_{(x_0, s_0)}, \chi_{(x_0, s_0)}) = 0$, cf. the proof of Theorem 3.2.

Before proving Theorem 3.7, we comment on the relation to Markov uniqueness. We first look at diffusion operators on $\mathbf{R}^1$, cf. the example above. There are three different cases :

I) If α has no zeros, then there is a unique symmetric sub–Markovian C^0 semigroup generated by an extension of $(\mathcal{L}, \mathcal{A})$, and this semigroup is L^2 ergodic.

II) If α has zeros, but the zeros are "good" in the sense that $1/\alpha$ is locally dx–integrable, then the Friedrichs extension of $(\mathcal{L}, \mathcal{A})$ still generates an ergodic semigroup. However, there is also a non–ergodic symmetric sub–Markovian C^0 semigroup generated by an extension of $(\mathcal{L}, \mathcal{A})$. In particular, $(\mathcal{L}, \mathcal{A})$ is not Markov unique.

III) If α has zeros at which $1/\alpha$ is not dx–integrable, then not even the semi-group generated by the Friedrichs extension of $(\mathcal{L}, \mathcal{A})$ is ergodic. If $1/\alpha$ is not dx–integrable at any zero of α, then Markov uniqueness holds again.

In particular, in the one–dimensional case, Markov uniqueness always holds if $\left(\hat{T}_t\right)_{t\geq0}$ is ergodic. In the multi–dimensional case, however, this is *not true*. If, for example, α vanishes only on some slit $I \subset \mathbb{R}^2$, and α converges to 0 rapidly enough near I, then Markov uniqueness does not hold, but ergodicity of $\left(\hat{T}_t\right)_{t\geq0}$ still holds, because the corresponding diffusion process can move around the slit. Nevertheless, there still seems to be a close relation between Markov uniqueness and some kind of local ergodicity.

PROOF OF THEOREM 3.7.   Let u be a function in $L^2(E; m)$. We first show that u is in $W^{1,2}(E, (a^{ij}); m)$ and satisfies $\hat{\mathcal{E}}(u, u) = 0$, if and only if u is constant dx–a.e. on each component of $\{x \in E;\; \alpha(x) > 0\}$. By Lemma 3.2 and 3.1, it is immediately clear that if u is constant on every component of $\{x \in E;\; \alpha(x) > 0\}$, then u is in $W^{1,2}(E, (a^{ij}); m)$ and satisfies $\hat{\partial}_i u = 0$ for all $1 \leq i \leq n$. Thus $\hat{\mathcal{E}}(u, u) = 0$.

Coversely, suppose u is in $W^{1,2}(E, (a^{ij}); m)$ and $\hat{\mathcal{E}}(u, u) = 0$, i.e., $\hat{\partial}_i u = 0$ m–a.e. for all $1 \leq i \leq n$. We can assume that u is bounded, otherwise we apply the following considerations to $(u \wedge c) \vee (-c)$ for every $c > 0$. Suppose there exists a connection component U of $\{x \in E;\; \alpha(x) > 0\}$ such that u is not constant dx–a.e. on U. Then there exists a cube $C = \prod_{i=1}^{n}[x_i, x_i + d]$, $x_i \in \mathbb{R}$, $d > 0$, such that $C \subset U$, and u is not constant dx–a.e. on C. In C, we can find two small cubes $C_1 = \prod_{i=1}^{n}[x_i + k_i\varepsilon_n, x_i + (k_i + 1)\varepsilon_n]$ and $C_2 = \prod_{i=1}^{n}[x_i + l_i\varepsilon_n, x_i + (l_i + 1)\varepsilon_n]$, $n \in \mathbb{N}$, $\varepsilon_n = 2^{-n}d$, $0 \leq k_i, l_i < 2^n$, such that $\int_{C_1} u\,dx \neq \int_{C_2} u\,dx$. We can even assume that the cubes C_1 and C_2 are in a row, i.e., there exists $1 \leq j \leq n$ such that $k_i = l_i$ for all $i \neq j$. Without loss of generality, we assume $j = 1$, whence

$$C_2 = [x_1 + l_1\varepsilon_n, x_1 + (l_1 + 1)\varepsilon_n] \times \prod_{i=2}^{n}[x_i + k_i\varepsilon_n, x_i + (k_i + 1)\varepsilon_n].$$

By Lemma 3.2 and 3.1, u has a modification $\tilde{u}$ on C such that $\tilde{u}(\cdot, y_2, y_3, \ldots, y_n)$ is absolutely continuous on $[x_1, x_1 + d]$ for all $y_i \in [x_i, x_i + d]$, $2 \leq i \leq n$, and

$$(\tilde{u}(\cdot, y_2, y_3, \ldots, y_n))'(y_1) = \left(\hat{\partial}_1 u\right)(y) = 0 \quad \text{for } dx\text{–a.e. } y \in C.$$

Integrating this inequality w.r.t. y yields

$$\tilde{u}(y_1, y_2, \ldots, y_n) = \tilde{u}(y_1 + (l_1 - k_1)\varepsilon_n, y_2, \ldots, y_n) \quad \text{for } dx\text{–a.e. } y \in C_1,$$

and hence $\int_{C_1} u\,dx = \int_{C_2} u\,dx$. This is a contradiction.

Since the semigroup $\left(\hat{T}_t\right)_{t\geq0}$ is L^2 ergodic if and only if the Dirichlet form $(\hat{\mathcal{E}}, W^{1,2}(E, (a^{ij}); m))$ is irreducible, the considerations above immediately prove

Assertion (i) of the theorem.

Now suppose the assumptions in Assertion (ii) are satisfied. We have to show that hence the Dirichlet form $(\mathcal{E}^{(0)}, H^{1,2}(E, (a^{ij}); m))$ is irreducible. Let $u \in H^{1,2}(E, (a^{ij}); m)$ such that $\mathcal{E}^{(0)}(u, u) = 0$. By the considerations above, we may assume that u is constant on every component of $\{x \in E;\ \alpha(x) > 0\}$. It only remains to show that the values taken on each component are the same. By the assumption, it suffices to show that u takes the same value on any two components U, V, for which there exists a C^1 diffeomorphism $\phi : (-1, 1) \times B \to E$ such that $\phi((-1, 0) \times B) \subset U$, $\phi((0, 1) \times B) \subset V$, and $\int_{-1}^{1} \int_B \alpha(\phi(s, y))^{-1} \, dy\, ds < \infty$. We fix two such components U, V, and a corresponding diffeomorphism ϕ. Suppose that $u(x) = u_1$ for all $x \in U$ and $u(x) = u_2$ for all $x \in V$. Since u is in $H^{1,2}(E, (a^{ij}); m)$, we can find a sequence $(f_k)_{k \in \mathbb{N}}$ of uniformly bounded smooth functions on E such that $f_n(x) \to u(x)$ as $n \to \infty$ for dx–a.e. x, and $\sup \mathcal{E}^{(0)}(f_n, f_n) < \infty$. For $n \in \mathbb{N}$ and $y \in B$, we have

$$|f_n(\phi(\varepsilon, y)) - f_n(\phi(-\varepsilon, y))| = \left| \int_{-\varepsilon}^{\varepsilon} \nabla f_n\left(\phi(s, y)\right) \cdot \frac{\partial \phi}{\partial s}(s, y)\, ds \right|$$

$$\leq K_1 \cdot \left(\int_{-1}^{1} \alpha(\phi(s,y)) \, |\nabla f_n(\phi(s,y))|^2 \, ds \right)^{1/2} \cdot \left(\int_{-\varepsilon}^{\varepsilon} \alpha(\phi(s,y))^{-1} \, ds \right)^{1/2}$$

for all $\varepsilon > 0$, where K_1 is a finite constant, i.e., independent of n, y and ε. Integrating w.r.t. y, we obtain

$$\int_B |f_n(\phi(\varepsilon, y)) - f_n(\phi(-\varepsilon, y))| \, dy$$

$$\leq K_1 \cdot \left(\int_{(-1,1) \times B} \alpha(\phi(x)) \, |\nabla f_n(\phi(x))|^2 \, dx \right)^{1/2} \left(\int_{(-\varepsilon, \varepsilon) \times B} \alpha(\phi(x))^{-1} \, dx \right)^{1/2}$$

$$\leq K_2 \cdot \mathcal{E}^{(0)}(f_n, f_n)^{1/2} \cdot \left(\int_{(-\varepsilon, \varepsilon) \times B} \alpha(\phi(x))^{-1} \, dx \right)^{1/2}$$

for all $\varepsilon > 0$ and $n \in \mathbb{N}$, where K_2 is a second constant. Hence

$$|u_1 - u_2| \cdot \int_B dy \leq \liminf_{n \to \infty} \int_B |f_n(\phi(\varepsilon, y)) - f_n(\phi(-\varepsilon, y))| \, dy$$

$$\leq K_2 \cdot \sup \mathcal{E}^{(0)}(f_n, f_n)^{1/2} \cdot \left(\int_{(-\varepsilon, \varepsilon) \times B} \alpha(\phi(x))^{-1} \, dx \right)^{1/2}$$

for ds–a.e. $\varepsilon > 0$. Since $1/(\alpha \circ \phi)$ is dx–integrable on $(-1, 1) \times B$, and the energies of f_n, $n \in \mathbb{N}$, are bounded, we obtain $u_1 = u_2$ as $\varepsilon \downarrow 0$. ∎

2) Ergodicity and extremality of symmetrizing measures

For a class of symmetric infinite dimensional diffusion operators on L^2 spaces, S. Albeverio, Y. Kondratiev and M. Röckner [AlbKoRö 97c] have shown that the weak Sobolev space is irreducible (i.e., the C^0 semigroup generated by the maximal Dirichlet extension of the operator is L^2 ergodic) if and only if the underlying measure is extremal in the convex set of all Gibbs measures. Here the Gibbs measures are precisely the symmetrizing probability measures for the operator defined on cylinder functions.

We will now show that a similar connection between irreducibility of $W^{1,2}(d)$ and extremality of the symmetrizing measure holds for arbitrary diffusion operators with symmetrizing probability measure. As a consequence, we point out that a *diffusion operator considered on an L^2 space w.r.t. a non-extremal symmetrizing measure can not be Markov unique, if the C^0 semigroup generated by the Friedrichs extension is ergodic.*

Let E be a set, and $\mathcal{B}$ a σ–algebra on E. Let $(\mathcal{L}, \mathcal{A})$ be a diffusion operator acting on the space $\mathcal{F}(E)$ of all measurable functions on E, i.e., Condition (i) and (ii) in Definition 1.5 hold pointwise instead of m–almost everywhere, with (L, A) replaced by $(\mathcal{L}, \mathcal{A})$. We assume that $\mathcal{A}$ consists of bounded functions that generate $\mathcal{B}$. For simplicity, we also assume that the constant function 1 is in the test function space $\mathcal{A}$. This is typically the case in the infinite dimensional applications we are interested in. In the finite dimensional situation studied in Subsection 1), we could also have considered the Markov uniqueness problem with the test function space $C_0^\infty(E)$ replaced by $C_0^\infty(E) \oplus \mathrm{span}\{1\}$. This would have made no essential difference for the considerations above.

A probability measure m on E is called **symmetrizing** for $(\mathcal{L}, \mathcal{A})$ if and only if $\mathcal{L}f$ is in $L^2(E; m)$ for all $f \in \mathcal{A}$, and

$$(3.34) \qquad \int f\, \mathcal{L}g\, dm \; = \; \int \mathcal{L}f\, g\, dm \qquad \text{for all } f, g \in \mathcal{A}.$$

We denote the space of all symmetrizing probability measures by $\mathcal{S}_1(\mathcal{L}, \mathcal{A})$.

REMARK. Suppose $\mathcal{L} = -\delta d$ for some generalized differential d taking values in a measurable co–tangent bundle $T'E$, and a generalized divergence operator δ. Let ∇ be the corresponding gradient on the tangent bundle (i.e., the dual of the co–tangent bundle), and let div denote the corresponding divergence on vector fields, cf. Appendix D, 6), for details. Then *a probability measure m is symmetrizing for $\mathcal{L}$ if and only if the differential d and the divergence operator δ are in duality w.r.t. m,* or, equivalently, *if and only if the integration by parts formula*

$$(3.35) \qquad \int Xf\, g\, dm \; = \; -\int f\, Xg\, dm \; + \; \int f g\, \mathrm{div}\, X\; dm$$

holds for all $f, g \in \mathcal{A}$ and all $X = \nabla h$, $h \in \mathcal{A}$.

Suppose moreover, that E is a Banach space, and H is a Hilbert space that is densely and continuously embedded into E. Then E' is densely and continuously embedded into H' as well. Let $j : H' \to H$ be the Riesz isometry. Suppose $T'_z E = H'$ for all $z \in E$, and $\mathcal{A} = \mathcal{F}C_b^\infty(E')$. In this situation, a measure m is called **a Gibbs measure** in [AlbKoRö 97c] and other articles, if and only if an integration by parts formula similar to (3.35) holds for every constant vector field $X(z) = j(\ell)$, $\ell \in E'$, i.e., $X = \nabla \ell$, cf. also Section e) in Chapter 5 below. The relation of this notion to the usual definition of Gibbs measures for lattice systems in classical statistical mechanics is discussed in [AlbKoRö 97c]. If m is a Gibbs measure in the sense just described, then (3.35) holds for all $X = \nabla h$, $h \in \mathcal{F}C_b^\infty(E')$, i.e., m is symmetrizing for the corresponding diffusion operator $(-\text{div}\,\nabla, \mathcal{F}C_b^\infty(E'))$. Conversely, if m is symmetrizing, then under very weak additional assumptions, m is also a Gibbs measure. Hence, in the situation described, *symmetrizing measures and Gibbs measures are essentially the same.*

We return to the general situation described above the remark. Since the functions in $\mathcal{A}$ generate the σ-algebra $\mathcal{B}$, (3.34) implies that the operator $(\mathcal{L}, \mathcal{A})$ respects m-classes, i.e., $\mathcal{L}f = 0$ m-a.e. for all $f \in \mathcal{A}$ such that $f = 0$ m-a.e. Hence $(\mathcal{L}, \mathcal{A})$ induces a symmetric diffusion operator on $L^2(E; m)$. The symmetry also implies that m is an invariant measure for $(\mathcal{L}, \mathcal{A})$, because $\mathcal{L}1 = 0$ by the diffusion property.

Let $(d, \mathcal{A})$ be a differential corresponding to $(\mathcal{L}, \mathcal{A})$, and let $H_0^{1,2}(d; m)$, $W^{1,2}(d; m)$, and $W_\infty^{1,2}(d; m)$ denote the corresponding strong and weak Sobolev spaces on $L^2(E; m)$, cf. Section b). The canonical quadratic forms on $H_0^{1,2}(d; m)$ and $W_\infty^{1,2}(d; m)$ are denoted by $\mathcal{E}$, $\hat{\mathcal{E}}$ respectively, i.e.,

$$\mathcal{E}(f, g) = \int (\bar{d}f, \bar{d}g)\, dm, \qquad f, g \in H_0^{1,2}(d; m),$$

is the Dirichlet form corresponding to the Friedrichs extension of the operator $(\mathcal{L}, \mathcal{A})$ on $L^2(E; m)$, and

$$\hat{\mathcal{E}}(f, g) = \int (\hat{d}f, \hat{d}g)\, dm, \qquad f, g \in W_\infty^{1,2}(d; m).$$

We make the following assumption :

Assumption : $(\hat{\mathcal{E}}, W_\infty^{1,2}(d; m))$ is a Dirichlet form.

The assumption is satisfied, if the weak differential $\hat{d}$ satisfies a chain rule on bounded functions in $W^{1,2}(d; m)$. As remarked before, I am not able to verify this in the general setting described here, but in essentially all concrete applications we are interested in, it is fulfilled, cf. Remark (ii) after Lemma 3.4, as well as Remark (vi) in Section c), 2) of Chapter 5. The assumption implies that the generator $\hat{L}$ of $(\hat{\mathcal{E}}, W_\infty^{1,2}(d; m))$ is the maximal Dirichlet extension of $(\mathcal{L}, \mathcal{A})$, cf. Corollary 3.1.

We can now state a generalization of the result from [AlbKoRö 97c] referred to at the beginning of this section. The proof is similar to that given in [AlbKoRö 97b, 97c]. Nevertheless, it is completely included below for the convenience of the reader.

Theorem 3.8 *Let m be a symmetrizing probability measure for $(\mathcal{L}, \mathcal{A})$ satisfying the assumptions above. Then the following assertions are equivalent :*

(i) The Dirichlet form $(\hat{\mathcal{E}}, W_\infty^{1,2}(d; m))$ is irreducible.

(ii) *Every positive bounded function* $u \in W^{1,2}(d; m)$ *such that* $\hat{d}u = 0$
  m–*a.e. is constant.*

(iii) m *is an extremal in the convex set* $\mathcal{S}_1(\mathcal{L}, \mathcal{A})$.

REMARK.   Suppose E is an open subset in $\mathbf{R}^n$, $\mathcal{A} = C_0^\infty(E)$, m is a not
necessarily finite measure such that $m = \varphi^2 \, dx$ for some $\varphi \in H_{loc}^{1,2}(E; dx)$, $\varphi >$
0 dx–a.e., and $(\mathcal{L}, \mathcal{A})$ is a symmetric diffusion operator on $L^2(E; m)$ of type
(3.6), such that the coefficients a^{ij} satisfy the assumptions from Section a), 2).
Then a result similar to Theorem 3.8 holds, but with the following modifications :

- The space $\mathcal{S}_1(\mathcal{L}, \mathcal{A})$ is replaced by the space $\mathcal{S}_{ac}(\mathcal{L}, \mathcal{A})$ of all (not necessarily finite) symmetrizing Radon measures that are locally absolutely
  continuous w.r.t. Lebesgue measure, or, equivalently, w.r.t. m. The absolute continuity condition is required only, because usually $\mathcal{L}f$, $f \in C_0^\infty(E)$,
  is defined only dx–a.e.

- The weak Sobolev space $W^{1,2}(E, (a^{ij}); m)$ ($= W_\infty^{1,2}(E, (a^{ij}); m)$) is
  replaced by the **extended weak Sobolev space**

$$W_{ext}^{1,2}(E, (a^{ij}); m) = \left\{ u \in W_{loc}^{1,2}(E, (a^{ij}); m); \int_E \sum_{i,j=1}^n a^{ij} \, \partial_i u \, \partial_j u \, dm < \infty \right\}$$

  The extended weak Sobolev space has to be used, because bounded functions of finite energy are not necessarily in $L^2(E; m)$ if m is not finite.

The proof of this finite dimensional counterpart to Theorem 3.8 is essentially
the same as that of the theorem. The use of the constant function 1 in the proof
of Theorem 3.8 below can be replaced by using functions in $C_0^\infty(E)$ that are
locally constant instead.

PROOF OF THEOREM 3.8.   (i) $\Rightarrow$ (ii) is obvious. Conversely, suppose that
(ii) holds, and let $u \in W_\infty^{1,2}(d; m)$ such that $\hat{\mathcal{E}}(u, u) = 0$. Let $n \in \mathbf{N}$. By the
contraction property for Dirichlet forms, the functions $u^+ \wedge n$ and $u^- \wedge n$ are
in $W_\infty^{1,2}(d; m)$ as well, and $\hat{\mathcal{E}}(u^+ \wedge n, u^+ \wedge n) = \hat{\mathcal{E}}(u^- \wedge n, u^- \wedge n) = 0$. Hence
$\hat{d}(u^+ \wedge n) = \hat{d}(u^- \wedge n) = 0$ m–a.e., whence $u^+ \wedge n$ and $u^- \wedge n$ are constant by
(iii). Since n has been chosen arbitrarily, u is constant as well. Thus (i) holds.

(ii) $\Rightarrow$ (iii) : Suppose m is not extremal in $\mathcal{S}_1(\mathcal{L}, \mathcal{A})$, i.e., $m = \alpha\mu + (1 - \alpha)\nu$ for
some $0 < \alpha < 1$ and $\mu, \nu \in \mathcal{S}_1(\mathcal{L}, \mathcal{A}) \setminus \{m\}$. Then μ is absolutely continuous
w.r.t. m and $d\mu/dm \le \alpha^{-1}$. Let $\rho := d\mu/dm$. Then

$$\int \rho \, d^*(f dg) \, dm = -\int (f \mathcal{L}g + \Gamma(f, g)) \, d\mu$$

$$= -\frac{1}{2} \int (f \mathcal{L}g - g \mathcal{L}f + \mathcal{L}(fg)) \, d\mu = 0$$

for all $f, g \in \mathcal{A}$, because μ is symmetrizing for $(\mathcal{L}, \mathcal{A})$, and $\mathcal{L}1 = 0$. Hence ρ is
a bounded function in $W^{1,2}(d; m)$, and $\hat{d}\rho = 0$. Since μ and m are probability

measures, and $\mu \neq m$, ρ is not constant. Hence (ii) does not hold if (iii) does not hold.

(iii) $\Rightarrow$ (ii) : Conversely, let ρ be a non–constant positive bounded function in $W^{1,2}(d; m)$ such that $\hat{d}\rho = 0$ m–a.e. W. l. g., we may assume $0 \leq \rho \leq 1$. Then, by the same equation as above, we obtain

$$(3.36) \quad 0 = \int \rho \, d^*(f \, dg) \, dm = -\frac{1}{2} \int (f \mathcal{L}g - g\mathcal{L}f + \mathcal{L}(fg)) \, \rho \, dm$$

for all $f, g \in \mathcal{A}$. By choosing first $f = 1$, we see that $\int \mathcal{L}g \, \rho \, dm = 0$ for all $g \in \mathcal{A}$. Now (3.36) yields $\int f \mathcal{L}g \, \rho \, dm = \int g \mathcal{L}f \, \rho \, dm$ for all $f, g \in \mathcal{A}$, i.e., $\rho \cdot m$ is symmetrizing for $(\mathcal{L}, \mathcal{A})$. We have

$$m = \rho \cdot m + (1 - \rho) \cdot m ,$$

and, since m is in $\mathcal{S}_1(\mathcal{L}, \mathcal{A})$, $(1 - \rho) \cdot m$ is symmetrizing as well. Now let

$$\alpha := \int \rho \, dm, \quad \mu := \alpha^{-1} \rho \cdot m, \quad \text{and} \quad \nu := (1 - \alpha)^{-1} (1 - \rho) \cdot m .$$

Since ρ is not m–a.e. constant, we have $0 < \alpha < 1$. Since $1 - \alpha = \int (1 - \rho) \, dm$, μ and ν are both *probability* measures, and thus in $\mathcal{S}_1(\mathcal{L}, \mathcal{A})$. Moreover,

$$m = \alpha \mu + (1 - \alpha)\nu ,$$

whence m is not extremal in $\mathcal{S}_1(\mathcal{L}, \mathcal{A})$. ∎.

3) Two kinds of non–uniqueness

A typical characteristic of several lattice systems from classical statistical mechanics is that they have a *phase transition* in the sense that for large temperatures, there exists a unique Gibbs measure, whereas for low temperatures, there are several Gibbs measures. We are interested in the case where uniqueness does not hold. As remarked above, the Gibbs measures are symmetrizing measures for the corresponding infinite dimensional diffusion operator defined on cylinder functions. Hence the symmetrizing measure for the operator is not unique. In particular, there are non–extremal symmetrizing measures. Let m be one of them. Then, by Theorem 3.8, one of the following assertions is true :

1) The C^0 semigroup generated by the Friedrichs extension of the operator on $L^2(E; m)$ is not $L^2(E; m)$ ergodic.

2) The operator is not Markov unique on $L^2(E; m)$.

In particular, if we can prove Markov uniqueness, then the Friedrichs semigroup can not be ergodic. If, on the other hand, we can prove ergodicity of the Friedrichs semigroup (– for example by proving a logarithmic Sobolev inequality for the corresponding Dirichlet form), then the operator is not Markov unique on $L^2(E; m)$. We now discuss the question, in which situations typically non–ergodicity of the Friedrichs semigroup, respectively non–Markov uniqueness occur.

The finite dimensional case

We first look once more at the corresponding problem for diffusion operators with finite dimensional state space. Here, by the results in Subsection 1), both cases can occur. If, with the notations from Subsection 1), the set $\{x \in E;\ \alpha(x) > 0\}$ is not connected, then the weak Sobolev space $W^{1,2}(E,\ (a^{ij});\ m)$ corresponding to the operator $(\mathcal{L},\ C_0^\infty(E))$ on $L^2(E;\ m)$ is not ergodic, i.e., there are several symmetrizing probability measures. In fact, every probability measure that is proportional to m on each fixed connection component of $\{x \in E;\ \alpha(x) > 0\}$ is symmetrizing for $(\mathcal{L},\ C_0^\infty(E))$. The extremal symmetrizing probability measures are proportional to m on one component, and vanish on the others. In particular, m is not extremal.

Now suppose α decays rapidly enough near its zeros. Then the diffusion process generated by the Friedrichs extension does not hit the singularity set $S(\alpha) := \{x \in E;\ \alpha(x) = 0\}$. Hence, by Corollary 3.4, even the operator $(\mathcal{L},\ C_0^\infty(E \backslash S(\alpha)))$ is Markov unique on $L^2(E;\ m)$, and the Friedrichs semigroup is not $L^2(E;\ m)$ ergodic. In fact, this behaviour occurs in the one–dimensional case if $1/\alpha$ is locally dx–integrable, cf. Theorem 3.2 and the example below Theorem 3.7. A sufficient condition for Markov uniqueness in the multi–dimensional case has been given in Sections e) and f).

Conversely, if α decays only slowly near a sufficiently large part of $S(\alpha)$, then Markov uniqueness does not hold. This is for example the case, if in the one–dimensional case, there exists a zero $s \in S(\alpha)$ such that $1/\alpha$ is not dx–integrable at s. In the multi-dimensional case, ergodicity of the Friedrichs semigroup holds, and Markov uniqueness does not hold, if the condition in Theorem 3.7 (ii) is satisfied. Of course, it is also possible that neither ergodicity of the Friedrichs semigroup nor Markov uniqueness occurs. This is for example the case, if the condition in Theorem 3.7 (ii) holds only for two adjacent components U, V of $E \backslash S(\alpha)$, whereas near the boundaries to all other components, α decays rapidly.

Infinite dimensional lattice systems

Let M be a Riemannian manifold, and let $d \in \mathbf{N}$. As in the finite dimensional case, non–uniqueness of the symmetrizing measure of a diffusion operator that describes the dynamics of a lattice system from classical statistical mechanics with state space $E = M^{\mathbf{Z}^d}$ can produce both non–ergodicity and/or non–uniqueness of the C^0 semigroup generated by the Friedrichs extension of the operator on $L^2(E;\ m)$, where m is a non–extremal symmetrizing measure.

If, however, the interaction is sufficiently smooth, bounded, and has, for example, finite range, then Markov uniqueness can always be expected. A corresponding uniqueness result has first been given in [AlbKoRö 95], [AlbKoRö 97c] respectively, cf. also Section e) in Chapter 5 below. Hence in the translation–invariant case, which is the most interesting from a physical point of view, we are usually in Situation 1), not 2).

Nevertheless, if the interactions are unbounded as $|i| \to \infty$ (where $i \in \mathbf{Z}^d$ is the lattice parameter), then Markov uniqueness can be destroyed, whereas the Friedrichs semigroup on $L^2(E\,;\,m)$ can be ergodic, even if m is a non-extremal symmetrizing measure. A corresponding example is given in Chapter 5, Section b). In this example, the state space E is not that of a lattice system, but $E = C([0,1] \to \mathbf{R})$. However, it is possible to replace the basic interval $[0,1]$ by a discrete set as, for example, $\{2^{-n};\, n \in \mathbf{N}\} \cup \{1 - 2^{-n};\, n \in \mathbf{N}\}$. In this way, a lattice system with unbounded interactions, which shows essentially the same behaviour, can be constructed.

Appendix D   The geometry of diffusion operators

Throughout this appendix, we fix a set E and a σ–algebra $\mathcal{B}$ on E. In many applications, we are confronted with the following situation : We are given a diffusion operator $\mathcal{L}$ acting on a vector space $\mathcal{A}$ consisting of some nice measurable functions on E. The operator $\mathcal{L}$ is, for example, defined as the generator of some diffusion process on E, which has been constructed stochastically. From Riemannian geometry we know, that if E is a smooth manifold, and $\mathcal{L}$ is a second order elliptic differential operator with smooth coefficients on E, then the analysis of $\mathcal{L}$ becomes the most natural, if we equip E with the unique Riemannian metric g such that $\mathcal{L} = \Delta_g + X$, where Δ_g denotes the Laplace–Beltrami operator w.r.t. g, and X is a smooth vector field on E. Can we proceed similarly if E is not assumed to be a smooth manifold, $\mathcal{L}$ is a general diffusion operator on E, and we do not impose any regularity assumptions on $\mathcal{L}$?

The purpose of this appendix is to demonstrate that under very weak assumptions, the answer is in principle "Yes". Moreover, we will consider a variety of examples of diffusion operators and their geometry. In particular, we look in detail at the geometry of measure–valued diffusions.

In Subsection 2) below, we show that *we can construct a unique generalized co–tangent bundle and generalized differential corresponding to the operator* $(\mathcal{L}, \mathcal{A})$. The uniqueness implies in particular that our co–tangent bundle and differential *can be identified with the corresponding classical objects whenever these exist*. A tangent bundle could also be constructed formally as the dual of the co–tangent bundle, but for our purposes, the co–tangent bundle is usually sufficient. The notions "generalized co–tangent bundle" and "generalized differential" are made precise in Subsection 1).

In Subsections 3)–5), we consider several classes of concrete examples. Here, we are typically already given a natural candidate for a differential, i.e., a linear operator d mapping test–functions in $\mathcal{A}$ to sections of some generalized vector bundle over E, such that d satisfies the product rule. The right metric on the bundle is then determined by the *Carré du champ* of the operator $(\mathcal{L}, \mathcal{A})$. The minimal complete bundle w.r.t. this metric such that df is a section for all

$f \in \mathcal{A}$ is a measurable co–tangent bundle for $(\mathcal{L}, \mathcal{A})$, and d is a corresponding generalized differential. In this way, we obtain for example co–tangent bundles corresponding to degenerate diffusion operators on $\mathbf{R}^n$ such that the dimension of the co–tangent spaces decreases at points of degeneracy (cf. Subsection 3), and co–tangent bundles on path and loop spaces over Riemannian manifolds (cf. Subsection 4), or on configuration spaces and more general spaces of measures (cf. Subsection 5), that are dual to tangent bundles which have been defined in previous articles by several authors, cf. e.g. [CrMal 94], [Mal 97], [AlbKoRö 97a,b], and [OvRöSchm 95]. To obtain a better understanding of geometries for measure-valued diffusions, we introduce the concepts of horizontal and vertical tangent resp. co–tangent bundles to spaces of measures.

Subsection 6) is a brief excursion to symmetrizing measures for diffusion operators of type $-\delta d$, where d is a generalized differential, and δ is a *generalized divergence operator*. We show that a probability measure m is symmetrizing for the operator $-\delta d$ if and only if d and δ are in duality w.r.t. m. Symmetrizing measures can be characterized by the equation $\delta^* m = 0$.

Finally, in Subsection 7), we apply our considerations to derive a *general representation theorem for diffusion operators* $(\mathcal{L}, \mathcal{A})$ acting on $L^2(E\,;\,m)$, where m is an invariant measure for the operator. We show that $\mathcal{L}$ can always be written as

$$\mathcal{L} \;=\; -d^* d \,+\, X\,,$$

where d is a differential, d^* is the adjoint w.r.t. the measure m, and X is an anti-symmetric derivation, i.e., the generalization of a vector field with m–divergence equal to 0. The operator $(\mathcal{L}, \mathcal{A})$ is symmetric if and only if X vanishes.

1) Generalized differentials

Recall that a collection $H = (H_z)_{z \in E}$ of Hilbert spaces together with a linear space $\mathcal{F}(E \to H)$ consisting of sections of H (i.e., maps ω from E to the disjoint union $\bigcup_{z \in E} H_z$ such that $\omega(z)$ is in H_z for every z) is called a **measurable field of Hilbert spaces**, if and only if

(i) A section ω of H is in $\mathcal{F}(E \to H)$ if and only if the map
$z \mapsto (\omega(z), \sigma(z))_{H_z}$ is measurable for every $\sigma \in \mathcal{F}(E \to H)$.

(ii) There exists a countable subset $\{e_i;\; i \in \mathbf{N}\}$ of $\mathcal{F}(E \to H)$ such that the span of $\{e_i(z); i \in \mathbf{N}\}$ is dense in H_z for every $z \in E$.

The sections in $\mathcal{F}(E \to H)$ are called **measurable**. If $\{e_i;\; i \in \mathbf{N}\}$ is a countable subset of $\mathcal{F}(E \to H)$ as in Condition (ii), then a section ω of H is measurable if and only if the functions $z \mapsto (\omega(z), e_i(z))_{H_z}$ are measurable for all $i \in \mathbf{N}$. See e.g. [Dix 69, Ch. II, Par. 1] for more details on measurable fields of Hilbert spaces.

Definition 3.6 *Let* $T'E = (T'_z E)_{z \in E}$ *be a measurable field of Hilbert spaces over* E, *and let* A *be an algebra of functions on* E *which generates the* σ*–algebra*

B. A map $d : A \to \mathcal{F}(E \to T'E)$ *is called a* (**generalized**) **differential**, *if and only if*

(i) $\{d_z f; f \in A\}$ *is dense in* $T_z'E$ *for every* $z \in E$.

(ii) d *is linear.*

(iii) $d(f \cdot g) = f \cdot dg + g \cdot df$ *for all* $f, g \in A$.

Here $d_z f$ *is used to denote the value of the section df at the point* $z \in E$. *The measurable field* $T'E$ *is called the corresponding measurable co–tangent bundle.*

REMARKS. (i) We use the notation $T'E$, although we will in general not introduce a corresponding tangent bundle TE, such that $T_z'E$ is the dual of $T_z E$ for all z. For various classes of examples, however, there exist natural tangent spaces, and $T_z'E$ can be identified with the dual, cf. Subsections 3)–5) below.
(ii) The first two conditions in Definition 3.6 imply that the sections df, $f \in A$, generate the measurable structure on $T'E$, i.e., a section ω of $T'E$ is measurable if and only if the functions $z \mapsto (\omega(z), d_z f)_{T_z'E}$ are measurable for all $f \in A$, cf. [Dix 69, Ch. II, Par. 1, Prop. 4, Démonstration].

In Subsections 3)–5) below, we consider several classes of examples of generalized differentials. Before, we show that to each diffusion operator on $(E, \mathcal{B})$, there correspond a unique (up to isometries of measurable fields) measurable co–tangent bundle, and a unique differential.

2) Measurable co–tangent bundles and differentials corresponding to diffusion operators

In Definition 1.5 in Appendix B, we have defined the notion of an *abstract diffusion operator* acting on m–classes of functions, where m is a σ–finite measure. In the same way, we call a linear operator $(\mathcal{L}, A)$ acting on the space $\mathcal{F}(E)$ an **abstract diffusion operator**, iff (i) and (ii) in Definition 1.5 hold (pointwise) with (L, A) replaced by $(\mathcal{L}, A)$.

From now on, we fix an (abstract) diffusion operator $(\mathcal{L}, A)$ on $\mathcal{F}(E)$. The corresponding Carré du champ operator $\Gamma : A \times A \to \mathcal{F}(E)$ is defined by

$$\Gamma(f, g) = \frac{1}{2} \left(\mathcal{L}(fg) - f\mathcal{L}g - g\mathcal{L}f \right).$$

The value of $\Gamma(f, g)$ at a point $z \in E$ is denoted by $\Gamma_z(f, g)$. In order to obtain a *measurable* co–tangent bundle corresponding to the operator $(\mathcal{L}, A)$, we need the following weak assumption :

(C 1) There exists a *countable* subset A_0 of A, such that for every $f \in A$ there is a sequence $(f_n)_{n \in \mathbb{N}}$ in A_0 satisfying $\lim_{n \to \infty} \Gamma_z(f - f_n, f - f_n) = 0$ for all $z \in E$.

The assumption can be easily verified for many interesting classes of examples, including those considered in Subsections 3)–5) below.

By an **isometry** between two measurable fields of Hilbert spaces $(H_z)_{z \in E}$ and $\left(\tilde{H}_z\right)_{z \in E}$, we mean a collection $\Phi = (\Phi_z)_{z \in E}$ of isometries $\Phi_z : H_z \to \tilde{H}_z$, such that a section ω of $(H_z)_{z \in E}$ is measurable if and only if the corresponding section $z \mapsto \Phi_z(\omega(z))$ of $\left(\tilde{H}_z\right)_{z \in E}$ is measurable.

Theorem 3.9 *Suppose (C 1) holds. Then :*

(i) There exist a measurable field of Hilbert spaces $T'E$, and a **differential** *$d : \mathcal{A} \to \mathcal{F}(E \to T'E)$ such that*

$$(3.37) \quad \Gamma_z(f, g) \;=\; (d_z f, d_z g)_{T'_z E} \qquad \textit{for all } f, g \in \mathcal{A} \textit{ and } z \in E.$$

(ii) The co–tangent bundle $T'E$ and the differential d in (i) are unique in the following sense : If $\tilde{T}'E$ is another measurable field of Hilbert spaces, and $\tilde{d} : \mathcal{A} \to \mathcal{F}(E \to \tilde{T}'E)$ is another differential such that (3.37) holds, then there exists an isometry $\Phi : T'E \to \tilde{T}'E$ such that $\tilde{d}_z f = \Phi_z(d_z f)$ for all $f \in \mathcal{A}$ and $z \in E$.

The uniqueness part of the theorem shows in particular, that the co–tangent bundle and the differential correponding to the operator $(\mathcal{L}, \mathcal{A})$ as introduced in the proof below *can be identified with the corresponding classical objects* whenever these exist, cf. also the examples in Subsections 3)–5).

REMARK. Our construction of a differential corresponding to the operator $(\mathcal{L}, \mathcal{A})$ in the proof of Theorem 3.9 is motivated partially by related considerations in the book of N. Bouleau and F. Hirsch, cf. [BouHi 91, Ch. V, Exercise 5.9].

The remaining part of this subsection contains the proof of Theorem 3.9. To construct the co–tangent bundle and the differential corresponding to $(\mathcal{L}, \mathcal{A})$, note that, for each $z \in E$, Γ_z is a positive definite symmetric bilinear form, i.e., a pseudo inner product on $\mathcal{A}$. Let

$$\ker \Gamma_z \; := \; \{ f \in \mathcal{A}; \; \Gamma_z(f, f) = 0 \}.$$

For $z \in E$, we define the Hilbert space $T'_z E$ as the *completion of $\mathcal{A}/\ker \Gamma_z$ w.r.t. the inner product induced by Γ_z.* Moreover, for $f \in \mathcal{A}$ and $z \in E$, let $d_z f$ be the *equivalence class in $T'_z E$ represented by the function f.* The map $z \mapsto d_z f$ is denoted by df. Then it is obvious that (3.37) holds.

So far, the spaces $T'_z E$, $z \in E$, are separate spaces, i.e., there is no "connection". Of course, since the coefficients of the operator $\mathcal{L}$ are not assumed to be smooth, we cannot expect $T'E$ to be a smooth vector bundle. However, under Condition (C 1), we can introduce a measurable structure. Let $\mathcal{F}(E \to T'E)$ be the space of all sections ω of $T'E$ such that the maps $z \mapsto (\omega(z), d_z f)_{T'_z E}$ are measurable for all $f \in \mathcal{A}$. By (3.37), df is in $\mathcal{F}(E \to T'E)$ for every $f \in \mathcal{A}$.

Now, let $\mathcal{A}_0$ be a countable subset of $\mathcal{A}$ such that (C 1) holds. Then for every $f \in \mathcal{A}$, there exists a sequence $(f_n)_{n \in \mathbb{N}}$ in $\mathcal{A}_0$ such that $\lim_{n \to \infty} d_z f_n = d_z f$ in $T'_z E$ for all $z \in E$, cf. (C 1) and (3.37). Hence the span of $\{d_z f; \ f \in \mathcal{A}_0\}$ is dense in $T'_z E$ for each z, and a section ω of $T'E$ is in $\mathcal{F}(E \to T'E)$ if and only if the maps $z \mapsto (\omega(z), d_z f)_{T'_z E}$ are measurable for all $f \in \mathcal{A}_0$. Since $\mathcal{A}_0$ is countable, this suffices to prove that $T'E$ is a *measurable field of Hilbert spaces* with measurable structure $\mathcal{F}(E \to T'E)$, cf. [Dix 69, Ch. II, Par. 1, 4., Prop. 4].

By construction, Conditions (i) and (ii) in Definition 3.6 hold. Moreover, the (pointwise) diffusion property of the operator $(\mathcal{L}, \mathcal{A})$ implies the product rule

$$\Gamma_z(fg, h) \ = \ f(z)\, \Gamma_z(g, h) \ + \ g(z)\, \Gamma_z(f, h)$$

for all f, g, $h \in \mathcal{A}$ and $z \in E$. Hence, by (3.37) and the density of $\{d_z h; \ h \in \mathcal{A}\}$ in $T'_z E$, we obtain the product rule

$$d_z\,(f \cdot g) \ = \ f\, d_z g \ + \ g\, d_z f \qquad \text{for all } f, g \in \mathcal{A} \text{ and } z \in E.$$

Thus d is a *generalized differential*. This proves the first assertion in Theorem 3.9.

To complete the proof of the theorem, it remains to show **uniqueness**. Suppose $\tilde{T}'E$ is another measurable field of Hilbert spaces, and $\tilde{d} : \mathcal{A} \to \mathcal{F}(E \to \tilde{T}'E)$ is another generalized differential such that (3.37) holds. Fix $z \in E$. Then the map $\tilde{d}_z : \mathcal{A} \to \tilde{T}'_z E$ preserves the metric, if $\mathcal{A}$ is endowed with the pseudo inner product Γ_z. Hence

$$\mathcal{A}/\ker \Gamma_z \ = \ \mathcal{A}/\ker \tilde{d}_z \ \cong \ \mathrm{im}\, \tilde{d}_z \ \subseteq \ \tilde{T}'_z E,$$

where "$\cong$" denotes the canonical isometry induced by $\tilde{d}_z$. Note that $T'_z E$ is the completion of $\mathcal{A}/\ker \Gamma_z$ w.r.t. Γ_z, and $\mathrm{im}\, \tilde{d}_z$ is dense in $\tilde{T}'_z E$, cf. Condition (i) in Definition 3.6. Hence the isometry above can be extended to a unique isometry

$$\Phi_z : \ T'_z E \ \to \ \tilde{T}'_z E .$$

By construction, $\Phi_z\,(d_z f) = \tilde{d}_z f$ for all $f \in \mathcal{A}$. Moreover, let ω be a measurable section of $T'E$. Then

$$(3.38) \qquad \Big(\Phi_z(\omega(z)),\, \tilde{d}_z f\Big)_{\tilde{T}'_z E} \ = \ (\omega(z),\, d_z f)_{T'_z E}$$

for all $z \in E$ and $f \in \mathcal{A}$. For $f \in \mathcal{A}$, the right hand side is a measurable function in z, whence by Remark (ii) under Definition 3.6, the section $z \mapsto \Phi_z(\omega(z))$ of $\tilde{T}'E$ is measurable as well. Similarly, if ω is an arbitrary section of $T'E$ such that $z \mapsto \Phi_z(\omega(z))$ is a measurable section of $\tilde{T}'E$, then, by (3.38), $z \mapsto (\omega(z), d_z f)_{T'_z E}$ is measurable for all $f \in \mathcal{A}$, whence ω is measurable. Therefore, $(\Phi_z)_{z \in E}$ is an isometry of the measurable fields of Hilbert spaces $T'E$ and $\tilde{T}'E$. This completes the proof of Theorem 3.9. ∎

3) Examples I : Diffusion operators on manifolds and vector spaces

(i) Smooth diffusion operators on Riemannian manifolds

Suppose E is a Riemannian manifold with Borel σ–algebra $\mathcal{B}$, the space $\mathcal{A}$ is either $C_0^\infty(E)$ or $C^\infty(E)$, and $\mathcal{L} = \Delta + X$, where Δ denotes the Laplace–Beltrami operator on E, and X is a smooth vector field. Let T^*E be the ordinary co–tangent bundle endowed with its natural metric. Then the ordinary differential $d : \mathcal{A} \to C^\infty(E \to T^*E)$ is a generalized differential for the operator $(\mathcal{L}, \mathcal{A})$. In particular,

$$\Gamma_z(f, g) = (df, dg)_{T_z^* E} \qquad \text{for all } f, g \in \mathcal{A} \text{ and } z \in E.$$

(ii) Measurable diffusion operators on $\mathbf{R}^n$

Suppose E is an open subset in $\mathbf{R}^n$, $n \in \mathbf{N}$, $\mathcal{B}$ is the Borel σ–algebra on E, $\mathcal{A} = C_0^\infty(E)$, and $\mathcal{L}$ is a second order differential operator of type

$$\mathcal{L}f = \sum_{i,j=1}^n a^{ij} \frac{\partial^2 f}{\partial x^i \partial x^j} + \sum_{i=1}^n \beta^i \frac{\partial f}{\partial x^i},$$

where a^{ij} and β^i, $1 \le i, j \le n$, are (measurable) functions on E, such that the matrix $(a^{ij}(z))$ is symmetric and positive definite (but possibly degenerate) for all $z \in E$. Then

$$\Gamma_z(f, g) = \sum_{i,j=1}^n a^{ij}(z) \frac{\partial f}{\partial x^i}(z) \frac{\partial g}{\partial x^j}(z) = a(d_z f, d_z g)(z)$$

for all $z \in E$ and $f, g \in \mathcal{A}$. Here $d_z : \mathcal{A} \to (\mathbf{R}^n)^*$ is the differential at z in the *ordinary sense*, and $a(\cdot, \cdot)(z)$ denotes the bilinear form on $(\mathbf{R}^n)^*$ which has the matrix $(a^{ij}(z))$ w.r.t. the canonical basis. Let $T'E$ be the measurable field of Hilbert spaces defined by

$$T'_z E = (\mathbf{R}^n)^* / \{ v \in (\mathbf{R}^n)^* ; a(v, v)(z) = 0 \}, \qquad z \in E,$$

with the metric induced by a. Then the map from $\mathcal{A}$ to $T'_z E$, $z \in E$, induced by the ordinary differential is a generalized differential for $(\mathcal{L}, \mathcal{A})$. On points of degeneracy, the dimension of the co–tangent space is less than n.

(iii) Infinite dimensional diffusion operators on linear spaces

Suppose E is a vector space, and K is a vector space of linear functionals on E. Let $\mathcal{B}$ be the σ–algebra generated by the functionals in K. By $C_b^\infty(\mathbf{R}^n)$, we denote the space of all smooth functions f on $\mathbf{R}^n$ such that f and all its derivatives are bounded. Let

$$
\begin{aligned}
\mathcal{A} \; &:= \; \mathcal{F}C_b^\infty(K) \\
&:= \; \{ z \mapsto f(\ell_1(z), \ldots, \ell_n(z)); \; n \in \mathbf{N}, \; f \in C_b^\infty(\mathbf{R}^n), \; \ell_1, \ldots, \ell_n \in K \}
\end{aligned}
$$

be the smooth cylinder functions based on functionals in K. Suppose $\mathcal{L}$ is a diffusion operator with domain $\mathcal{A}$ such that

$$
\mathcal{L}\left(f(\ell_1, \ldots, \ell_n) \right) = \sum_{i,j=1}^{n} a_{\ell_i, \ell_j} \frac{\partial^2 f}{\partial x_i \partial x_j} (\ell_1, \ldots, \ell_n) + \sum_{j=1}^{n} \beta_{\ell_j} \frac{\partial f}{\partial x_j} (\ell_1, \ldots, \ell_n)
$$

for all $n \in \mathbf{N}$, $f \in C_b^\infty(\mathbf{R}^n)$, and $\ell_1, \ldots, \ell_n \in K$, where $\beta : K \to \mathcal{F}(E)$, $\ell \mapsto \beta_\ell$, is a linear map, and $a : K \times K \to \mathcal{F}(E)$, $(\ell, \bar{\ell}) \mapsto a_{\ell\bar{\ell}}$ is a bilinear map such that $a(\cdot, \cdot)(z)$ is a (possibly degenerate) positive definite symmetric bilinear form on K for each $z \in E$. To ensure that Condition (C 1) holds, we assume that there exists a countable subset K_0 of K such that for every $\ell \in K$ there is a sequence $(\ell_n)_{n \in \mathbf{N}}$ in K_0 such that $\ell_n(z) \to \ell(z)$ as $n \to \infty$ and $\lim_{n \to \infty} a(\ell - \ell_n, \ell - \ell_n)(z) = 0$ for all $z \in E$. Let

$$
D_z F = \sum_{i=1}^{n} \frac{\partial f}{\partial x_i} (\ell_1(z), \ldots, \ell_n(z)) \, \ell_i
$$

denote the *ordinary differential* of a function $F = f(\ell_1, \ldots, \ell_n) \in \mathcal{F}C_b^\infty(K)$ at the point $z \in E$. Then

$$
\begin{aligned}
\Gamma_z(F, G) &= \sum_{i,j=1}^{n} a_{\ell_i, \ell_j}(z) \frac{\partial f}{\partial x_i} (\ell_1(z), \ldots, \ell_n(z)) \frac{\partial g}{\partial x_j} (\ell_1(z), \ldots, \ell_n(z)) \\
&= a\left(D_z F, D_z G \right)(z)
\end{aligned}
$$

for all $F, G \in \mathcal{F}C_b^\infty(K)$ and $z \in E$, where we have chosen $n \in \mathbf{N}$, $\ell_1, \ldots, \ell_n \in K$, and $f, g \in C_b^\infty(\mathbf{R}^n)$ such that $F = f(\ell_1, \ldots, \ell_n)$ and $G = g(\ell_1, \ldots, \ell_n)$.

If a is non–degenerate, i.e., $a_{\ell\ell}(z) > 0$ for all $z \in E$ and $\ell \in K \setminus \{0\}$, then D is a generalized differential for the operator $(\mathcal{L}, \mathcal{A})$ w.r.t. the co–tangent space $T_z'E$ defined as the completion of K w.r.t. the inner product $a(\cdot, \cdot)(z)$.

If a degenerates, then the co–tangent space $T_z'E$ corresponding to $(\mathcal{L}, \mathcal{A})$ is isometric to the completion of K modulo the kernel of $a(\cdot, \cdot)(z)$ w.r.t. $a(\cdot, \cdot)(z)$.

4) Examples II : Ornstein–Uhlenbeck geometries on path and loop spaces

We briefly show how to apply our considerations on path and loop spaces over Riemannian manifolds, cf. e.g. [Dri 95] for an overview and references to this active research area. Let M be a compact connected Riemannian manifold, and let x_0 be a fixed point in M. Suppose E is the space of all continuous paths $\omega : [0,1] \to M$ such that $\omega(0) = x_0$, or the space of all continuous loops $\omega : [0,1] \to M$ such that $\omega(0) = \omega(1) = x_0$. Let $\mathcal{B}$ be the Borel σ-algebra on E, and let

$$\mathcal{A} \; := \; \{ F : E \to \mathbf{R} ; \; F(\omega) = f(\omega(s_1), \dots, \omega(s_n)) \text{ for some } n \in \mathbf{N},$$
$$f \in C^\infty(M^n), \text{ and } s_1, \dots s_n \in (0,1) \}.$$

Let τ_s, $0 \le s \le 1$, denote a fixed version of the *stochastic parallel transport* along the paths of Brownian motion respectively the Brownian bridge on E, such that $\tau_s(\omega)$ is an isometry from $T_{x_0}M$ to $T_{\omega(s)}M$ for each $\omega \in E$. See e.g. [Dri 95] for the definition of the stochastic parallel transport. Suppose we are given a diffusion operator $\mathcal{L}$ with domain $\mathcal{A}$ on $\mathcal{F}(E)$ such that the corresponding Carré du champ operator is of type

$$\Gamma_\omega (F, \tilde{F}) \;=\; \sum_{i,j=1}^{n} G(s_i, s_j) \; \left(\tau_{s_i}^{-1}(\omega) \, (\nabla_i f) \, (\omega(s_1), \dots, \omega(s_n)), \right.$$

$$\left. \tau_{s_j}^{-1}(\omega) \, (\nabla_j \tilde{f}) \, (\omega(s_1), \dots, \omega(s_n)) \right)_{T_{x_0}M}$$

whenever $F(\omega) = f(\omega(s_1), \dots, \omega(s_n))$ and $\tilde{F}(\omega) = \tilde{f}(\omega(s_1), \dots, \omega(s_n))$ for some $n \in \mathbf{N}$, $f, \tilde{f} \in C^\infty(M^n)$, and $s_1, \dots, s_n \in (0,1)$. Here ∇_i denotes the gradient w.r.t. the i-th component, and $G(s,t) := \min(s,t)$ in the path space case, respectively $G(s,t) := \begin{cases} s(1-t) & \text{if } s \le t \\ t(1-s) & \text{if } s \ge t \end{cases}$ in the loop space case, is the Green's function for the operator $\frac{d^2}{ds^2}$ on $(0,1)$ with Dirichlet boundary conditions at 0, and Neumann resp. Dirichlet boundary conditions at 1.

The standard example of a diffusion operator with Carré du champ Γ is the **Ornstein–Uhlenbeck operator on E**. It is defined as the generator of the closure of the pre–Dirichlet form

$$\mathcal{E}(F, \tilde{F}) \;=\; \int \Gamma_\omega(F, \tilde{F}) \; P(d\omega), \qquad F, \tilde{F} \in \mathcal{A},$$

on $L^2(E; P)$, where P denotes the law of Brownian motion respectively the Brownian bridge on E, cf. e.g. [Eb 97].

We want to find a suitable expression for the differential DF of a cylinder function $F(\omega) = f(\omega(s_1), \dots, \omega(s_n))$ in $\mathcal{A}$. In any case, for each $\omega \in E$, and for each "sufficiently smooth" vector field X on M along ω, the *directional derivative* of F in direction X at point ω should exist, and be given by

$$(\partial_X F)(\omega) \;=\; \sum_{i=1}^{n} (d_i^M f)(\omega(s_1), \dots, \omega(s_n)) \; (X(s_i)),$$

where d_i^M denotes the ordinary differential on M w.r.t. the i-th component. Therefore, we define informally a "differential" D by

$$D_\omega F \;=\; \sum_{i=1}^{n} (d_i^M f)(\omega(s_1), \dots, \omega(s_n)) \; \delta_{s_i}$$

for F as above and $\omega \in E$, where δ_{s_i} is the Dirac distribution. We can view D_ω as a linear operator taking values in the *algebraic* direct sum of the spaces $T^*_{\omega(s)}M$, $0 < s < 1$. For the Carré du champ defined above, we obtain the representation

$$\Gamma_\omega(F, \tilde{F}) = \left(DF, D\tilde{F} \right)_\omega \qquad \text{for all } \omega \in E \text{ and } F, \tilde{F} \in \mathcal{A},$$

where the inner product $(\cdot , \cdot)_\omega$ on the algebraic direct sum is defined uniquely by

$$(\alpha \delta_s, \beta \delta_t)_\omega = G(s,t) \cdot ((\tau_s(\omega))^* \alpha, (\tau_t(\omega))^* \beta)_{T^*_{x_0}M}$$

for all $s, t \in (0,1)$, $\alpha \in T^*_{\omega(s)}M$, and $\beta \in T^*_{\omega(t)}M$. Here $(\tau_s(\omega))^* \alpha$ denotes the pull–back of the one–form α w.r.t. the isometry $\tau_s(\omega) : T_{x_0}M \to T_{\omega(s)}M$. Hence the completion of the algebraic direct sum $\bigoplus_{0<s<1} T^*_{\omega(s)}M$ w.r.t. $(\cdot , \cdot)_\omega$ serves as a co–tangent space for the operator $(\mathcal{L}, \mathcal{A})$, and the unique continuous extension of D is the corresponding generalized differential. The co–tangent bundle obtained in this way is in fact the dual bundle of the *tangent bundle* to E, which has been introduced e.g. in [CrMal 94].

5) Examples III :
Horizontal and vertical measure–valued diffusions

Let M be a Riemannian manifold. Suppose E is a space of measures over M. For example, E is the space $\mathcal{M}(M)$ of all (positive) finite measures on M, the space $\mathcal{M}_1(M)$ of all probability measures on M, or the **configuration space** Γ_M of all subsets γ of M such that $\gamma \cap B$ is finite for every bounded subset B of M. Γ_M can be viewed as a subset of the space $\mathcal{M}_p(M)$ consisting of all Radon measures on M which are countable sums of Dirac measures. A special case of a configuration space is the **Poisson space** consisting of all maps $\omega : (0,1) \to \mathbf{N}$ such that $\omega(x) = 0$ for small x, ω is increasing and right continuous, and has finitely many jumps of size one. This space can be identified with $\Gamma_{(0,1)}$.

Let $\mathcal{B}$ be the σ–algebra generated by the functionals $\mu \mapsto \int g\, d\mu$, $g \in C_0(M)$, and let

$$\mathcal{A} := \{ F : E \to \mathbf{R}; \ F(\mu) = f(\int g_1\, d\mu, \ldots, \int g_n\, d\mu)$$
$$\text{for some } n \in \mathbf{N}, \ f \in C_b^\infty(\mathbf{R}^n), \text{ and } g_1, \ldots, g_n \in C_0^\infty(M) \}.$$

There are in particular two types of carré du champ operators which correspond to generators of frequently considered measure–valued diffusions. The carré du champ Γ^v for **super–Brownian motion** and more general superprocesses on $\mathcal{M}(M)$ is given by

$$\Gamma_\mu^v(F, \tilde{F})$$
$$= \sum_{i,j=1}^n \int g_i\, g_j\, d\mu\ \frac{\partial f}{\partial x_i} (\int g_1\, d\mu, \ldots, \int g_n\, d\mu)\ \frac{\partial \tilde{f}}{\partial x_j} (\int g_1\, d\mu, \ldots, \int g_n\, d\mu)$$

whenever $F(\mu) = f(\int g_1 \, d\mu, \ldots, \int g_n \, d\mu)$ and $\tilde{F}(\mu) = \tilde{f}(\int g_1 \, d\mu, \ldots, \int g_n \, d\mu)$ for some $n \in \mathbf{N}$, $g_1, \ldots, g_n \in C_0^\infty(M)$, $f, \tilde{f} \in C_b^\infty(\mathbf{R}^n)$, and $f, \tilde{f} \in C_b^\infty(\mathbf{R}^n)$, cf. [Schi 97]. It can be written as

$$\Gamma_\mu^v (F, \tilde{F}) = \left(D_\mu^v F, D_\mu^v \tilde{F} \right)_{L^2(M;\mu)},$$

where the gradient $D^v F$ of a cylinder function F as above is given by

$$D_\mu^v F = \sum_{i=1}^n \frac{\partial f}{\partial x_i} (\int g_1 \, d\mu, \ldots, \int g_n \, d\mu) \; g_i \in L^2(M; \mu).$$

By identifying $L^2(M; \mu)$ with its dual, we can view D^v as a differential. In fact, it is a generalized differential corresponding to Γ^v w.r.t. the measurable co-tangent bundle $T'(\mathcal{M}(M))$ given by

$$T_\mu'(\mathcal{M}(M)) = \left(L^2(M; \mu) \right)' \cong L^2(M; \mu)$$

with measurable structure generated by the constant sections $\mu \mapsto f$, $f \in C_b(M)$ (i.e., a section ω is measurable if and only if $\mu \mapsto \int \omega(\mu) f \, d\mu$ is a measurable function for all $f \in C_b(M)$).

Similarly, the carré du champ $\tilde{\Gamma}^v$ for the **Fleming–Viot processes** on $\mathcal{M}_1(M)$ studied in [OvRöSchm 95], [OvRö 96], and [Schi 97] is a slight modification of Γ^v. For F, $\tilde{F} \in \mathcal{A}$ as above, it is given by

$$\tilde{\Gamma}_\mu^v (F, \tilde{F}) = \left(\tilde{D}_\mu^v F, \tilde{D}_\mu^v \tilde{F} \right)_{L^2(M;\mu)},$$

where

$$\tilde{D}_\mu^v F = \sum_{i=1}^n \frac{\partial f}{\partial x_i} (\int g_1 \, d\mu, \ldots, \int g_n \, d\mu) \; (g_i - \int g_i \, d\mu)$$

is the projection of $D_\mu^v F$ in $L^2(M; \mu)$ onto the functions with mean 0. Again, $\tilde{D}^v$ is a generalized differential for $\tilde{\Gamma}^v$ w.r.t. the measurable co-tangent bundle given by

$$T_\mu'(\mathcal{M}_1(M)) = \left\{ f \in L^2(M; \mu); \int f \, d\mu = 0 \right\}.$$

The dynamics corresponding to $\tilde{\Gamma}^v$ describes the motion in submanifolds of $\mathcal{M}(M)$ consisting of measures of constant mass.

The Brownian motion and more general, interacting particle systems on configuration spaces $E = \Gamma_M$ considered in [AlbKoRö 97 a and b] have a completely different carré du champ Γ^h. It is given by

$$\Gamma_\mu^h (F, \tilde{F}) = \sum_{i,j=1}^n \int \left(d^M g_i, d^M g_j \right)_{T_x M} \mu(dx) \; \frac{\partial f}{\partial x_i} (\int g_1 \, d\mu, \ldots, \int g_n \, d\mu)$$
$$\times \frac{\partial \tilde{f}}{\partial x_j} (\int g_1 \, d\mu, \ldots, \int g_n \, d\mu)$$

for F, $\tilde{F} \in \mathcal{A}$ as above, i.e.,

$$\Gamma_\mu^h (F, \tilde{F}) = \left(D_\mu^h F, D_\mu^h \tilde{F} \right)_{L^2(M \to T^*M; \mu)},$$

where $D^h F$ is defined by

$$D_\mu^h F = \sum_{i=1}^n \frac{\partial f}{\partial x_i} (\int g_1 \, d\mu, \ldots, \int g_n \, d\mu) \, d^M g_i \in L^2(M \to T^*M; \mu).$$

Here, d^M denotes the ordinary differential on M.

Hence D^h is a generalized differential corresponding to Γ^h w.r.t. the measurable co–tangent bundle $T'T_M$, where

$$T_\mu' \Gamma_M = L^2(M \to T^*M; \mu) \qquad \text{for every } \mu \in \Gamma_M.$$

Note that the range of D_μ^h is in fact the full space $L^2(M \to T^*M; \mu)$. The reason is that we have $L^2(M \to T^*M; \mu) \cong \bigoplus_{i \in I} T_{x_i}^* M$, and for every $(\omega_i)_{i \in I} \in \bigoplus_{i \in I} T_{x_i}^* M$, there exists a function $g \in C_0^\infty(M)$ such that $d_{x_i}^M g = \omega_i$ for all $i \in I$. If we would consider measures μ that are not in Γ_M, then $T_\mu' \Gamma_M$ had to be defined as the completion of $\{d^M g; \, g \in C_0^\infty(M)\}$ in $L^2(M \to T^*M; \mu)$, i.e., we would obtain $T_\mu' \Gamma_M \cong L^2(M \to T^*M; \mu) / \ker (d^M)^*$.

REMARKS. (i) S. Albeverio, Y. Kondratiev and M. Röckner introduced the space $L^2(M \to TM; \mu)$ as a tangent space to Γ_M at μ, cf. [AlbKoRö 97 and 97 a,b]. This space is in fact the dual space of the co–tangent space $T_\mu' \Gamma_M$ defined here.

(ii) It is also possible to take the differential D^v as a generalized differential corresponding to Γ^h. In this representation, the metric on the co–tangent space is given by

$$(g, h)_{T_\mu' E} = \left(d^M g, d^M h \right)_{L^2(M \to T^*M; \mu)} \qquad \text{for all } g, h \in C_0^\infty(M),$$

and the co–tangent space is the completion of $C_0^\infty(M)$ w.r.t. this metric, divided by $\{g \in C_0^\infty(M); \, d^M g = 0 \ \mu\text{-a.e.}\}$. Note that there is some similarity with the Cameron Martin type co–tangent spaces to path and loop spaces.

(iii) E. A. Carlen and E. Pardoux [CarPar 90] introduced a gradient operator on the **Poisson space** consisting of all maps $\omega : (0, 1) \to \mathbf{N}$ such that $\omega(x) = 0$ for small x, and ω is increasing and right continuous, and has finitely many jumps of size one. In particular, they defined a directional derivative in direction m, where m is a function in $L^2(0, 1; dt)$ with $\int_0^1 m(t) \, dt = 0$. Obviously, the Poisson space can be identified with $\Gamma_{(0,1)}$. It turns out that for $f \in \mathcal{A}$, the directional derivative defined by Carlen and Pardoux is precisely the directional derivative $(D_\mu^h F)(M)$ defined above, where $M \in L^2(0, 1; \mu) = L^2((0, 1) \to T(0, 1); \mu)$ is given by $M(s) = \int_0^s m(t) \, dt$. In this sense, Carlen and Pardoux consider essentially the geometry corresponding to D^h, but they use an alternative representation of the tangent space.

We finally comment on the meaning of the different geometries corresponding to Γ^h and Γ^v (resp. $\bar{\Gamma}^v$). Let $n \in \mathbf{N}$. For the moment, we consider the finite dimensional submanifold $\mathcal{M}^{(n)}(M)$ of $\mathcal{M}(M)$ consisting of all discrete measures μ of type $\mu = \sum_{i=1}^{n} c_i \, \delta_{x_i}$, $c_i > 0$, $x_i \in M$, $1 \leq i \leq n$, $x_i \neq x_j$ if $i \neq j$. Let M_Δ^n denote the open subset of all $(x_1, \ldots, x_n)$ in the product manifold M^n such that $x_i \neq x_j$ if $i \neq j$. Obviously,

$$\mathcal{M}^{(n)}(M) \; \cong \; M_\Delta^n \times (0, \infty)^n \; ,$$

whence the tangent bundle $T(\mathcal{M}^{(n)}(M))$ can be decomposed into two orthogonal components

$$
\begin{aligned}
T_\mu\left(\mathcal{M}^{(n)}(M)\right) &\cong \; T_{(x_1, \ldots, x_n)} M_\Delta^n \oplus T_{(c_1, \ldots, c_n)} (0, \infty)^n \\
&\cong \; L^2(M \to TM; \mu) \oplus L^2(M; \mu)
\end{aligned}
$$

if $\mu = \sum_{i=1}^{n} c_i \, \delta_{x_i}$. The tangent vectors in the first component are derivatives of curves in the submanifold obtained by variing the positions of the particles but not their masses. We call these tangent vectors **horizontal**. The tangent vectors in the second component are derivatives of curves in the submanifold obtained by variing the masses of the particles but not their positions. They are called **vertical**. Similarly, we obtain the decomposition

$$T_\mu^*\left(\mathcal{M}^{(n)}(M)\right) \; \cong \; L^2(M \to T^*M; \mu) \oplus L^2(M; \mu),$$

where we have identified $L^2(M; \mu)$ and its dual.

The differential D_μ^h has values in the first (horizontal) component of the cotangent space, whereas the differentials D_μ^v and $\bar{D}_\mu^v$ have values in the second (vertical) component. Hence a diffusion process on $\mathcal{M}^{(n)}(M)$ with carré du champ Γ^h consists of particles that move randomly on the manifold driven by independent Brownian motions, and, possibly, interact. In contrast to this, a diffusion process with carré du champ Γ^v or $\bar{\Gamma}^v$ consists of particles that have randomly fluctuating masses, where the fluctuation is driven by independent Brownian motions. The motion of these particles on the manifold is only due to deterministic interactions.

Now consider again the whole space $\mathcal{M}(M)$. Let μ be a measure in $\mathcal{M}(M)$ with a sufficiently smooth strictly positive density ρ w.r.t. the volume dx. Then the tangent space at μ can not be decomposed into a *disjoint* sum of horizontal and vertical tangent vectors. In fact, the directional derivative of a function $F \in \mathcal{A}$ in the direction of a "horizontal" tangent vector $X \in C_0^\infty(M \to TM) \subset L^2(M \to TM; \mu)$ is given by

$$
\begin{aligned}
\langle D_\mu^h F, X \rangle &= \sum_{i=1}^{n} \frac{\partial f}{\partial x_i} \left(\int g_1 \, d\mu, \ldots, \int g_n \, d\mu \right) \int X g_i \, \rho \, dx \\
&= -\sum_{i=1}^{n} \frac{\partial f}{\partial x_i} \left(\int g_1 \, d\mu, \ldots, \int g_n \, d\mu \right) \int g_i \, \frac{1}{\rho} \, \text{div} \, (\rho X) \, \rho \, dx
\end{aligned}
$$

if $F(\mu) = f(\int g_1 \, d\mu, \ldots, \int g_n \, d\mu)$ for some $n \in \mathbf{N}$, $f \in C_b^\infty(\mathbf{R}^n)$, and $g_1, \ldots, g_n \in C_0^\infty(M)$. Hence for all $F \in \mathcal{A}$, the directional derivative is the same as that in the direction of the "vertical" tangent vector $\check{X} := \operatorname{div}(\rho X)/\rho \in L^2(M; \mu)$. Thus every horizontal tangent vector at μ in $C_0^\infty(M \to TM)$ is also a vertical tangent vector, i.e., the "horizontal pre–tangent space" $C_0^\infty(M \to TM)$ at μ can be identified with the subspace $\{(1/\rho) \operatorname{div}(\rho X); X \in C_0^\infty(M \to TM)\}$ of the "vertical tangent space" $L^2(M; \mu)$ at μ. On the other hand, if M is compact and connected, then the only functions in $L^2(M; \mu)$ that are orthogonal to the subspace above, are the contant functions, whence every vertical tangent vector $f \in C_0^\infty(M)$ with $\int f \, d\mu = 0$ is also a horizontal tangent vector. Thus the "vertical pre–tangent space" $\{f \in C_0^\infty(M); \int f \, d\mu = 0\}$ w.r.t. $\bar{\Gamma}^v$ can be identified with a subspace of the horizontal tangent space $L^2(M \to TM; \mu)$. Hence for absolutely continuous measures μ, there is no essential difference between horizontal and vertical tangent vectors at μ, except that the full vertical tangent space to $\mathcal{M}(M)$ contains in addition the constant functions, which correspond to a variation of the total mass.

In spite of these considerations, it is useful to keep the concept of horizontal and vertical tangent vectors in mind, even when the measure μ, where the tangent space is considered, is absolutely continuous. The reason is that we use two entirely *different metrics* on the co–tangent bundle : The "horizontal metric" Γ^h and the "vertical metric" Γ^v resp. $\bar{\Gamma}^v$. Even for diffusion processes on absolutely continuous measures, it is useful to view a process corresponding to Γ^h as moving randomly in horizontal directions without creation of new mass, whereas a process corresponding to Γ^v or $\bar{\Gamma}^v$ can be viewed as a branching process with random creation and annihilation of mass. The total mass can nevertheless stay constant, provided the dynamics is determined by $\bar{\Gamma}^v$.

6) Divergence operators and symmetrizing measures

In this subsection, we look at a rather obvious but important characterization of symmetrizing measures for diffusion operators of type $-\delta d$, where d is a generalized differential, and δ is a generalized divergence operator. In particular, we show that the symmetrizing maesures for the operator $-\delta d$ are the solutions of the equation $\delta^* m = 0$, and that a measure m is symmetrizing if and only if δ and d are in duality w.r.t. m.

We consider the situation described at the beginning of Subsection 2), i.e., we assume that we are given a diffusion operator $(\mathcal{L}, \mathcal{A})$ acting on $\mathcal{F}(E)$ such that (C 1) holds. Let $T'E$ and $d : \mathcal{A} \to \mathcal{F}(E \to T'E)$ be a measurable co–tangent bundle and a differential associated with the operator $(\mathcal{L}, \mathcal{A})$ in the sense of Theorem 3.9. Let

$$\Omega(\mathcal{A}) := \operatorname{span}\{f \, dg; \ f, g \in \mathcal{A}\} \subseteq \mathcal{F}(E \to T'E).$$

Thinking of the elements in $\mathcal{A}$ as test functions, we may view the elements in $\Omega(\mathcal{A})$ as "test 1–forms". If, for example, E is a domain in $\mathbf{R}^n$, $\mathcal{A} = C_0^\infty(E)$, and d is the ordinary differential, then $\Omega(\mathcal{A}) = C_0^\infty(E \to (\mathbf{R}^n)^*)$.

We now assume in addition that we are given a **generalized divergence operator**

$$\delta \ : \ \Omega(\mathcal{A}) \subseteq \mathcal{F}(E \to T'E) \ \to \ \mathcal{F}(E)$$

corresponding to the operator $(\mathcal{L}, \mathcal{A})$, i.e.,

(i) δ is linear.

(ii) $\delta(f\omega) = f\,\delta\omega - (df, \omega)$ for all $f \in \mathcal{A}$ and $\omega \in \Omega(\mathcal{A})$.

(iii) $\mathcal{L}f = -\delta df$ for all $f \in \mathcal{A}$.

REMARKS. (i) The existence of a corresponding divergence operator is closely related to the symmetrizability of the operator $(\mathcal{L}, \mathcal{A})$. For non–symmetrizable operators, a representation of type $\mathcal{L} = -\delta d$ with an operator δ satisfying the product rule (ii) can not be expected.

(ii) A corresponding divergence operator is uniquely determined by $(\mathcal{L}, \mathcal{A})$, because

$$(3.39) \qquad \delta(f\,dg) = -f\,\mathcal{L}g - \Gamma(f,g) \qquad \text{for all } f, g \in \mathcal{A}$$

by (i) and (ii).

For simplicity, we also assume that the constant function 1 is in $\mathcal{A}$, and we restrict ourselves to finite symmetrizing measures. Let m be a finite measure on E. Recall that the **direct integral** of a measurable field $H = (H_z)_{z \in E}$ of Hilbert spaces w.r.t. m is the linear space consisting of all m–classes of m–square integrable sections of H. We denote the direct integral by $L^2(E \to H; m)$. Since $L^2(E; m)$ is separable, $L^2(E \to H; m)$ is also a separable Hilbert space with inner product

$$(\omega, \sigma)_{L^2(E \to H; m)} \ = \ \int (\omega(z), \sigma(z))_{H_z} \ m(dz),$$

cf. [Dix 69].

Now suppose that $\mathcal{L}f$ is in $L^1(E; m)$ for all $f \in \mathcal{A}$. Then $\Omega(\mathcal{A})$ is contained in $L^2(E \to T'E; m)$, because $\mathcal{A}$ consists of bounded functions, and

$$(df, df) \ = \ \Gamma(f, f) \ = \ \frac{1}{2}\mathcal{L}(f^2) - f\,\mathcal{L}f \ \in L^1(E; m)$$

for every $f \in \mathcal{A}$. Moreover, by (3.39), $\delta\omega$ is in $L^1(E; m)$ for all $\omega \in \Omega(\mathcal{A})$.

We also introduce formally a tangent bundle $TE = (T_z E)_{z \in E}$ as the dual bundle of $T'E = (T'_z E)_{z \in E}$, i.e., $T_z E$ is the dual space of $T'_z E$ for every $z \in E$. Clearly, TE is again a measurable field of Hilbert spaces with measurable structure induced by the isometry $j = (j_z)_{z \in E}$, where $j_z : T'_z E \to T_z E$ is the Riesz isometry for each z. We identify a vector field X, i.e., a section of TE, with the derivation operator $(X, \mathcal{A})$ on $\mathcal{F}(E)$ defined by

$$(Xf)(z) := \langle df(z), X(z) \rangle$$

for all $z \in E$ and $f \in \mathcal{A}$, where $\langle \cdot, \cdot \rangle$ is the dualisation between $T'_z E$ and $T_z E$. For a 1-form $\omega \in \mathcal{F}(E \to T'E)$, the corresponding vector field $z \mapsto j_z(\omega(z))$ in $\mathcal{F}(E \to TE)$ is denoted by $\omega^{\sharp}$. For $f \in \mathcal{A}$, let $\nabla f := (df)^{\sharp}$. Let

$$\mathcal{V}(\mathcal{A}) := \{ f \nabla g; \ f, g \in \mathcal{A} \} = \{ \omega^{\sharp}; \ \omega \in \Omega(\mathcal{A}) \}.$$

The divergence operator $\operatorname{div} : \mathcal{V}(\mathcal{A}) \to \mathcal{F}(E)$ on vector fields is defined by $\operatorname{div} \omega^{\sharp} = \delta \omega$ for all $\omega \in \Omega(\mathcal{A})$.

Theorem 3.10 *Let m be a finite measure on E such that $\mathcal{L}f$ is in $L^1(E; m)$ for all $f \in \mathcal{A}$, and the operator $(\mathcal{L}, \mathcal{A})$ respects m-classes. Then the following assertions are equivalent :*

(i) *m is a **symmetrizing measure** for $(\mathcal{L}, \mathcal{A})$, i.e.,*

$$\int f \mathcal{L}g \ dm = \int \mathcal{L}f \ g \ dm \qquad \text{for all } f, g \in \mathcal{A}.$$

(ii) $-\int \mathcal{L}f \ g \ dm = \int \Gamma(f, g) \ dm \qquad \text{for all } f, g \in \mathcal{A}.$

(iii) $\int \delta \omega \ dm = 0 \qquad \text{for all } \omega \in \Omega(\mathcal{A}).$

(iv) *The operators $(d, \mathcal{A})$ and $(\delta, \Omega(\mathcal{A}))$ are **in duality** w.r.t. the measure m, i.e.,*

$$\int (df, \omega) \ dm = \int f \ \delta \omega \ dm \qquad \text{for all } f \in \mathcal{A} \text{ and } \omega \in \Omega(\mathcal{A}).$$

(v) *The integration by parts formula*

$$\int Xf \ g \ dm = -\int f \ Xg \ dm + \int f \ g \ \operatorname{div} X \ dm$$

holds for all $f, g \in \mathcal{A}$ and all $X = \nabla h, \ h \in \mathcal{A}$.

REMARKS. (i) In compact notation, we can write Condition (iii) as $\delta^* m = 0$, where δ^* denotes the dual of the operator $(\delta, \Omega(\mathcal{A}))$ w.r.t. the dualisation between bounded functions and probability measures. Note the connection to the equation $(\delta d)^* m = 0$, which characterizes invariant measures of $(\mathcal{L}, \mathcal{A})$.

(ii) Condition (iv) means that the operator $(d, \mathcal{A})$ respects m-classes, and the adjoint d^*_m of the induced operator d_m from $L^2(E; m)$ to $L^2(E \to T'E; m)$ extends the divergence operator $(\delta, \Omega(\mathcal{A}))$.

PROOF OF THEOREM 3.10. By the diffusion property, $\mathcal{L}1 = 0$. Now suppose m is symmetrizing for $(\mathcal{L}, \mathcal{A})$. Then

$$\int \mathcal{L}f \ dm = \int f \ \mathcal{L}1 \ dm = 0 \qquad \text{for all } f \in \mathcal{A},$$

whence

$$\int \Gamma(f, g)\, dm \;=\; \Big(\int \mathcal{L}(fg)\, dm - \int f\,\mathcal{L}g\, dm - \int \mathcal{L}f\,g\, dm\Big)/2 \;=\; -\int f\,\mathcal{L}g\, dm$$

for all $f, g \in \mathcal{A}$, i.e., (ii) holds. Conversely, (ii) implies obviously the symmetry of $(\mathcal{L}, \mathcal{A})$. By (3.39), (iii) is just a rephrasing of (ii). Condition (iv) implies (iii) by choosing $f = 1$, and (iii) applied to $f\omega$ instead of ω yields the equation in (iv) by the product rule for δ. Finally, the integration by parts formula in (v) is precisely the equation (iv) with $\omega = g\, dh$, whence (iv) and (v) are equivalent. ∎

7) A representation theorem for diffusion operators on L^2 spaces

Let E be again an arbitrary set, and $\mathcal{B}$ a σ–algebra on E. So far, we have considered diffusion operators acting on measurable functions. We now fix a σ–finite measure m on $(E, \mathcal{B})$, and consider operators on $L^2(E; m)$. We assume that $L^2(E; m)$ is separable. Let $\mathcal{A}$ be an algebra of bounded m–square integrable functions, which is dense in $L^2(E; m)$. Suppose $\mathcal{L}$ and $\hat{\mathcal{L}}$ are two linear operators on $L^2(E; m)$ with domain $\mathcal{A}$, which are in duality w.r.t. each other, i.e.,

$$\int \mathcal{L}f\, g\, dm \;=\; \int f\,\hat{\mathcal{L}}g\, dm \qquad \text{for all } f, g \in \mathcal{A}.$$

We assume that $(\mathcal{L}, \mathcal{A})$ is a **diffusion operator**, and m is an **invariant measure** for $(\mathcal{L}, \mathcal{A})$, i.e., the range of $(\mathcal{L}, \mathcal{A})$ is in $L^1(E; m)$, and $\int \mathcal{L}f\, dm = 0$ for all $f \in \mathcal{A}$.

REMARK.   Let $\mathcal{L}$ be an arbitrary diffusion operator on $L^2(E; m)$ with domain $\mathcal{A}$. Suppose m is an invariant measure for $(\mathcal{L}, \mathcal{A})$. Then *there exists an operator* $(\hat{\mathcal{L}}, \mathcal{A})$ *such that* $\mathcal{L}$ *and* $\hat{\mathcal{L}}$ *are in duality w.r.t. each other if and only if the symmetric bilinear form* $(\mathcal{E}, \mathcal{A})$ *defined by*

$$(3.40) \qquad \mathcal{E}(f, g) \;=\; -\frac{1}{2}\left(\int \mathcal{L}f\, g\, dm + \int f\,\mathcal{L}g\, dm\right)$$

is **closable** *on* $L^2(E; m)$, i.e., $\lim_{n\to\infty} \mathcal{E}(f_n, f_n) = 0$ for every sequence $(f_n)_{n\in\mathbb{N}}$ in $\mathcal{A}$ such that $f_n \to 0$ in $L^2(E; m)$ and $\mathcal{E}(f_n - f_k, f_n - f_k) \to 0$ as $n, k \to \infty$. In fact, since $(\mathcal{L}, \mathcal{A})$ is a diffusion operator with invariant measure m, we have

$$\mathcal{E}(f, f) \;=\; \int \Gamma(f, f)\, dm \;\geq\; 0 \qquad \text{for all } f \in \mathcal{A},$$

where Γ denotes the Carré du champ for $\mathcal{L}$. If $\hat{\mathcal{L}}$ exists, then $(\mathcal{E}, \mathcal{A})$ is the bilinear form on $L^2(E; m)$ corresponding to the symmetric operator $(\mathcal{L}+\hat{\mathcal{L}})/2$, i.e.,

$$(3.41) \qquad \mathcal{E}(f, g) \;=\; -\int \frac{1}{2}(\mathcal{L}+\hat{\mathcal{L}})f\,g\, dm \qquad \text{for all } f, g \in \mathcal{A}.$$

Therefore, the form is closable, cf. e.g. [MaRö 92, Ch. I, Prop. 3.3]. Conversely, if $(\mathcal{E}, \mathcal{A})$ is closable on $L^2(E; m)$, and L^0 is the generator of the closure, then

$$\int \mathcal{L}f\, g\, dm = -2\,\mathcal{E}\,(f, g) - \int f\, \mathcal{L}g\, dm$$

$$= \int f\, (2L^0 - \mathcal{L})\, g\, dm$$

for all $f, g \in \mathcal{A}$, i.e., $\mathcal{L}$ is in duality with $\hat{\mathcal{L}} := 2L^0 - \mathcal{L}$ on $\mathcal{A}$.

Recall the definition of the direct integral $L^2(E \rightarrow T'E; m)$ from Subsection 6).

Definition 3.7 *a) A map $X : \mathcal{A} \rightarrow L^2(E; m)$ is called an L^2 derivation if and only if*

(i) X is linear.

(ii) $X(f \cdot g) = f \cdot Xg + g \cdot Xf$ m–a.e. for all $f, g \in \mathcal{A}$.

*b) Let $T'E = (T'_z E)_{z \in E}$ be a measurable field of Hilbert spaces over E. A map $d : \mathcal{A} \rightarrow L^2(E \rightarrow T'E; m)$ is called an L^2 **differential** (w.r.t. the co-tangent bundle $T'E$) if and only if*

(i) The span of $\{f\, dg;\ f, g \in \mathcal{A}\}$ is dense in $L^2(E \rightarrow T'E; m)$.

(ii) d is linear.

(iii) $d(f \cdot g) = f \cdot dg + g \cdot df$ for all $f, g \in \mathcal{A}$.

Suppose d is an L^2 differential w.r.t. the co–tangent bundle $T'E$, and let $f, g \in \mathcal{A}$. Then the m–class represented by the function $z \mapsto (d_z f, d_z g)_{T'_z E}$, where $d.f$ and $d.g$ are arbitrary versions of df and dg, is denoted by (df, dg).

Theorem 3.11
(i) Under the assumptions above, there exist a measurable field of Hilbert spaces $T'E = (T'_z E)_{z \in E}$, an L^2 differential $d : \mathcal{A} \rightarrow L^2(E \rightarrow T'E; m)$, and an anti-symmetric L^2 derivation $X : \mathcal{A} \rightarrow L^2(E; m)$ such that

$$(3.42) \qquad \mathcal{L}f = -d^*df + Xf \qquad \text{for all } f \in \mathcal{A}, \qquad i.e.,$$

$$\int \mathcal{L}f\, g\, dm = -\int (df, dg)\, dm + \int Xf\, g\, dm \qquad \text{for all } f, g \in \mathcal{A}.$$

(ii) The differential d and the derivation X in (i) are uniquely determined in the following sense : Suppose there exist another measurable field of Hilbert spaces $\tilde{T}'E = \left(\tilde{T}'_z E\right)_{z \in E}$, another differential $\tilde{d} : \mathcal{A} \rightarrow L^2(E \rightarrow \tilde{T}'E; m)$, and another derivation $\tilde{X} : \mathcal{A} \rightarrow L^2(E; m)$ such that (3.42) holds. Then $\tilde{X}f = Xf$ m–a.e. for all $f \in \mathcal{A}$. Moreover, there exists an m–measure zero set $N \subseteq E$ and an isometry $(\Phi_z)_{z \in E \setminus N}$ between the measurable fields $(T'_z E)_{z \in E \setminus N}$ and $\left(\tilde{T}'_z E\right)_{z \in E \setminus N}$ such that $\tilde{d}f(z) = \Phi_z(df(z))$ for m–a.e. z.

REMARKS. (i) The derivation X corresponding to $\mathcal{L}$ in the sense of Theorem 3.11 is given by

$$X f \;=\; \frac{1}{2}(\mathcal{L}f - \hat{\mathcal{L}}f) \qquad \text{for all } f \in \mathcal{A}.$$

In particular, X vanishes if and only if $\mathcal{L}$ is symmetric.

(ii) Every differential d corresponding to $\mathcal{L}$ in the sense of Theorem 3.11 satisfies

$$(3.43) \quad (\, df, dg\,) \;=\; \Gamma(f, g) \;=\; \frac{1}{2}(\mathcal{L}(fg) - f\mathcal{L}g - g\mathcal{L}f) \qquad m\text{–a.e.}$$

for all $f, g \in \mathcal{A}$, cf. the proof of the uniqueness part of Theorem 3.11 below. In particular, (df, dg) is in $L^2(E\,;\,m)$, because $\mathcal{A}$ consists of bounded functions.

(iii) Conversely, if d is an L^2 differential satisfying (3.43), and X is defined as in Remark (i), then (3.42) holds.

PROOF OF THEOREM 3.11. (i) Let $(\mathcal{E}, \mathcal{A})$ be the closable symmetric bilinear form on $L^2(E\,;\,m)$ defined by (3.40) respectively (3.41), and let $(\mathcal{E}, \bar{\mathcal{A}})$ denote the closure. Since m is an invariant measure for $(\mathcal{L}, \mathcal{A})$, we have

$$\mathcal{E}(f, g) \;=\; -\frac{1}{2}\int (\mathcal{L}f\, g + f\, \mathcal{L}g)\; dm \;=\; \int \Gamma(f, g)\; dm \quad \text{for all } f, g \in \mathcal{A}.$$

In particular, the Carré du champ operator Γ is a continuous bilinear map from $\mathcal{A} \times \mathcal{A}$ to $L^1(E\,;\,m)$, provided $\mathcal{A}$ is endowed with the inner product $\mathcal{E}_1(\,\cdot\,,\,\cdot\,) := \mathcal{E}(\,\cdot\,,\,\cdot\,) + (\,\cdot\,,\,\cdot\,)_{L^2(E\,;\,m)}$. It can be uniquely extended to a continuous bilinear map from $\bar{\mathcal{A}} \times \bar{\mathcal{A}}$ to $L^1(E\,;\,m)$, which we again denote by Γ. Clearly, $\Gamma(f, g) = \Gamma(g, f)$ and $\Gamma(f, f) \geq 0$ m–a.e. for all $f, g \in \bar{\mathcal{A}}$.

Let L^0 be the generator of the closed quadratic form $(\mathcal{E}, \bar{\mathcal{A}})$, i.e., L^0 is the Friedrichs extension of the operator $((\mathcal{L} + \hat{\mathcal{L}})/2, \mathcal{A})$. Since L^0 is a negative definite self-adjoint operator on $L^2(E\,;\,m)$, the resolvent $(1 - L^0)^{-1} : L^2(E\,;\,m) \to \bar{\mathcal{A}}$ exists, is continuous if $\bar{\mathcal{A}}$ is endowed with the $\mathcal{E}_1$–inner product, and has dense range. Thus $\bar{\mathcal{A}}$ is *separable* w.r.t. $\mathcal{E}_1$, because $L^2(E\,;\,m)$ is separable by assumption.

We fix linearly independent functions $f_n \in \bar{\mathcal{A}}$, $n \in \mathbf{N}$, such that the span of these functions is dense in $\bar{\mathcal{A}}$ w.r.t. $\mathcal{E}_1$. Let $\mathcal{A}_0 := \text{span}\{f_n;\, n \in \mathbf{N}\}$. We choose measurable m–versions $z \mapsto \Gamma_z(f_i, f_j)$ of $\Gamma(f_i, f_j)$, $i, j \in \mathbf{N}$, such that $\Gamma_z(f_i, f_i) \geq 0$ and $\Gamma_z(f_i, f_j) = \Gamma_z(f_j, f_i)$ for all $i, j \in \mathbf{N}$ and $z \in E$. For arbitrary $u, v \in \mathcal{A}_0$ such that $u = \sum_{i=1}^n \lambda_i f_i$ and $v = \sum_{i=1}^m \mu_i f_i$ with $n, m \in \mathbf{N}$ and $\lambda_i, \mu_i \in \mathbf{R}$, let

$$\Gamma_z(u, v) \;=\; \sum_{i=1}^n \sum_{j=1}^m \lambda_i \mu_j\, \Gamma_z(f_i, f_j)\,.$$

Clearly, $z \mapsto \Gamma_z(u, v)$ is a well–defined m–version of $\Gamma(u, v)$. Moreover, $\Gamma_z(u, v) = \Gamma_z(v, u)$ and $\Gamma_z(u, u) \geq 0$ hold for all $z \in E$ and $u, v \in \mathcal{A}_0$.

Now, a co–tangent bundle and a differential corresponding to Γ_z on $\mathcal{A}_0$ can be constructed similarly to the proof of Theorem 3.9 above : For $z \in E$, we define

$T'_z E$ as the completion of $\mathcal{A}_0/\ker \Gamma_z$ w.r.t. the inner product induced by Γ_z, and we denote the equivalence class in $T'_z E$ corresponding to a function $f \in \mathcal{A}_0$ by $d_z f$. Let $\mathcal{F}(E \to T'E)$ be the space of all sections ω of $T'E$ such that the maps $z \mapsto (\omega(z), d_z f)_{T'_z E}$ are measurable for all $f \in \mathcal{A}_0$. By [Dix 69, Ch. II, Par. 1, 4., Prop. 4], $T'E$ is a measurable field of Hilbert spaces with measurable structure $\mathcal{F}(E \to T'E)$.

For $f \in \mathcal{A}_0$, the map $df : z \mapsto d_z f$ is in $\mathcal{F}(E \to T'E)$. For $f, g \in \mathcal{A}_0$, the function $z \mapsto (d_z f, d_z g)_{T'_z E}$ is an m–version of $\Gamma(f, g)$, which is in $L^1(E; m)$. Hence

$$d \; : \; \mathcal{A}_0 \; \to \; L^2(E \to T'E; m)$$

is a linear operator that is continuous if $\mathcal{A}_0$ is endowed with the $\mathcal{E}_1$–norm. We denote the unique continuous extension to $\bar{\mathcal{A}}$ again by d. The following properties hold :

- By construction, d is linear and df is in $L^2(E \to T'E; m)$ for all $f \in \bar{\mathcal{A}}$.

- $(df, dg) = \Gamma(f, g)$ m–a.e.   for all $f, g \in \bar{\mathcal{A}}$.

- The span of $\{f \, dg; \; f, g \in \mathcal{A}\}$ is dense in $L^2(E \to T'E; m)$. In fact, since every section dg, $g \in \bar{\mathcal{A}}$, is the $L^2(E \to T'E; m)$ limit of sections of type dg_n, $n \in \mathbb{N}$, with $g_n \in \mathcal{A}$, it suffices to show that the span of $\{f \, dg; \; f \in \mathcal{A}, \; g \in \bar{\mathcal{A}}\}$ is dense in $L^2(E \to T'E; m)$. Suppose ω is a section in $L^2(E \to T'E; m)$ that is orthogonal to the span. Then

$$\int f \, (\omega, dg) \, dm \; = \; 0 \qquad \text{for all } f \in \mathcal{A} \text{ and } g \in \mathcal{A}_0.$$

Since every function in $L^\infty(E; m) \cap L^2(E; m)$ can be approximated m–a.e. by uniformly bounded functions in $\mathcal{A}$, and since $\mathcal{A}_0$ is countably generated,

$$(\omega(z), d_z g)_{T'_z E} = 0 \qquad \text{for all } g \in \mathcal{A}_0$$

holds for m–a.e. z. By construction of $T'E$, $\{d_z g; \; g \in \mathcal{A}_0\}$ is dense in $T'_z E$ for every z, whence $\omega = 0$ m–a.e. Thus the span of $\{f \, dg; \; f \in \mathcal{A}, \; g \in \bar{\mathcal{A}}\}$ is dense in $L^2(E \to T'E; m)$.

- The product rule $d(f \cdot g) = f \cdot dg + g \cdot df$ holds for all $f, g \in \bar{\mathcal{A}}$. In fact, for $\tilde{f}, \tilde{g} \in \mathcal{A}$ we have

$$\left(d(fg), \tilde{f} \, d\tilde{g} \right) \; = \; \tilde{f} \cdot \Gamma(fg, \tilde{g})$$
$$= \; \tilde{f} f \, \Gamma(g, \tilde{g}) + \tilde{f} g \, \Gamma(f, \tilde{g}) \; = \; \left(f \, dg + g \, df, \tilde{f} \, d\tilde{g} \right)$$

m–a.e. This implies the product rule for d, because the span of $\{\tilde{f} \, d\tilde{g}; \; \tilde{f}, \tilde{g} \in \mathcal{A}\}$ is dense in $L^2(E \to T'E; m)$.

In particular, we obtain

$$-\int \frac{1}{2}\left(\mathcal{L}+\hat{\mathcal{L}}\right) f\, g\, dm \;=\; \mathcal{E}\,(f,g) \;=\; \int (df, dg)\, dm \quad \text{for all } f,\, g \in \mathcal{A},$$

i.e.,

$$\frac{1}{2}\left(\mathcal{L}+\hat{\mathcal{L}}\right) f \;=\; -d^* df \qquad \text{for all } f \in \mathcal{A}.$$

Now let X be the operator $(\mathcal{L}-\hat{\mathcal{L}})/2$ with domain $\mathcal{A}$. Then X is anti–symmetric on $L^2(E;m)$, and

$$\mathcal{L}f \;=\; -d^* df + X f \qquad \text{for all } f \in \mathcal{A}.$$

For $f,\, g \in \mathcal{A}$ we have

$$\begin{aligned}
\mathcal{L}(fg) &= f\,\mathcal{L}g + g\,\mathcal{L}f + 2\Gamma\,(f,g), \qquad \text{and} \\
d^* d\,(fg) &= d^*\,(f\,dg) + d^*\,(g\,df) \\
&= f\, d^* dg + g\, d^* df - 2\,(df, dg)\,.
\end{aligned}$$

The product rule for the divergence operator d^* used in the last equation can be easily derived from the product rule for d. Adding both equations, we obtain

$$X\,(fg) \;=\; f\,Xg + g\,Xf \qquad \text{for every } f,\, g \in \mathcal{A},$$

i.e., X is a derivation. This completes the proof of the existence part of Theorem 3.11.

(ii) Let $\tilde{T}'E$, $\tilde{d}$ and $\tilde{X}$ be as in the assertion. For sections ω, σ of $\tilde{T}'E$, the function $z \mapsto (\omega(z), \sigma(z))_{\tilde{T}'_z E}$ is again denoted by (ω, σ). Since $\tilde{X}$ is a derivation, (3.42) implies

$$\begin{aligned}
\int f\,\Gamma(g,h)\, dm &= \frac{1}{2}\int f\,(\mathcal{L}(gh) - g\,\mathcal{L}h - h\,\mathcal{L}g)\, dm \\
&= -\frac{1}{2}\int \left\{(\tilde{d}f, \tilde{d}(gh)) - (\tilde{d}(fg), \tilde{d}h) - (\tilde{d}(fh), \tilde{d}g)\right\}\, dm \\
&= \int f\,(\tilde{d}g, \tilde{d}h)\, dm
\end{aligned}$$

for all $f,\, g,\, h \in \mathcal{A}$, whence

(3.44) $\Gamma\,(g,h) \;=\; (\tilde{d}g, \tilde{d}h)\qquad m\text{–a.e.}$

In particular, $(\tilde{d}g, \tilde{d}h) = (dg, dh)$ m–a.e. for all $g,\, h \in \mathcal{A}$, whence

$$\int \tilde{X}g\, h\, dm \;=\; \int \mathcal{L}g\, h\, dm + \int (\tilde{d}g, \tilde{d}h)\, dm \;=\; \int Xg\, h\, dm.$$

Hence $\tilde{X}g = Xg$ m–a.e. for all $g \in \mathcal{A}$.

Moreover, by (3.44), $\tilde{d} : \mathcal{A} \to L^2(E \to \tilde{T}'E; m)$ is continuous if $\mathcal{A}$ is endowed with the $\mathcal{E}_1$–norm. We denote the unique continuous extension to $\bar{\mathcal{A}}$ again by $\bar{d}$. Clearly, (3.44) holds for $g, h \in \bar{\mathcal{A}}$ as well.

We fix m–versions $z \mapsto \tilde{d}_z f_i$ of $\tilde{d} f_i$, $i \in \mathbf{N}$. For $g \in \mathcal{A}_0$ such that $g = \sum_{i=1}^n \lambda_i f_i$ with $n \in \mathbf{N}$ and $\lambda_i \in \mathbf{R}$, let $\tilde{d}_z g := \sum_{i=1}^n \lambda_i \tilde{d}_z f_i$, $z \in E$. By (3.44), there exists an m–measure zero set $N_1 \subseteq E$ such that

$$\Gamma_z(g, h) = \left(\tilde{d}_z g, \tilde{d}_z h\right)_{\tilde{T}'_z E} \qquad \text{for all } g, h \in \mathcal{A}_0 \text{ and } z \in E \setminus N_1 .$$

Hence for such z, $\tilde{d}_z$ induces a canonical isometry

$$(3.45) \qquad \mathcal{A}_0 / \ker \Gamma_z = \mathcal{A}_0 / \ker \tilde{d}_z \cong \tilde{d}_z(\mathcal{A}_0) \subseteq \tilde{T}'_z E .$$

Since $T'_z E$ is the completion of $\mathcal{A}_0 / \ker \Gamma_z$ w.r.t. Γ_z, the isometry extends to a unique isometry

$$\Phi_z : T'_z E \to U_z \subseteq \tilde{T}'_z E ,$$

where U_z denotes the closure of $\tilde{d}_z(\mathcal{A}_0)$ in $\tilde{T}'_z E$. Moreover, $U_z = \tilde{T}'_z E$ for m–a.e. $z \in E \setminus N_1$. In fact, suppose this would not be the case. Then we could find an m–square integrable section ω of $\tilde{T}'E$ such that $m(\{z \in E \setminus N_1; \omega(z) \text{ is not in } U_z\}) > 0$. However, such a section cannot exist, because the span of $\{f \tilde{d}g; f \in \mathcal{A}, g \in \mathcal{A}_0\}$ is dense in the span of $\{f \tilde{d}g; f \in \mathcal{A}, g \in \bar{\mathcal{A}}\}$ w.r.t. the $L^2(E \to \tilde{T}'E; m)$ norm by (3.44), and, therefore, dense in $L^2(E \to \tilde{T}'E; m)$ by Condition (i) in Definition 3.7, b). Hence there exists an m–measure zero set $N \subseteq E$ such that $N_1 \subseteq N$ and $U_z = \tilde{T}'_z E$ for all $z \in E \setminus N$. In particular, Φ_z is an isometry between $T'_z E$ and $\tilde{T}'_z E$ for all $z \in E \setminus N$. It can now be shown similarly as in the proof of Theorem 3.9 that Φ is an isometry between the measurable fields $(T'_z E)_{z \in E \setminus N}$ and $\left(\tilde{T}'_z E\right)_{z \in E \setminus N}$ such that $\tilde{d}_z f = \Phi_z(d_z f)$ for all $z \in E \setminus N$. This completes the proof of Theorem 3.11. ∎

Chapter 4

Probabilistic aspects of L^p and Markov uniqueness

In this chapter, we demonstrate how some analytic uniqueness results for finite dimensional diffusion operators can be explained from a probabilistic point of view. In Section a), we apply Feller's classification to both the boundaries and both sides of the singularities of a one–dimensional diffusion operator. Using the results obtained above, we discuss which kinds of uniqueness can occur depending on the Feller class the boundaries respectively singularity sides are in. We give intuitive explanations for the results.

In Section b), we briefly recall the relations between conservativity, ergodicity and Markov uniqueness proven in Chapter 3, and look at them from a probabilistic point of view. These relations can be viewed as multi–dimensional counterparts to the fact that a one–dimensional symmetric diffusion operator is Markov unique if and only if there is no regular boundary in Feller's sense, and no singularity that is regular from both sides.

The concluding Section c) is perhaps the most interesting one in this chapter. Here, we demonstrate, how it can be explained probabilistically that certain diffusion operators on domains in $\mathbf{R}^n$ are L^p unique for small p, but not for large p. Since the corresponding operators are strongly Markov unique, only one of the several C^0 semigroups generated by extensions of the operator on L^p for large p is the transition semigroup of an ordinary Markov process. However, some of the other semigroups can be viewed as transition semigroups of appropriate *particle systems*. The results in Section c) are stated without proof. Detailled proofs will be given in the forthcoming publication [Eb 99a].

REMARK. The relation between L^p and Markov uniqueness, and uniqueness of martingale problems is discussed in Section a) in Chapter 1.

a) Feller classification and uniqueness for singular one–dimensional diffusion operators

Feller [Fe 51] used his classification of the boundary points for an, in our sense, regular diffusion operator on a real interval to characterize different extensions of the operator that are generators. Implicitly, he already treated uniqueness problems for such operators. Also in [Wie 85], the relation between essential self-adjointness of one–dimensional generalized Schrödinger operators, and Feller's boundary classification has been pointed out. M. Takeda showed in [Ta 91] that a regular Sturm–Liouville diffusion operator (i.e., a regular one–dimensional diffusion operator considered w.r.t. its symmetrizing measure) is Markov unique if and only if both boundary points are regular in Feller's sense.

In this section, we apply Feller's classification to the boundaries *and* to both sides of each singularity of a *singular* Sturm–Liouville operator. We look at the relation between the different uniqueness results obtained in the preceeding chapters, and the refined Feller type classification for singular operators; i.e., we discuss which types of uniqueness (L^p uniqueness, strong Markov uniqueness, Markov uniqueness) can occur depending on the Feller class the boundaries and singularity sides are in. We then use the probabilistic interpretation of Feller's classification in terms of the behaviour of the corresponding diffusion processes near the boundaries resp. singularities (cf. e.g. [ItMKe 65]) to gain an intuitive understanding why certain types of uniqueness hold for a given diffusion operator, whereas other types of uniqueness do not hold.

1) A Feller type classification of boundaries and singularities

We consider the framework from Section e) in Chapter 2, i.e., (x_0, y_0), $-\infty \leq x_0 < y_0 \leq \infty$, is an interval, and $(\mathcal{L}, C_0^\infty(x_0, y_0))$ is a divergence form operator of type

$$(4.1) \qquad\qquad \mathcal{L} = \frac{1}{\rho}\frac{d}{dx}\left(\alpha\frac{d}{dx}\cdot\right),$$

where ρ is a continuos function on (x_0, y_0), α is absolutely continuous, ρ and α are strictly positive dx–a.e., and the coefficients α/ρ and α'/ρ of the operator in non–divergence form are at least in $L^2_{\text{loc}}(x_0, y_0\,;\,\rho\,dx)$. In particular, $(\mathcal{L}, C_0^\infty(x_0, y_0))$ is a densely defined *symmetric* operator on $L^2(x_0, y_0\,;\,\rho\,dx)$. For simplicity, we also assume that the *singularity set*

$$S := \{\, s \in (x_0, y_0)\,;\, \alpha(s) = 0\,\}$$

is finite. Hence the set $(x_0, y_0) \setminus S$ can be decomposed into finitely many components (s_i, s_{i+1}), $0 \leq i \leq k$, $x_0 = s_0 < s_1 < \ldots < s_k < s_{k+1} = y_0$, where $s_1, \ldots, s_k$ are the elements in S. On each of the components, $\mathcal{L}$ is a regular Sturm–Liouville operator.

REMARK. Note that every *regular* diffusion operator $(\mathcal{L}, C_0^\infty(x_0, y_0))$ of type

$$\mathcal{L} = a\,\frac{d^2}{dx^2} + \beta\,\frac{d}{dx}\;,$$

can be rewritten in the form (4.1) with

$$\alpha := \exp \int_{z_0}^{\bullet} (\beta/a)\,dx, \qquad \rho := (1/a)\,\exp \int_{z_0}^{\bullet} (\beta/a)\,dx,$$

wher z_0 is an arbitrary fixed point in (x_0, y_0). The measure $\rho\,dx$ is the **speed measure**, and the function $1/\alpha$ is the **scale function** of the operator $\mathcal{L}$, respectively of the corresponding diffusion process.

Definition 4.1 (Feller type classification)
(i) Let $z_0 \in (x_0, y_0)$ such that $\alpha(x) > 0$ for all $x \in (z_0, y_0)$. We call y_0 an **exit boundary** *for $\mathcal{L}$, iff*

$$(4.2) \qquad \int_{z_0}^{y_0} \int_{z_0}^{y} \rho(x)\,dx\,\,(\alpha(y))^{-1}\,dy\; < \; \infty.$$

Similarly, we call y_0 an **entrance boundary** *for $\mathcal{L}$, iff*

$$(4.3) \qquad \int_{z_0}^{y_0} \int_{z_0}^{y} (\alpha(x))^{-1}\,dx\,\,\rho(y)\,dy\; < \; \infty.$$

The definitions for x_0 are analogous. A boundary which is both an exit and an entrance boundary is called **regular**, *a boundary which is neither exit nor entrance is called* **natural**.

(ii) Let $s \in S$. We call $s-$ respectively $s+$ an **exit point** *(resp. an* **entrance point**, *resp.* **regular**, *resp.* **natural**), *iff s is an exit (resp. entrance, resp. regular, resp. natural) boundary of the interval (x_0, s) respectively (s, y_0).*

REMARKS. (i) The definitions of exit and entrance boundaries do not depend on the choice of z_0.

(ii) Our notations are in accordance with those of Itô/McKean [ItMKe 65], which differ slightly from the original notations. In a large part of the literature, a boudary is called an exit respectively entrance boundary only if it is an exit/no entrance, resp. an entrance/no exit boundary according to our notation. An exit boundary in our sense is then called accessible, cf. e.g. [Man 68].

(iii) The classification describes the behaviour of the diffusion process generated by $\mathcal{L}$ near the boundary resp. singularity. Roughly speaking, an exit point is hit by the diffusion process corresponding to $\mathcal{L}$ with strictly positive probability in finite time, whereas the diffusion can be started at an entrance point in such a way that it immediately enters the corresponding interval; cf. [ItMKe 65] for a more precise probabilistic interpretation of exit and entrance boundaries.

Note that in a neighbourhood of each singularity, the function ρ is bounded by assumption. Hence the following lemma is an immediate consequence of the definition of exit and entrance points :

Lemma 4.1 *Let $s \in S$. Then $s-$ is either regular, natural, or entrance/ no exit. More precisely, let $z_0 \in (x_0, s)$ such that $\alpha(x) > 0$ for all $x \in [z_0, s)$. Then*

(i) *$s-$ is regular if and only if $1/\alpha$ is dx–integrable on (z_0, s).*

(ii) *$s-$ is entrance/no exit, if and only if $1/\alpha$ is not dx–integrable on (z_0, s), but*
$$\int_{z_0}^{s} \int_{z_0}^{y} (\alpha(x))^{-1} \, dx \; \rho(y) \, dy \; < \; \infty.$$

(iii) *$s-$ is natural if and only if the double integral above is infinite.*

Analogue assertions hold for $s+$.

REMARK.   The same simplified classification holds for boundary points, provided ρ is dx–integrable at the boundary. In particular, exit/no entrance boundaries don't exist if the measure $\rho \, dx$ is finite.

2) Uniqueness depending on Feller class

We now summarize the uniqueness results for singular Sturm–Liouville operators proven in this work, and we point out the relations to the Feller type classification. Let $(\mathcal{L}, C_0^\infty(x_0, y_0))$ be as above.

Markov uniqueness

The operator $(\mathcal{L}, C_0^\infty(x_0, y_0))$ is *Markov unique* on $L^2(x_0, y_0 ; \rho \, dx)$, *if and only if there is no regular boundary, and no singularity which is regular from both sides*, cf. Section d) in Chapter 3 above.

L^1 uniqueness

By Remark (ii) under Theorem 2.4 above, and by Lemma 4.1, the operator $(\mathcal{L}, C_0^\infty(x_0, y_0))$ is $L^1(x_0, y_0 , \rho \, dx)$ *unique, if and only if there is no exit boundary, and no singularity $s \in S$ which is regular from both sides.*

L^p uniqueness for $p > 1$

Let $p \in (1, \infty)$ such that the range of the operator $(\mathcal{L}, C_0^\infty(x_0, y_0))$ is contained in $L^p(x_0, y_0 ; \rho \, dx)$, and let $q \in (1, \infty)$ such that $\frac{1}{p} + \frac{1}{q} = 1$. Recall that the $L^p(\rho \, dx)$ limit point case is said to hold at y_0, respectively at $s-$, $s \in S$, if and only if $\int^{\bullet}_{z_0} (1/\alpha) \, dx$ is not in $L^q(z_0, y_0 ; \rho \, dx)$ resp. $L^q(z_0, s ; \rho \, dx)$ for any $z_0 \in (x_0, y_0)$ resp. $z_0 \in (x_0, s)$. The definition for x_0 and $s+$ is analogous. By Theorem 2.4, the operator $(\mathcal{L}, C_0^\infty(x_0, y_0))$ is $L^p(x_0, y_0 ; \rho \, dx)$ unique if and only if the $L^p(\rho \, dx)$ limit point case holds at both boundaries, and at $s-$ or $s+$ for each singularity $s \in S$.

Claim : For each side of a singularity, and for each boundary, there exists a **critical value** $p_0 \in [1, \infty]$, such that the $L^p(\rho\,dx)$ limit point case holds for $1 < p < p_0$, but not for $p_0 < p < \infty$. At regular points, $p_0 = 1$, at natural and exit/no entrance points, $p_0 = \infty$, and at entrance/no exit points, p_0 can have any value in $[1, \infty]$.

PROOF OF THE CLAIM. We prove the claim for y_0, the reasoning for x_0 and the singularity sides is analogous. Let $z_0 \in (x_0, y_0)$ such that α is strictly positive on $[z_0, y_0)$.
(i) Suppose y_0 is not an entrance boundary. If the measure $\rho\,dx$ is finite in a neighbourhood of y_0, then $\int_{z_0}^{\bullet}(1/\alpha)\,dx$ is not dx–inegrable at y_0. Hence $\int_{z_0}^{\bullet}(1/\alpha)\,dx$ is not in $L^q(z_0, y_0 \, ; \, \rho\,dx)$ for any $q \in (1, \infty)$, i.e., the $L^p(\rho\,dx)$ limit point case holds for all $p \in (1, \infty)$. The same is true if $\rho\,dx$ is infinite at y_0, since $\int_{z_0}^{\bullet}(1/\alpha)\,dx$ is strictly positive and increasing on (z_0, y_0).
(ii) Suppose y_0 is regular. Then both ρ and $1/\alpha$ are dx–integrable at y_0. Hence the $L^p(\rho\,dx)$ limit point case does not hold for any $p > 1$.
(iii) Suppose y_0 is an entrance/no exit point. Then the measure $\rho\,dx$ is finite, and the function $\int_{z_0}^{\bullet}(1/\alpha)\,dx$ is in $L^1(z_0, y_0 \, ; \, \rho\,dx)$ but not in $L^\infty(z_0, y_0 \, ; \, \rho\,dx)$. Hence there exists $p_0 \in [1, \infty]$, such that the $L^p(\rho\,dx)$ limit point case holds for $1 < p < p_0$, but not for $p_0 < p < \infty$. ∎

In particular, the claim implies :

Corollary 4.1 *(i) If for each $s \in S$, $s-$ or $s+$ is natural, and neither x_0 nor y_0 is an entrance boundary, then the operator $(\mathcal{L}, C_0^\infty(x_0, y_0))$ is $L^p(x_0, y_0 \, ; \, \rho\,dx)$ unique for every $p > 1$ such that the range of $\mathcal{L}$ is in L^p.*
(ii) If there exists a regular boundary, or a singularity $s \in S$ such that $s-$ and $s+$ are regular, then $(\mathcal{L}, C_0^\infty(x_0, y_0))$ is not $L^p(x_0, y_0 \, ; \, \rho\,dx)$ unique for any $p > 1$.

The in some sense most interesting case, however, is that, where none of the boundaries is regular, and none of the singularities is regular from both sides, but there exists a singularity which is entrance from both sides and/or a boundary which is entrance. In this case, all we can say in general is that $L^p(x_0, y_0 \, ; \, \rho\,dx)$ uniqueness holds for $1 < p < p_0(\mathcal{L})$ but not for $p > p_0(\mathcal{L})$ with

$$p_0(\mathcal{L}) \;:=\; \min\,\big(\, p_0(x_0),\, p_0(y_0),\, \max\,(p_0(s_1-), p_0(s_1+)),\, \ldots$$
$$\ldots,\, \max\,(p_0(s_k-), p_0(s_k+))\,\big).$$

Here $p_0(t)$ denotes the critical value of the singularity side resp. boundary point t, nad $s_1, \ldots, s_k$ are the elements in S.

Strong Markov uniqueness

Since we did not prove any direct results on strong Markov uniqueness in this work, all we can say is, that the operator $(\mathcal{L}, C_0^\infty(x_0, y_0))$ is strongly Markov unique if it is defined on $L^p(x_0, y_0 \, ; \, \rho\,dx)$ and L^p unique for some $p \in [1, \infty)$, and that it is not strongly Markov unique if it is not Markov unique, cf. Section e) in Chapter 1 above. In the cases, where the operator is Markov unique, but not $L^p(x_0, y_0 \,; \rho\,dx)$ unique for any $p \in [1, \infty)$, it is not clear whether strong Markov uniqueness holds.

REMARK. We point out that by the results listed above, **Markov uniqueness of the operator ($\mathcal{L}$, $C_0^\infty(x_0, y_0)$) is equivalent to $L^1(\rho\,dx)$ uniqueness provided there is no exit boundary.** In particular, this is the case if the measure $\rho\,dx$ is finite. Hence, for singular one–dimensional symmetric diffusion operators, there is no difference between Markov uniqueness and L^1 uniqueness, if the diffusion process generated by the Friedrichs extension of the operator is **conservative**.

The results for the following two special cases are summarized in Table 4.1 and 4.2 below :

1) $\mathcal{L}$ is a regular diffusion operator, i.e., $S = \emptyset$.

2) Both boundaries are natural, and S consists of a single point s.

Obviously, the same results as listed in the tables hold if x_0 and y_0, respectively $s-$ and $s+$ are interchanged.

3) Intuitive explanations for one–dimensional uniqueness results

We fix an operator ($\mathcal{L}$, $C_0^\infty(x_0, y_0)$) as before. We will now demonstrate, how the probabilistic interpretation of Feller's classification can help us to understand *intuitively* the uniqueness results listed in Table 4.1 and 4.2. "Intuitively" means also that, in contrast to the other parts of this work, this subsection does not contain "rigorous" results and complete proofs.

Regular case

We first look at the results in the regular case, see Table 4.1. We distinguish four cases:

(i) *At least one boundary is regular.* The diffusion process can both reach and get away from the regular boundary in finite time. Thus it can be *reflected* at the regular boundary, whence we can expect at least two different reversible solutions of the martingale problem for the operator ($\mathcal{L}$, $C_0^\infty(x_0, y_0)$) : A process with absorption at the regular boundary, and a reflected process. This explains why the operator ($\mathcal{L}$, $C_0^\infty(x_0, y_0)$) is *not even Markov unique* on $L^2(x_0, y_0 \,; \rho\,dx)$ in this case.

(ii) *One boundary is exit/no entrance, the other is entrance/no exit.* In this case, the only possible way to obtain a Markov process solving the martingale problem for ($\mathcal{L}$, $C_0^\infty(x_0, y_0)$), which is different from that with absorption at the exit boundary, is to let the process jump to the entrance boundary when it

Table 4.1: Uniqueness of the operator $(\mathcal{L}, C_0^\infty(x_0, y_0))$ in the regular case $(S = \emptyset)$.

x_0	y_0	$L^p(\rho\,dx)$ uniqueness for $p > 1$	$L^1(\rho\,dx)$ uniqueness	strong Markov uniqueness	Markov uniqueness
regular	arbitrary	–	–	–	–
ex./no entr.	ex./no entr.	+	–	+	+
ex./no entr.	entr./no ex.	+/–	–	+/?	+
entr./no ex.	entr./no ex.	+/–	+	+	+
entr./no ex.	natural	+/–	+	+	+
natural	natural	+	+	+	+

+ :   Uniqueness always holds.

– :   Uniqueness never holds.

+/– :   There exists $p_0 \in [1, \infty]$ such that L^p uniqueness holds for $1 < p < p_0$, but not for $p > p_0$.

+/? :   Strong Markov uniqueness holds if p_0 as above does not equal 1, otherwise not clear.

Table 4.2: Uniqueness of the operator $(\mathcal{L}, C_0^\infty(x_0, y_0))$ provided both boundaries are natural, and $S = \{s\}$.

$s-$	$s+$	$L^p(\rho\,dx)$ uniqueness for $p > 1$	$L^1(\rho\,dx)$ uniqueness	strong Markov uniqueness	Markov uniqueness
regular	regular	–	–	–	–
regular	entr./no ex.	+/–	+	+	+
entr./no ex.	entr./no ex.	+/–	+	+	+
natural	arbitrary	+	+	+	+

reaches the exit boundary. However, a process constructed in this way will not be reversible, since it jumps from the exit boundary to the entrance boundary, but not vice versa. This gives us an intuitive explanation why *Markov uniqueness* holds.

It is not clear, whether we can construct a Markov process as described above in such a way that the process started with initial distribution $\rho\,dx$ is sub–stationary. Therefore, the probabilistic picture does not give a direct hint whether strong Markov uniqueness holds on $L^1(x_0, y_0 ; \rho\,dx)$.

However, it should always be possible to modify the absorbed diffusion by creating new particles at the entrance boundary, and letting them move into the interval. In this way, we can construct a semigroup, which is not sub–Markovian, but bounded w.r.t. the L^∞ norm, and different from the transition semigroup of the absorbed diffusion, cf. Section c) below. This semigroup is not the transition semigroup of an ordinary Markov process, but of a particle system, which can be viewed as a diffusion with creation of new particles. It *depends on the behaviour of the transition function near the entrance boundary* whether the constructed semigroup is bounded w.r.t. the $L^p(x_0, y_0 ; \rho\,dx)$ norm for $p \in [1, \infty)$, see Section c). If it is not, we can hope for $L^p(x_0, y_0 ; \rho\,dx)$ *uniqueness*, otherwise not. This helps to understand why $L^p(x_0, y_0 ; \rho\,dx)$ uniqueness only holds for some p.

(iii) *There is no exit boundary, and at least one entrance boundary.* Then there is a canonical reversible diffusion process solving the martingale problem for the operator $(\mathcal{L}, C_0^\infty(x_0, y_0))$, which does not hit the boundaries. The only possibility to give a probabilistic construction of a second semigroup generated by an extension of the operator $(\mathcal{L}, C_0^\infty(x_0, y_0))$, seems to be to pour in new particles at the entrance boundary, as described in the considerations for Case (ii). This explains why *strong Markov uniqueness* holds, whereas it *depends on the behaviour of the transition function near the entrance boundaries* whether $L^p(x_0, y_0 ; \rho\,dx)$ *uniqueness* holds for $p \in [1, \infty)$.

(iv) *There is no entrance boundary.* Then there seems to be no possibility of constructing probabilistically a semigroup generated by an extension of $(\mathcal{L}, C_0^\infty(x_0, y_0))$, which is different from the transition semigroup of the canonical reversible diffusion solving the corresponding martingale problem. This is consistent with the fact that *all types of uniqueness hold* in this case (–except, however, L^1 uniqueness, which does not hold if there is an exit boundary).

Singular case

We now consider the singular case with two natural boundaries, and a single singularity at s. Here the results summarized in Table 4.2 show, that the operator $(\mathcal{L}, C_0^\infty(x_0, y_0))$ is not even Markov unique if both $s-$ and $s+$ are regular, and that it is strongly Markov unique but not necessarily $L^p(x_0, y_0 ; \rho\,dx)$ unique for $p > 1$, if both $s-$ and $s+$ are entrance, but at least one side of the singularity is not exit. Only in the case where $s-$ or $s+$ is natural happens what one might naively expect in the other cases as well : All kinds of uniqueness hold. How can we intuitively explain these observations ?

We distinguish three cases :

(i) $s-$ and $s+$ are regular. In this case, there exist two reversible diffusion processes solving the martingale problem for the operator $(\mathcal{L}, C_0^\infty(x_0, y_0))$: One, which passes through the singularity s, and another one, which is reflected at s. If s would not be a singularity, then the generator of the reflected process would not extend $(\mathcal{L}, C_0^\infty(x_0, y_0))$. For example, the generator of one-dimensional Brownian motion with reflection at a point $x \in \mathbf{R}$ has only those functions f in its domain, which satisfy the Neumann condition $f'(x) = 0$. However, the corresponding Neumann condition for the operator $\mathcal{L}$ at s is $\alpha(s) f'(s) = 0$. This condition is satisfied for all functions f in $C_0^\infty(x_0, y_0)$, because $\alpha(s) = 0$. Thus, there are indeed two different reversible diffusion processes with generators extending $(\mathcal{L}, C_0^\infty(x_0, y_0))$, which explains why *not even Markov uniqueness* holds.

(ii) $s-$ and $s+$ are entrance, and $s-$ or $s+$ is not exit. In this case, a reversible diffusion process passing through the singularity s does not exist, since s can not be reached from the non-exit side by diffusions solving the martingale problem for $(\mathcal{L}, C_0^\infty(x_0, y_0))$. The only reversible solution of the martingale problem is the non-ergodic diffusion which runs separately on (x_0, s) and (s, y_0), and is reflected at the regular side of s, if there exists one. Thus we can "explain", why *Markov uniqueness* holds.

If one side of the singularity is exit, then there is another conservative diffusion process solving the martingale problem, which passes through the singularity from the exit side to the entrance/no exit side. However, this process cannot be reversible, since it can get from one side to the other, but not vice versa. For the same reason, it cannot be sub–stationary if the initial distribution has a strictly positive density both on (x_0, s) and (s, y_0). This helps to understand why *strong Markov uniqueness* holds. Moreover, it gives us an idea why L^p *uniqueness does not necessarily hold* : The second process has a transition semigroup which is, in particular, L^∞ contractive. For large p, it might hence be a C^0 semigroup on L^p, whence there might be several extensions of $(\mathcal{L}, C_0^\infty(x_0, y_0))$ which generate C^0 semigroups, provided p is large enough.

In the case where both sides of the singularity are entrance/no exit, it is a little more difficult to see how L^p uniqueness could be "destroyed". We first give a new interpretation of the process passing through s in the case considered above, where one side of s is exit. Instead of particles passing through s, we may think of a particle which is killed at the exit side of s when it hits s, and a new particle on the entrance/no exit side of s, which is created instead. In the case we consider now, there is no exit side, whence particles cannot be killed without changing the generator. However, instead of killing a particle, a *negatively charged* particle can be created at one side of the singularity, which has the same effect on the transition semigroup. Indeed, it is possible to construct several C^0 semigroups on $L^p(x_0, y_0; \rho\, dx)$ with generators that extend

$(\mathcal{L}, C_0^\infty(x_0, y_0))$ as transition semigroups of systems of charged particles, provided the canonical diffusion process associated with $\mathcal{L}$ behaves in an approriate way near the singularity s, see the results in Section c). Here, negatively and positively charged particles are created with the same probability rate depending on the state of the system, but the negatively charged particles always move to one side of s, whereas the positively charged particles move to the other side. The corresponding semigroups are not even positivity preserving, cf. Section c) for details.

(iii) $s-$ *or* $s+$ *is natural.* In this case, constructions as above are impossible. One can pour in new particles at the non–natural side of s (if there exists one), but this changes the generator. Creating both positively and negatively charged particles with the same probability rate at the non–natural side does not change the generator, but it does not change the transition semigroup either. This gives us an idea, why all kinds of uniqueness hold.

b) Conservativity, ergodicity and Markov uniqueness

As remarked in Section a), one–dimensional symmetric diffusion operators are Markov unique if and only if there is no regular boundary and no singularity which is regular from both sides. There are similar relations in higher dimensions, which have partially already been clarified in Sections f) and g) of Chapter 3. At this place, we just briefly summarize the corresponding results from Chapter 3, and look at them from a probabilistic point of view.

One aspect of the one–dimensional result quoted above is that Markov uniqueness holds if there is no "bad" singularity and no exit boundary. The higher–dimensional counterpart is Corollary 3.4 : *Conservativity of the diffusion process generated by the Friedrichs extension implies Markov uniqueness if the degeneracy is "controllable".* Probabilistically, this is not surprising, because conservativity means that the diffusion process does not reach a boundary where it could be absorbed or reflected.

More interesting is the rôle of the singularities. Let $(\mathcal{L}, C_0^\infty(x_0, y_0))$ be a one–dimensional diffusion operator with a singularity s that is regular from both sides. Then, as pointed out in the last section, there exist (at least) two different processes generated by extensions of the operator, one passing through s, and another one that is reflected at s. In contrast to the first process, the second process can not be ergodic. In fact, it is not difficult to show that the process passing through s is generated by the Friedrichs extension of the operator $(\mathcal{L}, C_0^\infty(x_0, y_0))$, and the non–ergodic process is generated by the maximal Dirichlet extension.

Again, we have a similar phenomenon in higher dimensions, which is, however, more involved. We have remarked in Chapter 3, Section g), that in many multi–dimensional (and even infinite dimensional) situations, where Markov uniqueness does not hold, the semigroup generated by the Friedrichs extension of the operator is ergodic, but the semigroup generated by the maximal Dirichlet extension is not. An infinite dimensional example for this situation is given in Section b) of Chapter 5. However, in contrast to the one–dimensional case, ergodicity of the maximal semigroup plus conservativity of the Friedrichs semigroup do not imply Markov uniqueness in $\mathbf{R}^n$, $n \geq 2$. The reason is simply that the diffusion process generated by the maximal extension can move around the singularity set S, if S is for example a slit in $\mathbf{R}^2$. I suspect, that a kind of *"local ergodicity"* of the maximal process (i.e., ergodicity of the process restricted to each ball in the state space with reflection at the ball's boundary) plus conservativity of the minimal process do still imply Markov uniqueness in the finite dimensional case.

c) Probabilistic explanations for some uniqueness results on L^p

The results in Chapter 2 show that for many singular one–dimensional diffusion operators, L^p uniqueness holds for small p, but it does not hold for large p, cf. also Section a) in this chapter. For example, for $1 \leq p < n$, the operator $\left(\frac{d^2}{dx^2} + \frac{n-1}{x} \frac{d}{dx}, C_0^\infty(\mathbf{R}) \right)$ is $L^p(\mathbf{R}; x^{n-1}dx)$ unique if and only if $p \leq n/2$, cf. Theorem 2.4. A related phenomenon, although caused by a boundary and not a singularity, is that the Laplacian on $C_0^\infty(\mathbf{R}^n \setminus \{0\})$, $n \in \mathbf{N}$, is $L^p(\mathbf{R}^n; dx)$ unique if and only if $p \leq n/2$. This well-known fact can be verified by similar arguments as used in the proof of the lemma in Section d), 2), in Chapter 2.

We demonstrate now how to give a probabalistic explanation for such analytic L^p uniqueness results. At first glance, one would not expect such an explanation, because the operators considered are often L^p unique for small p, and thus strongly Markov unique. Hence only one of the several semigroups generated by the operator on L^p for large p is the transition semigroup of an ordinary Markov process. However, by considering *particle systems* instead of ordinary Markov processes, we can also give a stochastic interpretation for some of the other semigroups. In many cases, the transition semigroups of the corresponding particle systems induce a semigroup of *bounded* operators on L^p precisely for those p where L^p uniqueness does not hold. We will apply the particle systems approach to derive a sufficient condition for *non–L^p uniqueness* of general diffusion operators on finite dimensional state spaces.

In this work, we just sketch the basic ideas. The complete proofs, which are not too difficult, but very technical, will be given in a follow–up article [Eb 99a], where also further examples will be considered.

1) The basic idea

Consider the operator $(\Delta, C_0^\infty(\mathbf{R}^n \setminus \{0\}))$ on $L^p(\mathbf{R}^n; dx)$, $n \geq 2$, $p \in [1, \infty)$. We know that the transition function of Brownian motion induces a C^0 semigroup on $L^p(\mathbf{R}^n; dx)$ with a generator that extends $(\Delta, C_0^\infty(\mathbf{R}^n \setminus \{0\}))$. Since Brownian motion does not hit the point 0, there is no way to disturb this process in order to obtain another process such that its generator is the same on $C_0^\infty(\mathbf{R}^n \setminus \{0\})$. In fact, if we would let a Brownian particle jump to 0 after some random time, then this would not only change the generator at 0, but also at the position in $\mathbf{R}^n \setminus \{0\}$ where the jump started. However, we can consider the following stochastic process : After a random time depending on the Brownian path, a new particle (a "child" of the original Brownian particle) is created at 0. Afterwards, both the original particle and the child fulfil a Brownian motion, and after some random times depending on their paths, each of the two particles again creates a child which starts moving at 0, and so on.

A particle system as described can be realized as a stochastic process $(X_t, (P_\mu)_{\mu \in \mathcal{M}_p(\mathbf{R}^n)})$ on the space $\mathcal{M}_p(\mathbf{R}^n)$ of finite sums of dirac measures on $\mathbf{R}^n$. If the branching mechanism is chosen appropriately, then $(X_t, (P_\mu))$ is a Markov process, and, moreover, the operators p_t, $t \geq 0$, given by

$$(4.4) \qquad (p_t f)(x) = E_{\delta_x}\left[\int f \, dX_t\right],$$

form a semigroup on the bounded functions on $\mathbf{R}^n$.

Now suppose that $(p_t)_{t \geq 0}$ induces a C^0 semigroup $(T_t)_{t \geq 0}$ on $L^p(\mathbf{R}^n; dx)$. Then one should expect that the generator of $(T_t)_{t \geq 0}$ coincides with that of the Brownian semigroup on $C_0^\infty(\mathbf{R}^n \setminus \{0\})$, because the new particles are always born at 0, whereas the initial particle moves on undisturbed. Since obviously, $(p_t)_{t \geq 0}$ strictly dominates the Brownian semigroup from above, the operator $(\Delta, C_0^\infty(\mathbf{R}^n \setminus \{0\}))$ is not $L^p(\mathbf{R}^n; dx)$ unique in this case.

These considerations can be made rigorous, cf. Theorem 4.1 below. The semigroup $(p_t)_{t \geq 0}$ induces a C^0 semigroup on $L^p(\mathbf{R}^n; dx)$ if and only if 0 is an $L^p(\mathbf{R}^n; dx)$ entrance boundary (cf. Definition 4.2 below) for Brownian motion, which is the case exactly for $p > n/2$, see Example (i) below. By a similar argument, we also obtain that the operator $\left(\frac{d^2}{dx^2} + \frac{n-1}{x}\frac{d}{dx}, C_0^\infty(0, \infty)\right)$ is not $L^p(\mathbf{R}; x^{n-1} dx)$ unique for $p > n/2$, $n \geq 2$. The non$-L^p(\mathbf{R}; x^{n-1} dx)$ uniqueness of the operator $\left(\frac{d^2}{dx^2} + \frac{n-1}{x}\frac{d}{dx}, C_0^\infty(\mathbf{R})\right)$ for $p > n/2$, $n \geq 2$, however, cannot be explained in this way, since here the functions in the domain of the operator do not vanish at 0. A C^0 semigroup $(T_t)_{t \geq 0}$ constructed as above would have a generator that differs from $\frac{d^2}{dx^2} + \frac{n-1}{x}\frac{d}{dx}$ on functions $f \in C_0^\infty(\mathbf{R})$ by a term depending on $f(0)$. To construct different C^0 semigroups generated by $\left(\frac{d^2}{dx^2} + \frac{n-1}{x}\frac{d}{dx}, C_0^\infty(\mathbf{R})\right)$, we have to use systems of charged particles, i.e., Markov processes on the space of all finite sums and differences of dirac measures. If we create positive and negative particles at 0 with the same probability depending on the path of the parent particle, then the generator will not change on functions that are continuous at 0, cf. [Eb 99a] for details. Nevertheless, if we let the positive particles diffuse to the right, but the negative particles to the left, then the semigroup given by (4.4) will differ from the semigroup $\left(p_t^{(0)}\right)_{t \geq 0}$ of the ordinary diffusion process corresponding to the operator $\left(\frac{d^2}{dx^2} + \frac{n-1}{x} \frac{d}{dx}, C_0^\infty(\mathbf{R})\right)$. In fact, for $t > 0$ we then have $p_t f > p_t^{(0)} f$ if $f : \mathbf{R} \to [0, \infty)$ is strictly positive on $(0, \infty)$, and vanishes on $(-\infty, 0)$. Similarly, $p_t f < p_t^{(0)} f$ if f is strictly positive on $(-\infty, 0)$ but vanishes on $(0, \infty)$.

2) Results and examples

To state the results in detail, let E be a topological space. Suppose we are given a conservative diffusion process $(X_t^{(0)}, (P_x^{(0)})_{x \in E})$ on E with transition function $p_t^{(0)}$, $t \geq 0$. Let m be a σ-finite sub-stationary measure for $(p_t^{(0)})_{t \geq 0}$, i.e.,

$$\int p_t^{(0)} f \, dm \leq \int f \, dm \quad \text{for all positive functions } f \text{ on } E.$$

We assume that the continuous functions are dense in $L^p(E; m)$ for all $1 \leq p < \infty$. Then the semigroup $(p_t^{(0)})_{t \geq 0}$ induces a C^0 semigroup of contractions $(T_t^{(0)})_{t \geq 0}$ on each $L^p(E; m)$, $1 \leq p < \infty$. Now let $B \subseteq E$ be a measurable subset such that $m(B) = 0$. B should be viewed as some kind of **boundary** of $E \setminus B$.

Definition 4.2 *Let $1 < p < \infty$. We call B an $L^p(E; m)$ entrance boundary for the diffusion process $(X_t^{(0)}, (P_x^{(0)})_{x \in E})$, iff there exist a probability measure ν on B and $\varepsilon > 0$ such that the finite measure*

$$(4.5) \qquad \nu_\varepsilon := \int_0^\varepsilon \int p_s^{(0)}(y, \cdot) \ \nu(dy) \ ds$$

is absolutely continuous w.r.t. m, and

$$\frac{d\nu_\varepsilon}{dm} \in L^q(E; m), \qquad \text{where } \frac{1}{p} + \frac{1}{q} = 1.$$

Let $\mathcal{A}$ be a linear space of functions in $L^p(E; m)$ which converge to 0 at the boundary B fast enough in the following sense :

$$(4.6) \qquad \lim_{t \downarrow 0} \sup_{y \in B} p_t^{(0)} f(y) = 0 \quad \text{for all } f \in \mathcal{A}.$$

Let $(\mathcal{L}, \mathcal{A})$ be a linear operator on functions in E such that the range of $(\mathcal{L}, \mathcal{A})$ is contained in $L^p(E; m)$, and suppose that the generator $L^{(0)}$ of $(T_t^{(0)})_{t \geq 0}$ extends $(\mathcal{L}, \mathcal{A})$. The following theorem is the first main result of this section :

Theorem 4.1
Let $1 < p < \infty$. Suppose B is an $L^p(E\,;\,m)$ entrance boundary for $(X_t^{(0)},\ (P_x^{(0)}))$. Then the operator $(\mathcal{L},\ \mathcal{A})$ is not $L^p(E\,;\,m)$ unique.

The proof will be given in a follow–up article [Eb 99a]. Actually, we can explicitly construct and describe various L^p extensions of $(\mathcal{L},\ \mathcal{A})$, if B is an $L^p(E\,;\,m)$ entrance boundary : Let ν be a probability measure on B satisfying the condition in the definition of "$L^p(E\,;\,m)$ entrance boundary", and let $k^+,\ k^- : E \times \mathcal{B}(B) \to [0,\infty)$ be two positive kernels such that

$$k^+(x,\ \cdot) \ \leq \ g(x) \cdot \nu \quad \text{and} \quad k^-(x,\ \cdot) \ \leq \ g(x) \cdot \nu \quad \text{for all } x \in E$$

for some bounded function $g \in L^p(E\,;\,m)$. Then we can construct a C^0 semigroup $(T_t)_{t \geq 0}$ on $L^p(E\,;\,m)$ such that every bounded function f in the domain of the generator $L^{(0)}$ of $(T_t^{(0)})_{t \geq 0}$ which satisfies

$$(4.7) \qquad\qquad \lim_{t \downarrow 0} \sup_{y \in B} |(p_t^{(0)}f)(y) - f(y)| \ = \ 0$$

is in the domain of the generator L of $(T_t)_{t \geq 0}$, and

$$(4.8) \qquad Lf \ = \ L^{(0)}f \ + \ \int_B f(y)\, k^+(\,\cdot\,, dy) \ - \ \int_B f(y)\, k^-(\,\cdot\,, dy).$$

The stochastic process corresponding to the semigroup $(T_t)_{t \geq 0}$ is a particle system on E which can be described roughly as follows : Suppose first that k^- vanishes, and $k^+(x, dy) = g(x) \cdot q(x, dy)$, where q is a probability kernel from E to B, and g is a bounded positive function on E. All particles move independently according to the law of motion of the diffusion $(X_t^{(0)},\ (P_x^{(0)}))$. In an infinitesimal time–interval $[t, t+dt]$, a particle α, which at time t is at position X_t^α creates a child with probability $g(X_t^\alpha)\,dt$. The child is born at a random point on the boundary, which is distributed as $q(X_t^\alpha, dy)$. After their birth, the children move independently according to the same law of motion and branching mechanism as the other particles. A rigorous construction can be given similarly to Dynkin's construction of particle systems that approximate super processes, cf. [Dy 91] and [Schi 91].

If k^- does not vanish then we have to consider a system of particles which are either positively or negatively charged. Let g be a bounded positive function on E, and let q^+ and q^- be positive kernels from E to B such that $q^+ + q^-$ is a probability kernel, and $k^+(x, dy) = g(x)\, q^+(x, dy)$, $k^-(x, dy) = g(x)\, q^-(x, dy)$. All (positively and negatively charged) particles move and create children as before. If X_t^α is the position of a particle at the birth time of one of its children, then the child has the same charge with probability $q^+(X_t^\alpha, B)$, and the opposite charge with probability $q^-(X_t^\alpha, B)$. If the child has the same charge, then it starts at a random point on the boundary which is distributed as $q^+(X_t^\alpha, dy)$, otherwise the distribution of its place of birth is $q^-(X_t^\alpha, dy)$

EXAMPLES. (i) Suppose that $E = \mathbf{R}^n$, $n \geq 2$, $(X_t^{(0)}, (P_x^{(0)}))$ is Brownian motion, m is Lebesgue measure, and $B \subset \mathbf{R}^n$ is a linear subspace of dimension $n_B \leq n - 2$. Note that B is a polar set for Brownian motion. An easy explicit calculation shows that B is an $L^p(\mathbf{R}^n; dx)$ entrance boundary for Brownian motion if $p > (n - n_B)/2$. Hence we can prove probabilistically that for $p > (n - n_B)/2$, the operator $(\Delta, C_0^\infty(\mathbf{R}^n \setminus B))$ is not $L^p(\mathbf{R}^n; dx)$ unique. It can be shown analytically that the obtained condition for non-L^p uniqueness of $(\Delta, C_0^\infty(\mathbf{R}^n \setminus B))$ is *sharp*.

(ii) The considerations for the first example remain true, if the boundary B is any smooth submanifold $B \subset \mathbf{R}^n$ of dimension $\dim(B) \leq n - 2$.

So far, we have considered the uniqueness problem on functions that vanish on a boundary. We can also use the particle systems approach to study the influence of singularities of the coefficients of the generator on L^p uniqueness. For simplicity, we restrict ourselves to diffusion operators on $C_0^\infty(\mathbf{R}^1)$ that have a drift which changes its direction at 0 :

Fix $1 < p < \infty$, a function $\rho \in L_{\text{loc}}^1(\mathbf{R}; dx)$, and functions a, $b \in L_{\text{loc}}^p(\mathbf{R}; \rho\,dx)$, $a \geq 0$ dx–a.e. Let $\mathcal{L}$ be the diffusion operator on $\mathbf{R}$ given by

$$\mathcal{L}f = a\,f'' + b\,f'.$$

We consider the $L^p(\mathbf{R}; \rho\,dx)$ uniqueness problem for $(\mathcal{L}, C_0^\infty(\mathbf{R}))$. Assume that there exist *conservative* diffusion processes $(X_t^{(+)}, (P_x^{(+)})_{x \in [0,\infty)})$ and $(X_t^{(-)}, (P_x^{(-)})_{x \in (-\infty,0]})$ on $[0,\infty)$, $(-\infty, 0]$ respectively, such that $\rho\,dx$ is a sub-invariant measure for both diffusions, and the induced C^0 semigroups $(T_t^{(+)})_{t \geq 0}$ and $(T_t^{(-)})_{t \geq 0}$ on $L^p(0,\infty; \rho\,dx)$, $L^p(-\infty, 0; \rho\,dx)$ respectively, have generators that extend the operator $\mathcal{L}$ defined on the restrictions of functions in $C_0^\infty(\mathbf{R})$ to $[0,\infty)$, $(-\infty, 0]$ respectively. Note that for both diffusions, 0 is not an exit point. Hence, roughly speaking, $\mathcal{L}$ corresponds to a diffusion which drifts to the right right of 0, and to the left left of 0. The two diffusions can be viewed as parts of one diffusion $(X_t^{(0)}, (P_x^{(0)})_{x \in E})$ on the disjoint union $E := (-\infty, 0] \cup [0, \infty)$. We identify $L^p(E; \rho\,dx)$ and $L^p(\mathbf{R}; \rho\,dx)$. The transition semigroup $(p_t^{(0)})_{t \geq 0}$ of $(X_t^{(0)}, (P_x^{(0)})_{x \in E})$ induces a C^0 semigroup $(T_t^{(0)})_{t \geq 0}$ on $L^p(\mathbf{R}; \rho\,dx)$ such that $T_t^{(0)} f(x) = T_t^{(+)} f(x)$ for $x > 0$ and $T_t^{(0)} f(x) = T_t^{(-)} f(x)$ for $x < 0$ dx–a.e. for all $t \geq 0$ and $f \in L^p(\mathbf{R}; \rho\,dx)$. In particular, let $f : \mathbf{R} \to \mathbf{R}$ be a function such that both $f|_{(-\infty,0)}$ and $f|_{(0,\infty)}$ are restrictions of functions in $C_0^\infty(\mathbf{R})$. Then f is in the domain of the generator $L^{(0)}$ of $(T_t^{(0)})_{t \geq 0}$, and

(4.9) $L^{(0)}f = \mathcal{L}f$ dx–a.e. on $(-\infty, 0) \cup (0, \infty)$.

By applying the particle systems approach to $(X_t^{(0)}, (P_x^{(0)})_{x \in E})$, we can prove :

Theorem 4.2 *Suppose 0 is an $L^p(0,\infty; \rho\,dx)$ entrance boundary for $(X_t^{(+)}, (P_x^{(+)})_{x \in [0,\infty)})$, and an $L^p(-\infty, 0; \rho\,dx)$ entrance boundary for $(X_t^{(-)}, (P_x^{(-)})_{x \in (-\infty,0]})$. Then $(\mathcal{L}, C_0^\infty(\mathbf{R}))$ is not $L^p(\mathbf{R}; \rho\,dx)$ unique.*

EXAMPLE.   (iii) Fix $n \geq 2$. We consider the operator $\mathcal{L}f = (f'' + (n-1)x^{-1}f')/2$. Let $1 < p < n$. Then $1/x$ is in $L^p_{\text{loc}}(\mathbf{R}; x^{n-1}\,dx)$, whence $(\mathcal{L}, C_0^\infty(\mathbf{R}))$ is a densely defined operator on $L^p(\mathbf{R}; x^{n-1}\,dx)$. The associated diffusion on $[0, \infty)$, $(-\infty, 0]$ respectively, is a Bessel process (resp. a negative Bessel process) with parameter $\nu = (n-2)/2$, and $m := x^{n-1}\,dx$ is a symmetrizing measure for this process. Again, an easy calculation shows that for $p > n/2$, 0 is an $L^p(0, \infty; m)$ entrance boundary for the Bessel process. Hence by the theorem, the operator $(\mathcal{L}, C_0^\infty(\mathbf{R}))$ is not $L^p(\mathbf{R}; m)$ unique if $p > n/2$. Again, this condition is sharp, cf. Theorem 2.4.

Chapter 5

First steps in infinite dimensions

The uniqueness problem for infinite dimensional diffusion operators is still understood very insufficiently in the sense that there are several important types of infinite dimensional diffusion operators for which it is not known whether uniqueness holds or not. Typical examples are operators arising in quantum field theory, cf. e.g. [Alb 97, 6.4.2], or Ornstein–Uhlenbeck operators on path and loop spaces over Riemanninan manifolds, cf. e.g. [Eb 97].

There are various research articles where essential self–adjointness has been shown for some special operators, e.g. Ornstein–Uhlenbeck type operators on submanifolds of Wiener space [Ai 93] and on group–valued path spaces [Aco 94], or for lattice systems from classical statistical mechanics, cf. e.g. [AlbKoRö 95a, 95b]. Moreover, essential self–adjointness respectively Markov uniqueness has been proven for (in some sense) "small" perturbations of such operators, cf. e.g. [Shi 95], [RöZha 92], and [Eb 93, Satz 35]. In [KoTsy 93], essential self–adjointness of symmetric infinite–dimensional Dirichlet operators of type $\Delta +$ $(\beta, \nabla \cdot)$ (i.e., generators of gradient Dirichlet forms) has been shown under smoothness assumptions on β. In many applications, however, neither these smoothness assumptions are satisfied, nor are the operators one is interested in small perturbations of "smooth" operators.

There are two approaches to the uniqueness problem for infinite dimensional diffusion operators that go essentially beyond results of the type described above. First, there is a projective limit approach to prove Markov uniqueness of infinite dimensional symmetric Dirichlet operators, cf. the article [AlbRöZha 92] and Section c) of this chapter, and see also [AlbHK 77] and [Ta 87] for related previous results. Secondly, in [AlbKoRö 95] and several subsequent articles [AlbKoRö 95a, 95b, PaYoo 97], an approximative criterion for essential self–adjointness of infinite dimensional symmetric Dirichlet operators has been derived and applied, cf. Remark (iii) in Section d) below.

Results on existence and uniqueness for non–symmetric infinite dimensional diffusion operators might also be rather important for applications, e.g., to measure–valued diffusions, infinite dimensional stochastic differential equations, and stochastic partial differential equations. Nevertheless, the only uniqueness result for such operators that I know is a small perturbation result in [St 96].

The aim of this chapter is to present a simple unified approach, which covers several important uniqueness results known so far, and which also applies in non–symmetric cases. We restrict ourselves to the flat case, i.e., we only consider diffusion operators on linear spaces such that the "metric" generated by the second order coefficients of the operator is flat. However, our framework and the proofs of our results are formulated in such a way that the underlying geometric structure becomes clear. Hopefully, this should make it possible to carry over the results to non–flat cases. In particular, we give a new analytic proof for a non–symmetric generalization of the essential self–adjointness result in [AlbKoRö 95]. This proof is based on a variant of the Bochner technique in analysis on manifolds, which clearly applies in non–flat cases as well.

The organization of this chapter is as follows :
After describing our framework in Section a), we consider two examples of non–unique infinite dimensional diffusion operators defined on cylinder functions in Section b). Both are variants of the generator of Funaki's random motion of strings, considered on different L^2 spaces. Whereas in the first example, uniqueness can still be achieved by defining the operator on a larger space of cylinder functions, this is not possible in the second example. In contrast to corresponding finite dimensional counterexamples, the non–uniqueness is not caused by an exit boundary or a singularity of the operator coefficients. Nevertheless, there is still a relation to ergodicity : In contrast to the minimal semigroup, the maximal semigroup is not ergodic.

In Section c), we give a generalization of the projective limit approach to Markov uniqueness of infinite dimensional diffusion operators. Here we do not restrict ourselves to the flat case. The range of applications of this approach is limited, but it can be used to study "small perturbations" of "nice" operators, and it provides some theoretical insight.

Section d) is the central part of this chapter. Here, we give an approximative approach to L^p uniqueness for (not necessarily symmetric) infinite dimensional diffusion operators, which *unifies the projective limit approach to Markov uniqueness and the approximative approach from [AlbKoRö 95]* described above, cf. Theorems 5.2 and 5.3 below. We do not exactly recover the results in these articles, but to the applications considered so far, our results seem to apply equally well. Moreover, in contrast to [AlbRöZha 92], Theorem 5.2 below yields not only Markov uniqueness but L^p uniqueness for $p \in [1, 2)$, and it also applies in the *non–symmetric case.* Theorem 5.3 below gives a simple and purely analytic proof of results similar to those in [AlbKoRö 95], and generalizes the latter results to the non–symmetric case. Our methods do not only yield uniqueness, but also existence on L^p, cf. Corollary 5.3 below.

In Section e), we apply the results obtained so far to lattice systems from classical statistical mechanics. These offer an easily accessible class of examples, which reveal advantages and disadvantages of the different approaches.

The concluding Sections f) and g) deal with "small" (i.e., H–valued) perturbations of nice diffusion operators. Here, we can prove L^p uniqueness for $p < 2$ under very weak assumptions. The general perturbation result, and its application to perturbations of operators with linear drift including infinite dimensional generalized Schrödinger operators are presented in Section f). In Section g), we consider applications to *stochastic quantization in finite volume*, *perturbations of Ornstein–Uhlenbeck operators on Wiener space*, and *Brownian strings in a velocity field*.

a) Infinite dimensional diffusion operators on linear spaces

In this section, we describe a framework, which covers many of the examples of infinite dimensional diffusion operators defined on cylinder functions we are interested in. However, as mentioned before, we restrict ourselves to the *flat case* throughout this chapter except Section c), i.e., we assume that the state space E is a vector space, and the finite dimensional projections of the diffusion operator $\mathcal{L}$ considered have constant non–degenerate diffusion matrices. This means that for approriate normalized projections $\Pi : E \to \mathbf{R}^n$, $n \in \mathbf{N}$, we have

$$\mathcal{L}(f \circ \Pi) = (\Delta f) \circ \Pi + \beta_\Pi \cdot ((\nabla f) \circ \Pi)$$

for all sufficiently smooth functions f on $\mathbf{R}^n$, where $\beta_\Pi : E \to \mathbf{R}^n$ is a measurable function.

1) State space and cylinder functions

Let E be a vector space, and K a vector space of linear functionals on E. By $C_b^\infty(\mathbf{R}^n)$ we denote the space of all smooth functions f on $\mathbf{R}^n$ such that f and all its derivatives are bounded. Let

$$\mathcal{F}C_b^\infty(K) := \{ f(\ell_1, \ldots, \ell_n) ; \ n \in \mathbf{N}, \ f \in C_b^\infty(\mathbf{R}^n), \ \ell_1, \ldots \ell_n \in K \}$$

be the smooth cylinder functions on E based on K. We assume that E is endowed with the σ–algebra generated by the functionals in K. This is for example the case, if E is a Banach space endowed with its Borel σ–algebra, and K is a dense subspace of the topological dual E'. Fix $p \in [1, \infty)$, and a probability measure m on E such that the marginals $m \circ (\ell_1, \ldots, \ell_n)^{-1}$ have full support on $\mathbf{R}^n$ for all linearly independent functionals $\ell_1, \ldots, \ell_n \in K$, $n \in \mathbf{N}$. Then two different functions in $\mathcal{F}C_b^\infty(K)$ represent two different m–classes. The space $\mathcal{F}C_b^\infty(K)$ is dense in $L^p(E; m)$.

2) Diffusion operators on cylinder functions and associated co–tangent spaces

Suppose that we are given a diffusion operator $\mathcal{L}$ with domain $\mathcal{F}C_b^\infty(K)$ on $L^p(E\,;\,m)$ such that

$$\mathcal{L}\left(f(\ell_1,\dots,\ell_n)\right) = \sum_{i,j=1}^{n} a_{\ell_i,\ell_j}\frac{\partial^2 f}{\partial x_i \partial x_j}(\ell_1,\dots,\ell_n) + \sum_{j=1}^{n}\beta_{\ell_j}\frac{\partial f}{\partial x_j}(\ell_1,\dots,\ell_n)$$

for all $n \in \mathbf{N}$, $f \in C_b^\infty(\mathbf{R}^n)$, and $\ell_1,\dots,\ell_n \in K$.

Note that, by the chain rule, one can show that any diffusion operator with domain $\mathcal{F}C_b^\infty(K)$ on $L^p(E\,;\,m)$ has a representation as above, where $a : K \times K \to L^p(E\,;\,m)$, $(\ell,\tilde{\ell}) \mapsto a_{\ell,\tilde{\ell}}$, is a symmetric bilinear map, and $\beta : K \to L^p(E\,;\,m)$ is linear. Since we restrict ourselves to the flat case, we assume, however :

- $a : K \times K \to \mathbf{R}$ is a non–degenerate, positive definite symmetric bilinear form on K.

- $\beta : K \to L^p(E\,;\,m)$ is a linear map.

Hence a is an inner product on K. We call the Hilbert space H' obtained by completing K w.r.t. $a(\,\cdot\,,\,\cdot\,)^{1/2}$ the "co–tangent space" associated with the diffusion operator $(\mathcal{L},\mathcal{F}C_b^\infty(K))$, cf. also Example (iii) in Subsection 3) of Appendix D above. The dual $H := H''$ of H' is called the corresponding "tangent space".

REMARKS. (i) Suppose $E = \mathbf{R}^n$, g is a constant but non–trivial metric on E, $K = (\mathbf{R}^n)^*$, and $\mathcal{L}$ is the Laplace–Beltrami operator on (E,g). Then a is the natural inner product on the ordinary co-tangent space T^*E, i.e., the matrix $a(e_i^*,e_j^*)$, $1 \le i,j \le n$, is the inverse of the matrix $g(e_i,e_j)$, $1 \le i,j \le n$, where e_i and e_i^*, $1 \le i \le n$, denote the canonical bases on $\mathbf{R}^n$ resp. $(\mathbf{R}^n)^*$. Hence the Hilbert space H' is the ordinary co–tangent space with standard inner product, and the Hilbert space H is the ordinary tangent space with the metric g.

(ii) Note that K can indeed be viewed as a "restricted co–tangent space" to E, since for a function $F = f(\ell_1,\dots,\ell_n)$ in $\mathcal{F}C_b^\infty(K)$, the differential

$$(DF)(\omega) = \sum_{i=1}^{n}\frac{\partial f}{\partial x_i}(\ell_1(\omega),\dots,\ell_n(\omega))\,\ell_i$$

is a function from E to K.

3) General framework

We fix E, K, m and p as above. From now on, we assume moreover, that we are given a Hilbert space H ("the *tangent space to* E"), such that $E \cap H$ is dense in H, and the linear functionals in K are continuous on $E \cap H$ w.r.t. the norm on H. We can then extend the functionals in K to continuous linear functionals on H, i.e., K is a subspace of the dual H'. Let $j : H' \to H$ denote the Riesz isomorphism, i.e.,

$$(j(\ell), h)_H = \ell(h) \qquad \text{for all } \ell \in H' \text{ and } h \in H.$$

We equip H' with the natural inner product, whence j becomes an isometry. We have :

$$K \subseteq H' \xrightarrow{\ j\ } H$$

REMARK. (iii) In many applications (though not in all), E is a topological vector space, and H is a densely and continuously embedded subspace of E. In this case, we have the following situation :

$$K \subseteq E' \subseteq H' \xrightarrow{\ j\ } H \subseteq E,$$

where the embeddings $E' \subseteq H'$ and $H \subseteq E$ are dense and continuous. We call this the **standard framework**, since it has been used in many previous articles, cf. e.g. [AlbRö 90]. There are, however, examples which do not fit into the standard framework, since the natural tangent space H is larger than the natural state space E, see e.g. Example (iv) in Subsection 6) below. Usually, these examples can nevertheless be included in the standard framework by enlarging the state space E. However, it seems more natural to modify the framework as described above.

In Subsection 2), we have shown how to construct spaces H' and H as above corresponding to a given diffusion operator on $\mathcal{F}C_b^\infty(K)$. Conversely, we now define diffusion operators corresponding to the given "tangent space" H. Let $\beta : K \to L^p(E; m)$, $\ell \mapsto \beta_\ell$, be a linear map. Note that if β would be continuous w.r.t. $(\cdot, \cdot)_{H'}$, then it would correspond to a "vector field" in $L^p(E \to H; m)$, but we don't assume any continuity here. Nevertheless, it is convenient to think of β as an L^p vector field on E. We consider the diffusion operator $(\mathcal{L}, \mathcal{F}C_b^\infty(K))$ on $L^p(E; m)$ defined by

$$\boxed{\mathcal{L}F = \Delta_H F + \langle \beta, DF \rangle}$$

Here $\langle \cdot, \cdot \rangle$ is the dualisation between linear functionals on K and elements in K, Δ_H denotes the cylindrical Laplacian w.r.t. the metric $(\cdot, \cdot)_H$ (respectively w.r.t. the metric $(\cdot, \cdot)_{H'}$ on the cotangent space), and DF is the differential, i.e.,

$$(\Delta_H F)(\omega) = \sum_{i,j=1}^n (\ell_i, \ell_j)_{H'} \frac{\partial^2 f}{\partial x_i \partial x_j}(\ell_1(\omega), \dots, \ell_n(\omega)), \quad \text{and}$$

$$(DF)(\omega) = \sum_{i=1}^n \frac{\partial f}{\partial x_i}(\ell_1(\omega), \dots, \ell_n(\omega)) \, \ell_i \qquad \text{for all } \omega \in E,$$

whenever $F = f(\ell_1, \ldots, \ell_n)$ for some $n \in \mathbf{N}$, $f \in C_b^\infty(\mathbf{R}^n)$, and $\ell_1, \ldots, \ell_n \in K$. Explicitly, $\mathcal{L}$ is hence given by
(5.1)

$$\mathcal{L}\left(f(\ell_1,\ldots,\ell_n)\right) = \sum_{i,j=1}^{n} (\ell_i, \ell_j)_{H'} \frac{\partial^2 f}{\partial x_i \partial x_j}(\ell_1,\ldots,\ell_n) + \sum_{j=1}^{n} \beta_{\ell_j} \frac{\partial f}{\partial x_j}(\ell_1,\ldots,\ell_n)$$

In the next two subsections, we consider two alternative representations of the operator $\mathcal{L}$: Firstly, an intrinsic representation in terms of the tangent space H (instead of the cotangent space H'), and secondly a divergence form representation.

4) Directional derivatives, H–gradient, and H–representation of the operator

As remarked above, K is contained in H'. We introduce generalized directional derivatives, i.e., for a cylinder function $F = f(\ell_1, \ldots, \ell_n)$ in $\mathcal{F}C_b^\infty(K)$, and $h \in H$, we define the function $\partial_h F$ in $\mathcal{F}C_b^\infty(K)$ by

$$(\partial_h F)(\omega) := \sum_{i=1}^{n} \frac{\partial f}{\partial x_i}(\ell_1(\omega), \ldots, \ell_n(\omega)) \, \ell_i(h) \qquad \text{for all } \omega \in E.$$

Note that for $\omega \in E$ and $h \in H \cap E$, $(\partial_h F)(\omega)$ is the ordinary directional derivative at ω of the function F in direction h. The functional $h \mapsto (\partial_h F)(\omega)$ is the element in H' corresponding to the functional $(DF)(\omega)$ in K, i.e., it is the unique continuous extension of $(DF)(\omega)$ from $E \cap H$ to H.

Lemma Let $F \in \mathcal{F}C_b^\infty(K)$. Then

$$\Delta_H F = \operatorname{tr}_H D^2 F, \qquad i.e.,$$

for every complete orthonormal system $\{e_k ; k \in \mathbf{N}\}$ in H, the sum $\sum_{k=1}^{N} \partial_{e_k} \partial_{e_k} F$ converges pointwise and in $L^p(E; m)$ to $\Delta_H F$ as $N \to \infty$.

PROOF. If $F = f(\ell_1, \ldots, \ell_n)$ for some $n \in \mathbf{N}$, $f \in C_b^\infty(\mathbf{R}^n)$, and $\ell_1, \ldots, \ell_n \in K$, and $\{e_k ; k \in \mathbf{N}\}$ is a complete orthonormal system in H, then

$$\sum_{k=1}^{N} \partial_{e_k} \partial_{e_k} F = \sum_{k=1}^{N} \sum_{\lambda=1}^{n} \sum_{\nu=1}^{n} \frac{\partial^2 f}{\partial x_\lambda \partial x_\nu}(\ell_1, \ldots, \ell_n) \, \ell_\lambda(e_k) \, \ell_\nu(e_k)$$

$$= \sum_{\lambda=1}^{n} \sum_{\nu=1}^{n} \frac{\partial^2 f}{\partial x_\lambda \partial x_\nu}(\ell_1, \ldots, \ell_n) \sum_{k=1}^{N} (j(\ell_\lambda), e_k)_H \, (j(\ell_\nu), e_k)_H$$

$$\xrightarrow{N \uparrow \infty} \sum_{\lambda=1}^{n} \sum_{\nu=1}^{n} (j(\ell_\lambda), j(\ell_\nu))_H \frac{\partial^2 f}{\partial x_\lambda \partial x_\nu}(\ell_1, \ldots, \ell_n)$$

$$= \sum_{\lambda=1}^{n} \sum_{\nu=1}^{n} (\ell_\lambda, \ell_\nu)_{H'} \frac{\partial^2 f}{\partial x_\lambda \partial x_\nu}(\ell_1, \ldots, \ell_n)$$

pointwise and in $L^p(E\,;\,m)$. ∎

We define the **H-gradient** $\nabla F \in L^\infty(E \to H\,;\,m)$ of a cylinder function $F \in \mathcal{F}C_b^\infty(K)$ by

$$(\nabla F)\,(\omega) \;=\; j\,(DF(\omega)), \qquad \omega \in E,$$

where we have identified the functional $(DF)\,(\omega)$ in K with the corresponding functional in H'. Hence

$$(h,\,(\nabla F)\,(\omega))_H \;=\; (\partial_h F)\,(\omega) \qquad \text{for all } h \in H \text{ and } \omega \in E.$$

REMARK.   If (E, m) is the Wiener space, and H is the Cameron–Martin subspace of E, then ∇F is the **Malliavin gradient** of F.

Now suppose for the moment, that there exists a vector field β^H in $L^p(E \to H\,;\,m)$ such that $\beta_\ell = \ell \circ \beta^H$ $(= (j(\ell),\,\beta^H)_H\,)$ for all $\ell \in K$. Then we have the following representation of the operator $\mathcal{L}$:

$$(5.2) \qquad \mathcal{L}F \;=\; \mathrm{tr}_H\,D^2 F + (\beta^H,\,\nabla F)_H \qquad \text{for all } F \in \mathcal{F}C_b^\infty(K).$$

In most of the interesting applications, there does not exist an H-*valued* vector field β^H as above. Then (5.2) only holds in a generalized sense. Rigorously, we still have

$$(5.3) \qquad\qquad \mathcal{L}F \;=\; \mathrm{tr}_H\,D^2 F + \langle \beta,\,DF \rangle.$$

5) Integration by parts identity and divergence form representation of the operator

From now on, we make the following assumption on the measure m :

(IP) There exist functions $\beta_\ell^m \in L^p(E\,;\,m)$, $\ell \in K$, such that the integration by parts identity

$$(5.4) \qquad\qquad \int \partial_{j(\ell)} F\,dm \;=\; -\int \beta_\ell^m\,F\,dm$$

holds for all $F \in \mathcal{F}C_b^\infty(K)$ and $\ell \in K$.

Suppose $F = f(\ell_1, \dots, \ell_n)$ for some $n \in \mathbf{N}$, $f \in C_b^\infty(\mathbf{R}^n)$, and $\ell_1, \dots, \ell_n \in K$. Since $\partial_{j(\ell)} F = (DF)\,(j(\ell)) = (\ell, DF)_{H'}$, we can rewrite (5.4) in the more explicit form

$$(5.5) \quad \int \sum_{i=1}^n (\ell,\ell_i)_H\,\frac{\partial f}{\partial x_i}\,(\ell_1, \dots, \ell_n)\,dm \;=\; -\int \beta_\ell^m\,f\,(\ell_1, \dots, \ell_n)\,dm.$$

Integration by parts identities if type (IP) are known to hold for many measures on infinite dimensional spaces one is interested in in applications, cf. the examples below and in the subsequent sections, and see also [MaRö 92] and [RöSchm 95].

REMARKS.   (iv) If (IP) holds then the map $\beta^m : K \to L^p(E\,;\,m)$, $\ell \mapsto \beta_\ell^m$ is linear.

(v) Obviously, $\beta_{j(\ell)}$ is the divergence w.r.t. m of the constant vector field $\omega \mapsto j(\ell)$ on E.

(vi) We point out, that the map β^m **does not only depend on the measure m, but also on the embedding j, i.e., on the choice of H** respectively H'. In particular, β^m is in general not the logarithmic derivative Z^m of the measure m, but $\beta^m = Z^m \circ j$. In the next subsection, we look at some simple examples, which may help to clarify the difference between β^m and the logarithmic derivative of m.

We now use (IP) to rewrite $\mathcal{L}$ in divergence form. Let

$$(5.6) \qquad\qquad b := \beta - \beta^m,$$

i.e., $b_\ell = \beta_\ell - \beta_\ell^m \in L^p(E\,;\,m)$ for all $\ell \in K$. Note that if $\ell_1, \dots, \ell_n \in K$ are orthonormal w.r.t. $(\,\cdot\,,\,\cdot\,)_{H'}$, and $\ell = \ell_i$ for some $1 \le i \le n$, then (5.5) reduces to

$$(5.7) \qquad \int \frac{\partial f}{\partial x_i} \,(\ell_1,\dots,\ell_n)\; dm \;=\; -\int \beta_{\ell_i}^m\, f\,(\ell_1,\dots,\ell_n)\; dm.$$

Now fix cylinder functions $F, G \in \mathcal{F}C_b^\infty(K)$. We can represent F and G as $F = f(\ell_1,\dots,\ell_n)$, $G = g(\ell_1,\dots,\ell_n)$ for some $n \in \mathbf{N}$, $f, g \in C_b^\infty(\mathbf{R}^n)$, and orthonormal functionals $\ell_1, \dots, \ell_n \in K$. We then have

$$\mathcal{L}F \;=\; (\Delta f)\,(\ell_1,\dots,\ell_n) \;+\; \sum_{i=1}^n \beta_{\ell_i} \frac{\partial f}{\partial x_i}\,(\ell_1,\dots,\ell_n),$$

whence, by (5.7),

$$\int \mathcal{L}F\,G\; dm \;=\; -\int (\nabla f \cdot \nabla g)\,(\ell_1,\dots,\ell_n)\; dm$$

$$(5.8) \qquad\qquad + \int \sum_{i=1}^n b_{\ell_i} \frac{\partial f}{\partial x_i}\,(\ell_1,\dots,\ell_n)\; g\,(\ell_1,\dots,\ell_n)\; dm$$

$$= \;-\int (DF, DG)_{H'}\; dm + \int \langle b, DF \rangle\; G\; dm.$$

This yields the **divergence form representation**

$$(5.9) \qquad\qquad \boxed{\mathcal{L}F \;=\; -D^*DF + \langle b, DF \rangle,}$$

where D^* denotes the adjoint of D considered as an operator from $L^p(E\,;\,m)$ to $L^p(E \to H'\,;\,m)$. Note that D^* depends on the choice of the inner product $(\,\cdot\,,\,\cdot\,)_{H'}$.

REMARK. (vii) Suppose there exists a vector field $b^H \in L^p(E \to H; m)$, such that

$$b_\ell = \ell \circ b^H \quad (= (j(\ell), b^H)_H) \quad \text{for all } \ell \in K.$$

Then

$$(5.10) \quad \mathcal{L}F = -D^* DF + (b^H, \nabla F)_H = -\nabla^* \nabla F + (b^H, \nabla F)_H$$

for all $F \in \mathcal{F}C_b^\infty(K)$, where ∇^* denotes the adjoint of the densely defined linear operator $\nabla : \mathcal{F}C_b^\infty(K) \subset L^p(E; m) \to L^p(E \to H; m)$. In particular, (5.10) holds with $b^H = 0$ in the **symmetric case**. Here

$$\mathcal{L}F = -\nabla^* \nabla F = -D^* DF = \Delta_H F + \langle \beta^m, DF \rangle$$

for all $F \in \mathcal{F}C_b^\infty(K)$. In non–symmetric cases, however, it is somehow restrictive to assume that b is H–valued. If b is H–valued, then $\mathcal{L}$ can be viewed as a (in some sense) **"small"** perturbation of the symmetric operator $-\nabla^* \nabla$, cf. also Section f) below. It is the aim of this chapter to include the small perturbation case, but also to go a few steps beyond.

6) Examples

(i) $\mathbf{R}^n$ with non–Euclidean metric

Suppose that $E = H = \mathbf{R}^n$, and H carries an inner product with coefficients $g_{ij} = (e_i, e_j)_H$, where $\{e_i; 1 \leq i \leq n\}$ is the canonical basis in $\mathbf{R}^n$. Then the inner product on H' $(= (\mathbf{R}^n)^*)$ induced by the isomorphism j is given by

$$\left(\ell, \tilde{\ell} \right)_{H'} = \sum_{i,j=1}^n a_{ij} \, \ell(e_i) \, \tilde{\ell}(e_j),$$

where (a_{ij}) is the inverse of the matrix (g_{ij}). Suppose $m = \rho \, dx$ for a strictly positive differentiable function ρ on $\mathbf{R}^n$. Then the logarithmic derivative Z_h^m of m in direction $h \in \mathbf{R}^n$ is given by

$$Z_h^m = \sum_{i=1}^n \frac{1}{\rho} \frac{\partial \rho}{\partial x_i} h_i, \quad \text{whereas}$$

$$\beta_\ell^m = \sum_{i,j=1}^n a_{ij} \frac{1}{\rho} \frac{\partial \rho}{\partial x_j} \ell(e_i) \quad \text{for all } \ell \in (\mathbf{R}^n)^*.$$

(ii) Lattice systems

We fix a countable set Λ, e.g., $\Lambda = \mathbf{Z}^n$ for some $n \in \mathbf{N}$. Let

$$E := \mathbf{R}^\Lambda = \{ x = (x^i)_{i \in \Lambda}; \ x^i \in \mathbf{R} \text{ for all } i \}$$

endowed with the product σ–algebra, and let $K := \mathrm{span}\left\{x \mapsto x^i ; i \in \Lambda\right\}$. As the tangent space H, we take the space $\ell^2(\Lambda) := \left\{x \in \mathbf{R}^\Lambda; \sum_{i\in\Lambda}(x^i)^2 < \infty\right\}$ with the canonical inner product. The metric on H' is then given by

$$\left(\ell, \tilde{\ell}\right)_{H'} = \sum_{i\in\Lambda} \ell(e_i)\, \tilde{\ell}(e_i),$$

where $\{e_i ; i \in \Lambda\}$ denotes the canonical basis in E. Note that to verify (IP), it is sufficient to show that there are functions $\beta^m_{(i)} \in L^p(E ; m)$, $i \in \Lambda$, such that

$$\int \frac{\partial f}{\partial x^{i_k}}\left(x^{i_1},\dots,x^{i_n}\right) m(dx) = \int \beta^m_{(i_k)}(x)\, f(x^{i_1},\dots,x^{i_n})\, m(dx)$$

for all $n \in \mathbf{N}$, $f \in C^\infty_b(\mathbf{R}^n)$, $i_1,\dots,i_n \in \Lambda$, and $1 \le k \le n$. Measures satisfying integration by parts identities of this type are obtained as *Gibbs measures* — diffusion operators of interest are the generators of of the corresponding stochastic dynamics, cf. Section e) below.

(iii) Wiener space with Cameron–Martin metric

Fix $n \in \mathbf{N}$. Suppose E is the space of all continuous paths $\omega : [0,1] \to \mathbf{R}^n$ starting at 0, and m is Wiener measure on E. Let H be the Cameron–Martin space consisting of all absolutely continuous functions h in E such that $\int_0^1 |h'(s)|^2\, ds < \infty$, endowed with inner product

$$(h_1, h_2)_H = \int_0^1 h'_1(s) \cdot h'_2(s)\, ds.$$

Then, by the Cameron-Martin theorem, the integration by parts identity

$$(5.11) \qquad \int \partial_h F\, dm = -\int Z^m_h F\, dm$$

holds for all $h \in H$ and $F \in \mathcal{F}C^\infty_b(E')$, where $-Z^m_h$ is the stochastic integral $\int_0^1 h'(s)\, dW_s$ w.r.t. Brownian motion on (E, m), which is in $\bigcap_{1\le p<\infty} L^p(E ; m)$, cf. e.g. [MaRö 92, Ch. II, Thm. 3.11]. Moreover,

$$(5.12) \qquad \beta^m_\ell = Z^m_{j(\ell)} = -\ell \qquad m\text{–a.e.} \qquad \text{for all } \ell \in E'.$$

Heuristically, this is obvious, since

$$\text{`` } Z^m_{j(\ell)}(\omega) = (j(\ell), \omega)_H = \ell(\omega) \text{ ''.}$$

Rigorously, one easily verifies that for $\ell \in E'$ given by $\ell(\omega) = \int_0^1 \omega(s)\, dg(s)$ for some function g of bounded variation such that $g(1) = 0$, $h := j(\ell)$ satisfies $h' = g$, whence

$$\ell = \int_0^1 W_s\, dg(s) = -\int_0^1 g(s)\, dW_s = Z^m_h \qquad m\text{–a.e.}$$

By (5.12), the symmetric operator $\Delta_H + \langle \beta^m, D\cdot\rangle$ is the **Ornstein–Uhlenbeck operator** on E.

(iv) Wiener space with L^2 metric

Fix $n \in \mathbb{N}$. As in Example (iii), let (E, m) be the Wiener space over $\mathbb{R}^n$, but now we choose $H = L^2([0,1] \to \mathbb{R}^n; ds)$. We assume that K is a dense subspace of E' such that $j(K)$ is contained in the Cameron–Martin space H_{CM}, for example,

$$K = \left\{ \omega \mapsto \int_0^1 h(s)\,\omega(s)\,ds\,; \ h \in H_{CM} \right\}.$$

Again, (5.11) holds for all $h \in j(K)$, where $-Z_h^m$ is the stochastic integral $\int_0^1 h'(s)\,dW_s$. Hence Z^m is the same as above, but β^m is different. In fact, choose $h \in C_0^\infty(0,1)$, and let $\ell := j^{-1}(h)$, i.e., $\ell(\omega) = \int_0^1 h\,\omega\,ds$ for all $\omega \in E$. Then

$$\beta_\ell^m = Z_h^m = -\int_0^1 h'(s)\,dW_s = \int_0^1 h''(s)\,W_s\,ds \qquad m\text{--a.e.},$$

i.e.,

$$\beta_{j^{-1}(h)}^m = j^{-1}(h'') \qquad m\text{--a.e.}$$

The corresponding symmetric operator $\Delta_H + \langle \beta^m, D \cdot \rangle$ is the generator of **Funaki's random motions of strings**, cf. Section b) below.

7) Divergence bound and dissipativity

As in the finite–dimensional case, we will make the following assumption throughout this chapter :

(A 1) There exists $\alpha \geq 0$ such that

$$\int \langle b, DF \rangle \, dm \leq \alpha \int F \, dm \qquad \text{for all positive } F \in \mathcal{F}C_b^\infty(K).$$

Setting $G = 1$ in (5.8), we see that (A 1) holds if and only if m is a sub–invariant measure for the operator $(\mathcal{L} - \alpha, \mathcal{F}C_b^\infty(K))$. In particular, $(\mathcal{L} - \frac{\alpha}{p}, \mathcal{F}C_b^\infty(K))$ is dissipative on $L^p(E; m)$ in this case, cf. Lemma 1.8 in Appendix B.

8) Finite dimensional projections

In the uniqueness and existence results in Section d) below, we assume that we are given a sequence $(\Lambda_N)_{N \in \mathbb{N}}$ of finite sets, and finite dimensional projections $\Pi_N : E \to \mathbb{R}^{\Lambda_N}$, $N \in \mathbb{N}$, such that the following conditions hold :

• The spaces

$$K_N := \left\{ \ell_N \circ \Pi_N\,; \ \ell_N : \mathbb{R}^{\Lambda_N} \to \mathbb{R} \text{ linear} \right\}, \qquad N \in \mathbb{N}$$

are subspaces of K.

- The maps Π_N, $N \in \mathbf{N}$, generate the σ–algebra on E, i.e.,

$$\sigma\left(K\right) \; = \; \sigma\left(\Pi_N \, ; \, N \in \mathbf{N}\right) \; = \; \sigma\left(\bigcup_{N \in \mathbf{N}} K_N\right).$$

- The sequence $(K_N)_{N \in \mathbf{N}}$ is increasing (or, equivalently, the families $\{\Pi_N^{-1}(A) \, ; \, A \subseteq \mathbf{R}^{\Lambda_N} \text{ measurable}\}$ of all Π_N–cylinder sets increase with N).

- $\left(\Pi_N^i, \Pi_N^j\right)_{H'} = \delta_{ij}$ for all $N \in \mathbf{N}$ and $i, j \in \Lambda_N$. Here Π_N^i denotes the i–th component of the $\mathbf{R}^{\Lambda_N}$–valued map Π_N. Π_N^i is a functional in K.

If the conditions hold, then the space $\mathcal{F}C_b^\infty(\bigcup_{N \in \mathbf{N}} K_N)$ of all smooth cylinder functions based on Π_N, $N \in \mathbf{N}$, is dense in $L^p(E; m)$. For $N \in \mathbf{N}$ and $f \in C_b^\infty(\mathbf{R}^{\Lambda_N})$ we have

$$(5.13) \qquad \mathcal{L}\left(f \circ \Pi_N\right) \; = \; (\Delta f) \circ \Pi_N \; + \; \beta_N \cdot \left((\nabla f) \circ \Pi_N\right),$$

where $\beta_N := \left(\beta_{\Pi_N^i}\right)_{i \in \Lambda_N} \in L^p(E \to \mathbf{R}^{\Lambda_N} ; m)$, and the "dot" denotes the Euclidean inner product on $\mathbf{R}^{\Lambda_N}$.

We finally comment on possible choices for the sequence Π_N, $N \in \mathbf{N}$, of finite dimensional projections. Suppose $(i_N)_{N \in \mathbf{N}}$ is a strictly increasing sequence of positive integers, and $\{e_i, \, i \in \mathbf{N}\}$ is a complete orthonormal system of H, such that e_i is in $j(K)$ for each i. Note that such a system always exists if K is a dense subspace of H', since then $j(K)$ is dense in H. Let ℓ_i, $i \in \mathbf{N}$, be the functionals in K such that $j(\ell_i) = e_i$. For $N \in \mathbf{N}$, let $\Lambda_N := \{1, \ldots, i_N\}$, and let $\Pi_N : E \to \mathbf{R}^{\Lambda_N} \cong \mathbf{R}^{i_N}$ be the projection given by

$$\Pi_N\left(\omega\right) \; = \; \left(\ell_1(\omega), \ldots, \ell_{i_N}(\omega)\right) \qquad \text{for all } \omega \in E.$$

Then all assumptions listed above are satisfied. Note also that, for $h \in H$, $\ell_i(h) = (e_i, h)_H$, whence

$$\Pi_N\left(h\right) \; = \; \left((e_1, h)_H, \ldots, (e_{i_N}, h)_H\right).$$

Thus Π_N is the unique continuous extension to E of the orthogonal projection onto $\text{span}\{e_1, \ldots, e_{i_N}\}$ in H.

EXAMPLE.   For the lattice systems considered in Example (ii) above, a canonical choice for Π_N, $N \in \mathbf{N}$, would be

$$\Pi_N\left(x\right) \; = \; \left(x^i\right)_{i \in \Lambda_N},$$

where $(\Lambda_N)_{N \in \mathbf{N}}$ is an increasing sequence of finite subsets of Λ such that $\bigcup_{N \in \mathbf{N}} \Lambda_N = \Lambda$. For example, $\Lambda_N = \{-N, -N+1, \ldots, N-1, N\}^n$ if $\Lambda = \mathbf{Z}^n$.

REMARK.   (viii) Sometimes, it is appropriate to choose the projections Π_N in a slightly different way. Suppose, for example, E is a space of continuous paths $\omega : [0,1] \to \mathbf{R}^n$, and $K = \{\omega \mapsto \omega^i(s); 0 \le s \le 1, 1 \le i \le n\}$. Then a natural choice for Π_N, $N \in \mathbf{N}$, would be

$$\Lambda_N = \{(j \cdot 2^{-n}, i); 0 \le j \le 2^N, 1 \le i \le n\},$$
$$\Pi_N(\omega) = (\omega^i(s))_{(s,i) \in \Lambda_N}.$$

However, these projections usually don't satisfy the orthonormality condition

$$\left(\Pi_N^{(s,i)}, \Pi_N^{(s',i')} \right)_{H'} = \delta_{ss'} \delta_{ii'}.$$

For example, for the Ornstein–Uhlenbeck operator on the Wiener space (cf. Example (iii) above), $j(\Pi_N^{(s,i)})$ is the function $t \mapsto (s \wedge t) \cdot e_i$ in H, whence

$$\left(\Pi_N^{(s,i)}, \Pi_N^{(s',i')} \right)_{H'} = \left(j(\Pi_N^{(s,i)}), j(\Pi_N^{(s',i')}) \right)_H$$
$$= e_i \cdot e_i' \int_0^1 \frac{d}{dt}(s \wedge t) \frac{d}{dt}(s' \wedge t)\, dt = \delta_{ii'} \cdot (s \wedge s').$$

By orthonormalization of $\left\{ \Pi_N^{(s,i)}; (s,i) \in \Lambda_N \right\}$ w.r.t. $(\cdot,\cdot)_{H'}$, we can construct a normalized projection $\tilde{\Pi}_N$ from Π_N.

b) The generator of the Brownian string : Two examples for non–uniqueness on L^2

In this section, we study two simple but instructive examples of a non–unique symmetric, infinite dimensional diffusion operator on different L^2 spaces. The operator appears as the generator of the random motions of an elastic string considered by T. Funaki [Fun 82, 83]. These random motions are $C([0,1] \to \mathbf{R}^d)$–valued diffusion processes which solve the stochastic evolution equation

$$(5.14) \qquad dX_t = dW_t + \frac{d^2}{ds^2} X_t\, dt$$

for a cylindrical Brownian motion $(W_t)_{t \ge 0}$ on $L^2([0,1] \to \mathbf{R}^d; ds)$, respectively the corresponding stochastic partial differential equation

$$(5.15) \qquad \frac{\partial X}{\partial t} = \frac{\partial W}{\partial t} + \frac{\partial^2 X}{\partial s^2},$$

where $\frac{\partial W}{\partial t}$ is space–time white noise. For notational simplicity, we restrict ourselves to the case $d = 1$, but the case $d > 1$ can be treated analogously. Let $E := C([0,1] \to \mathbf{R})$. It is not difficult to see that a diffusion process solves the equation (5.14) in the distributioinal sense[1] if and only if, on cylinder functions of type

$$F(\omega) = f\left(\int_0^1 g_1(s)\,\omega(s)\,ds, \ldots, \int_0^1 g_n(s)\,\omega(s)\,ds \right), \qquad \omega \in E,$$

with $n \in \mathbf{N}$, $g_1, \ldots, g_n \in C_0^\infty(0,1)$ and $f \in C_b^\infty(\mathbf{R}^n)$, the generator $\mathcal{L}$ of the diffusion process is given by

$$(\mathcal{L}F)(\omega) = \sum_{i,j=1}^n \int_0^1 g_i\, g_j\, ds\ \frac{\partial^2 f}{\partial x_i \partial x_j} \left(\int_0^1 g_1\,\omega\,ds, \ldots, \int_0^1 g_n\,\omega\,ds \right)$$

$$(5.16) \qquad + \sum_{i=1}^n \int_0^1 g_i''\,\omega\,ds\ \frac{\partial f}{\partial x_i} \left(\int_0^1 g_1\,\omega\,ds, \ldots, \int_0^1 g_n\,\omega\,ds \right).$$

In the sequel, we will denote the functional $\omega \mapsto \int_0^1 g\omega\,ds$ on E respectively $L^2(0,1;ds)$, $g \in L^2(0,1;ds)$, by g^*. Instead of $L^2(0,1;ds)$ we briefly write $L^2(0,1)$, or just L^2. For a subspace V of $L^2(0,1)$, let $\mathcal{F}C_b^\infty(V)$ denote the space of all cylinder functions of type $f(g_1^*, \ldots, g_n^*)$ with $n \in \mathbf{N}$, $f \in C_b^\infty(\mathbf{R}^n)$, and $g_1, \ldots, g_n \in V$. In compact notation,

$$(5.17) \qquad\qquad \mathcal{L}F = \Delta_{L^2}F + \langle \beta, DF \rangle$$

for all $F \in \mathcal{F}C_b^\infty(C_0^\infty(0,1))$, where

$$\beta_{g^*}(\omega) = (g'', \omega)_{L^2} \qquad \text{for all } g \in C_0^\infty(0,1) \text{ and } \omega \in E.$$

Note that β_{g^*} is a stochastic integral w.r.t. Brownian motion :

$$(5.18) \qquad \beta_{g^*} = -\int_0^1 g'(s)\,dW_s \qquad P_x\text{-a.s. for all } x \in \mathbf{R},$$

where P_x is the law on E of one–dimensional Brownian motion starting at x, and $W_s(\omega) = \omega(s)$, $0 \le s \le 1$.

In Subsection 1) and 2), we discuss the non–uniqueness of the corresponding diffusion operator on two different L^2 spaces. In the preceeding subsections 3) and 4), we identify the Friedrichs extension and the maximal Markovian extension of the operator explicitly. We also describe the corresponding diffusion processes.

---

[1] An E-valued continuous stochastic process $(X_t)_{t \ge 0}$ defined on a probability space $(\hat{\Omega}, \hat{\mathcal{F}}, \hat{P})$ is called a solution of (5.14) in the distributional sense if and only if the $(C_0^\infty(0,1))'$ valued process $(W_t)_{t \ge 0}$ defined by

$$W_t(g) = (X_t, g)_{L^2} - (X_0, g)_{L^2} - \int_0^t \left(X_\tau, g'' \right)_{L^2} d\tau, \quad g \in C_0^\infty(0,1),$$

is an $L^2(0,1;ds)$ Wiener process, i.e., the process $t \mapsto (W_t(g_1), \ldots, W_t(g_n))$ is a standard Brownian motion in $\mathbf{R}^n$ for all $n \in \mathbf{N}$ and any test–functions $g_1, \ldots, g_n \in C_0^\infty(0,1)$ that are orthonormal w.r.t. the $L^2(0,1;ds)$ inner product.

1) A first example for non–uniqueness

The martingale problem for the operator $(\mathcal{L}, \mathcal{F}C_b^\infty(C_0^\infty(0,1)))$ is not unique. In fact, the solution of the stochastic evolution equation (5.14), respectively the SPDE (5.15), is not uniquely determined by its initial values, but we also have to prescribe boundary values. There exist various random motions of strings, e.g. the random string with fixed end–points, or that with freely moving end–points, cf. [Fun 82, 83]. Both are diffusion processes solving the martingale problem for $(\mathcal{L}, \mathcal{F}C_b^\infty(C_0^\infty(0,1)))$.

Similarly, it turns out that the operator $(\mathcal{L}, \mathcal{F}C_b^\infty(C_0^\infty(0,1)))$ is symmetric but **not Markov unique** on $L^2(E; P_0)$. In fact, let $H_0^{1,2}(0,1)$ denote the Sobolev space of all absolutely continuous functions $h : [0,1] \to \mathbf{R}$ such that h' is square integrable and $h(0) = h(1) = 0$. Let $P_{x,y}$ be the law of the Brownian bridge from x to y, $x, y \in \mathbf{R}$. We have :

Lemma 5.1 *(i) The operator* $(\mathcal{L}, \mathcal{F}C_b^\infty(C_0^\infty(0,1)))$ *is symmetric w.r.t. each probability measure P of type*

$$P = \int P_{x,y} \; \sigma\,(dx dy) \,,$$

where σ is an arbitrary probability measure on $\mathbf{R}^2$.

(ii) For each probability measure P as in (i), the space $\mathcal{F}C_b^\infty(H_0^{1,2}(0,1))$ is contained in the domain of the closure $\bar{\mathcal{L}}$ of the operator $(\mathcal{L}, \mathcal{F}C_b^\infty(C_0^\infty(0,1)))$ on $L^2(E; P)$, and

$$\bar{\mathcal{L}}F = \Delta_{L^2} F + \langle \beta, DF \rangle \qquad P\text{-a.s.}$$

for all $F \in \mathcal{F}C_b^\infty(H_0^{1,2}(0,1))$, where $\beta : \{g^*; g \in H_0^{1,2}(0,1)\} \to L^2(E; P)$ is defined by

$$\beta_{g^*} := -\int_0^1 g'(s) \; dW_s \qquad P\text{-a.s.}$$

Lemma 5.2 *The operator* $(\mathcal{L}, \mathcal{F}C_b^\infty(C_0^\infty(0,1)))$ *is not Markov unique on $L^2(E; P_0)$.*

In particular, the operator is not $L^p(E; P_0)$ unique for any $p \in [1, \infty)$, cf. Section e) in Chapter 1.

REMARKS. (i) Note that the non–Markov uniqueness of the operator $(\mathcal{L}, \mathcal{F}C_b^\infty(C_0^\infty(0,1)))$ does not directly follow from the non–uniqueness of the corresponding martingale problem, because different solutions of the martingale problem may have very different symmetrizing measures.

(ii) Uniqueness of the operator $(\mathcal{L}, \mathcal{F}C_b^\infty(C_0^\infty(0,1)))$ on $L^2(E; P)$ for some symmetrizing probability measure P can only be expected, if the semigroup generated by the Friedrichs extension is not ergodic, or if $P = P_{x,y}$ for some $x, y \in \mathbf{R}$, i.e., P is an extremal in the convex set of all symmetrizing measures — cf. the proof of Lemma 5.2, and see also Lemma 5.3 below, where we look at a second example of a probability measure P such that $(\mathcal{L}, \mathcal{F}C_b^\infty(C_0^\infty(0,1)))$ is symmetric but not Markov unique on $L^2(E; P)$.

(iii) In Subsections 2) and 3), we calculate the Friedrichs extension and the maximal Markovian extension of the operator $(\mathcal{L}, \mathcal{F}C_b^\infty(C_0^\infty(0,1)))$ on $L^2(E; P_0)$ explicitly. It turns out that $\mathcal{F}C_b^\infty(H_{CM})$ is an operator core for the Friedrichs extension, where H_{CM} denotes the Cameron–Martin space. The corresponding diffusion process solves the equation (5.14) for the Laplacian Δ $(= \frac{d^2}{ds^2})$ on $L^2(0,1; ds)$ with boundary conditions $h(0) = 0$ and $h'(1) = 0$. The diffusion process corresonding to the maximal Markovian extension solves (5.14) for the Laplacian on $L^2(0,1; ds)$ with Dirichlet boundary conditions. Note that this is converse to extensions of symmetric diffusion operators on finite dimensional domains, where the Friedrichs extension satisfies Dirichlet boundary conditions, and the maximal Markovian extension satisfies Neumann boundary conditions. If we would analyze the operator $\mathcal{L}$ on $L^2(E; P_{dx})$ instead of $L^2(E; P_0)$, then the Friedrichs extension would correspond to Neumann boundary conditions at *both* boundaries, and the maximal Markovian extension to Dirichlet boundary conditions. Since we do the analysis on $L^2(E; P_0)$, the Neumann conditions for the Friedrichs extension appear only at one boundary.

PROOF OF LEMMA 5.1. (i) This is a consequence of the Cameron–Martin theorem. We first show that $P_{0,0}$ is a symmetrizing measure. Indeed, suppose we endow $H_0^{1,2}(0,1)$ with the inner product

$$\left(h, \tilde{h} \right)_{CM} := \int_0^1 h'(s)\, \tilde{h}'(s)\, ds .$$

Then $(E, H_0^{1,2}(0,1), P_{0,0})$ is an **abstract Wiener space**, i.e., $H_0^{1,2}(0,1)$ is densely and continuously embedded into E, and $P_{0,0}$ is a mean zero Gaussian measure on E such that each $\ell \in E'$ is $N(0, \|J(\ell)\|_{CM}^2)$ distributed, where $J : (H_0^{1,2}(0,1))' \to H_0^{1,2}(0,1)$ is the Riesz isometry, cf. e.g. [MaRö 92, Ch. II, Sect. 3 c)]. Hence the general version of the Cameron–Martin theorem implies that the integration by parts identity

$$E_{0,0}[\partial_h F] = E_{0,0}\left[F \int_0^1 h'(s)\, dW_s \right]$$

holds for all $h \in H_0^{1,2}(0,1)$ and $F \in \mathcal{F}C_b^\infty(E')$, see e.g. [MaRö 92, Ch. II, Thm. 3.11]. By (5.17) and (5.18), it is now easy to conclude (cf. Section a), 5)), that

$$-E_{0,0}[F\, \mathcal{L}G] = E_{0,0}[(\nabla F, \nabla G)_{L^2}] \quad \text{for all } F, G \in \mathcal{F}C_b^\infty(C_0^\infty(0,1)),$$

where the L^2 gradient ∇F of a cylinder function $F = f(g_1^*, \ldots, g_n^*)$ is given by

$$(\nabla F)(\omega) = \sum_{i=1}^n g_i\, \frac{\partial f}{\partial x_i}(g_1^*(\omega), \ldots, g_n^*(\omega)) \quad \text{for all } \omega \in E.$$

This proves the symmetry w.r.t. $P_{0,0}$.

Now fix arbitrary $x, y \in \mathbf{R}$. Let $\varphi_{x,y}(s) := (1-s)x + sy$, $0 \leq s \leq 1$. It is well-known that $P_{x,y}$ is the image of $P_{0,0}$ under the translation $\omega \mapsto \omega + \varphi_{x,y}$ on E. Thus, for $h \in H_0^{1,2}(0,1)$ and $F \in \mathcal{F}C_b^\infty(E')$,

$$E_{x,y}[\partial_h F] = E_{0,0}[(\partial_h F)(\varphi_{x,y} + \bullet)] = E_{0,0}[\partial_h(F(\varphi_{x,y} + \bullet))]$$

$$= E_{0,0}[F(\varphi_{x,y} + \bullet) \int_0^1 h'(s)\, dW_s] = E_{x,y}[F \int_0^1 h'(s)\, dW_s],$$

because $\int_0^1 h'(s)\, d\varphi_{x,y}(s) = (y-x) \cdot \int_0^1 h'(s)\, ds = 0$. Now, we again obtain

$$-E_{x,y}[F\,\mathcal{L}G] = E_{x,y}[(\nabla F, \nabla G)_{L^2}] \qquad \text{for all } F, G \in \mathcal{F}C_b^\infty(C_0^\infty(0,1)).$$

Since this holds for every $x, y \in \mathbf{R}$, it implies the first assertion.

(ii) Let P be a probability measure as in (i). Fix $F \in \mathcal{F}C_b^\infty(H_0^{1,2}(0,1))$, $F = f(g_1^*, \ldots, g_n^*)$ for some $n \in \mathbf{N}$, $f \in C_b^\infty(\mathbf{R}^n)$, and $g_1, \ldots, g_n \in H_0^{1,2}(0,1)$. Let $g_1^{(k)}, \ldots, g_n^{(k)}$, $k \in \mathbf{N}$, be a sequence of functions in $C_0^\infty(0,1)$ such that $g_i^{(k)} \to g_i$ and $\left(g_i^{(k)}\right)' \to g_i'$ in $L^2(0,1)$ as $k \to \infty$ for all $1 \leq i \leq n$. For $k \in \mathbf{N}$ let $F_k := f((g_1^{(k)})^*, \ldots, (g_n^{(k)})^*)$. Then for $1 \leq i \leq n$ and $k \in \mathbf{N}$, $\beta_{(g_i^{(k)})^*}$ is the stochastic integral $-\int_0^1 (g_i^{(k)})'(s)\, dW_s$, which converges to $-\int_0^1 g_i'(s)\, dW_s$ in $L^2(E; P)$ as $k \to \infty$, since $(g_i^{(k)})' \to g_i'$ in $L^2(0,1)$. Now, it is not difficult to see that $F_k \to F$ in $L^2(E; P)$, and

$$\mathcal{L}F_k \to \Delta_{L^2}F + \langle \beta, DF \rangle \qquad \text{in } L^2(E; P)$$

as $k \to \infty$. Hence F is in the domain of the closure $\bar{\mathcal{L}}$ on $L^2(E; P)$, and $\bar{\mathcal{L}}F = \Delta_{L^2}F + \langle \beta, DF \rangle$. ∎

PROOF OF LEMMA 5.2. By Lemma 5.1, P_0 is a symmetrizing measure for the operator $(\mathcal{L}, \mathcal{F}C_b^\infty(C_0^\infty(0,1)))$, but it is not an extremal point in the set of all symmetrizing measures. Hence the C^0 semigroup generated by the maximal Markovian extension of $(\mathcal{L}, \mathcal{F}C_b^\infty(C_0^\infty(0,1)))$ is not ergodic, cf. Theorem 3.8. On the other hand, the C^0 semigroup generated by the Friedrichs extension of the operator is ergodic, because the Friedrichs extension has a spectral gap. Indeed, for $F \in \mathcal{F}C_b^\infty(C_0^\infty(0,1))$ such that $F = f(g_1^*, \ldots, g_n^*)$ for some $n \in \mathbf{N}$, $f \in C_b^\infty(\mathbf{R}^n)$, and functions $g_1, \ldots, g_n \in C_0^\infty(0,1)$ that are orthonormal in $L^2(0,1)$, we have

$$-E_0[F\,\mathcal{L}F] = E_0[(\nabla F, \nabla F)_{L^2}] = E_0[|\nabla f|^2 (g_1^*, \ldots, g_n^*)]$$

$$\geq E_0\left[\sum_{i,j=1}^n \int_0^1 \int_0^1 (s \wedge t)\, g_i(s)\, g_j(t)\, dt\, ds\, \frac{\partial f}{\partial x_i}(g_1^*, \ldots, g_n^*)\, \frac{\partial f}{\partial x_j}(g_1^*, \ldots, g_n^*)\right]$$

$$= E_0[|\nabla_M F|_{CM}^2] \geq E_0[(F - E_0[F])^2],$$

where E_0 denotes expectation w.r.t. P_0, ∇_M is the Malliavin gradient, and $|\cdot|_{\text{CM}}$ is the usual norm on the Cameron–Martin space. Thus the Friedrichs extension and the maximal Markovian extension of $(\mathcal{L}, \mathcal{F}C_b^\infty(C_0^\infty(0,1)))$ on $L^2(E; P_0)$ do not coincide, i.e., the operator is not Markov–unique. ∎

As obvious as the non–Markov uniqueness of the operator above is, as surprising it is from a finite dimensional point of view. In fact, we have shown in Section f) in Chapter 3 that a symmetric diffusion operator defined on $C_0^\infty(E)$, where E is a domain in $\mathbf{R}^n$, is *always Markov unique provided the diffusion matrix is non–degenerate, and the diffusion process generated by the Friedrichs extension is conservative*. In our example, $\mathcal{L}1 = 0$, whence *the semigroup* $(T_t)_{t \geq 0}$ *generated by the Friedrichs extension is conservative, i.e.,* $T_t 1 = 1$ *for all* $t \geq 0$, *but nevertheless the operator is not Markov unique.* Hence we are confronted with a *new infinite dimensional phenomenon which causes non–uniqueness*.

2) A refined example for non–uniqueness

The example presented above provoques the following objection :

Objection :   Non–uniqueness only occurs because we consider the "wrong" cylinder functions. The space $\mathcal{F}C_b^\infty(H_{\text{CM}})$ is an operator core for the Friedrichs extension.

One should bear in mind, that non–uniqueness *always* occurs, because we consider the operator on the "wrong" test–functions (which do not form a core). Nevertheless, this does not refute the objection above. In fact, we will prove in Section f) below, that for an infinite dimensional diffusion operator $\mathcal{L}$ of type $\Delta_H + \langle \beta, D \cdot \rangle$ with *linear drift* β determined by a linear operator (A, V) on the "tangent space" H, L^p uniqueness for $p < 2$ holds under very weak conditions, provided we consider $\mathcal{L}$ on cylinder functions of type $f((g_1, \bullet)_H, \dots, (g_n, \bullet)_H)$ with $g_1, \dots, g_n \in V$, and A is (for example) essentially self–adjoint on V. Note that in the example from Subsection 1), the operator $(\frac{d^2}{ds^2}, C_0^\infty(0,1))$ on $L^2(0,1)$ is indeed not essentially self–adjoint, whereas the space of all C^2 functions $g : [0,1] \to \mathbf{R}$ such that $g(0) = 0$ and $g'(0) = 0$ is a domain of essential self–adjointness, which is contained in H_{CM}.

Now suppose again that $E = C([0,1] \to \mathbf{R})$, and $(\mathcal{L}, \mathcal{F}C_b^\infty(C_0^\infty(0,1)))$ is the diffusion operator defined by (5.16). The considerations above suggest, that we should perhaps consider a weakened uniqueness problem of the following type :

Problem :   Suppose P is a probability measure on E such that the operator $(\mathcal{L}, \mathcal{F}C_b^\infty(C_0^\infty(0,1)))$ is symmetric on $L^2(E; P)$. Does there exist a subspace $K \subseteq E'$ such that $\mathcal{F}C_b^\infty(K)$ is a core (or, less restrictively, a domain of Markov uniqueness) for the Friedrichs extension of $(\mathcal{L}, \mathcal{F}C_b^\infty(C_0^\infty(0,1)))$?

As remarked above, the answer is "Yes" if $P = P_0$. We will now look at a symmetrizing measure for which the answer is "No". Let $T : E \to E$ denote the *time reversal*, i.e.,

$$(T\omega)(s) = \omega(1-s) \qquad \text{for all } s \in [0,1] \text{ and } \omega \in E.$$

We define a probability measure $\bar{P}_0$ on E by

$$\bar{P}_0 := \frac{1}{2}\left(P_0 + P_0 \circ T^{-1}\right).$$

Clearly, $\bar{P}_0$ is a mixture of Brownian bridge measures, whence it is symmetrizing for the operator $(\mathcal{L}, \mathcal{F}C_b^\infty(C_0^\infty(0,1)))$, cf. Lemma 5.1. Let L^0 denote the *Friedrichs extension* of this operator on $L^2(E; \bar{P}_0)$.

Lemma 5.3 *(i) A functional $\ell \in E'$ is contained in the domain of L^0 if and only if $\ell = h^*$ for some $h \in H_0^{1,2}(0,1)$.*
(ii) The operator $(L^0, \mathcal{F}C_b^\infty(H_0^{1,2}(0,1)))$ is not Markov unique on $L^2(E; \bar{P}_0)$.

Hence there does not exist a subspace $K \subseteq E'$ such that K is contained in the domain of the Friedrichs extension L^0, and $(L^0, \mathcal{F}C_b^\infty(K))$ is Markov unique.

PROOF. (i) By Lemma 5.1 (ii), it is not difficult to see that the functionals h^*, $h \in H_0^{1,2}(0,1)$, are even contained in the domain of the closure of $(\mathcal{L}, \mathcal{F}C_b^\infty(C_0^\infty(0,1)))$ on $L^2(E; \bar{P}_0)$. Thus they are contained in the domain of L^0, which proves the "if"–part of the assertion.

Now let ℓ be an arbitrary functional in E' that is in the domain of L^0. We first show $\ell = g^*$ for some $g \in L^2(0,1)$. The functional ℓ is in particular contained in the domain of the closure $\bar{\mathcal{E}}$ of the quadratic form $(\mathcal{E}, \mathcal{F}C_b^\infty(C_0^\infty(0,1)))$ on $L^2(E; \bar{P}_0)$,

$$\mathcal{E}(F, G) = -\bar{E}_0[F \mathcal{L} G] = \bar{E}_0[(\nabla F, \nabla G)_{L^2}],$$

where $\bar{E}_0$ is expectation w.r.t. $\bar{P}_0$, and ∇ denotes the L^2 gradient. We have

$$\bar{\mathcal{E}}(F, G) = \bar{E}_0[(\bar{\nabla} F, \bar{\nabla} G)_{L^2}]$$

for all F, G in the domain of $\bar{\mathcal{E}}$, where $\bar{\nabla}$ is the closure of the operator $\nabla : L^2(E; \bar{P}_0) \to L^2(E \to L^2(0,1); \bar{P}_0)$. Now note that for $g \in C_0^\infty(0,1)$, the ordinary directional derivative $\partial_g \ell$ exists, and is equal to $\ell(h)$. Moreover,

$$\bar{E}_0[\partial_g \ell\, G] = -\bar{E}_0[\ell\,(\partial_g G + \beta_{g^*} \cdot G)] = \bar{E}_0[(g, \bar{\nabla} \ell)_{L^2}\, G]$$

for all $G \in \mathcal{F}C_b^\infty(C_0^\infty(0,1))$, where $\beta_{g^*} = -\int_0^1 g'(s)\, dW_s$ $\bar{P}_0$–a.s. Here the left equation holds directly by the Cameron-Martin theorem, cf. the proof of Lemma 5.1 (i), and the right equation holds because the integration by parts identity for the gradient $(\nabla, \mathcal{F}C_b^\infty(C_0^\infty(0,1)))$ extends by continuity to an integration by parts identity for $\bar{\nabla}$. Thus

$$(g, \bar{\nabla} \ell)_{L^2} = \partial_g \ell = \ell(g)$$

for all $g \in C_0^\infty(0,1)$, whence ℓ is continuous on $C_0^\infty(0,1)$ w.r.t. the $L^2(0,1)$ norm. We obtain $\ell = h^*$ for some $h \in L^2(0,1)$, and $\bar{\nabla}\ell = h$.

Next, we use that ℓ is not only contained in the domain of the form $\bar{\mathcal{E}}$, but even in the domain of its generator L^0. Thus

$$(5.19) \qquad \bar{E}_0[\partial_h G] = \bar{E}_0[(\bar{\nabla}\ell, \bar{\nabla}G)_{L^2}] = \bar{E}_0[L^0\ell \, G]$$

for all $G \in \mathcal{F}C_b^\infty(C_0^\infty(0,1))$.

The outer equation in (5.19) is an integration by parts formula for the measure $\bar{P}_0$. We finally show that this formula implies that h is in $H_0^{1,2}(0,1)$. Obviously, this completes the proof of Assertion (i).

Suppose first that h is not in $H^{1,2}(0,1)$. Since the embedding of $H^{1,2}(0,1)$ into $L^2(0,1)$ is Hilbert–Schmidt, there exists an intermediate (separable) Hilbert space H_h, $H^{1,2}(0,1) \subset H_h \subset L^2(0,1)$, such that the embedding of $H^{1,2}(0,1)$ into H_h is Hilbert–Schmidt, the embedding of H_h into $L^2(0,1)$ is continuous, and h is not in H_h. Let $\tilde{H} := H_h \oplus \mathrm{span}\{h\}$. We define an inner product on $\tilde{H}$ in the obvious way. We then have

$$H^{1,2}(0,1) \subset H_h \subset \tilde{H} \subsetneq L^2(0,1),$$

where the first embedding is Hilbert–Schmidt, and the other embeddings are continuous. We extend the measures P_0 and $\bar{P}_0$ trivially to $L^2(0,1)$. Since the embedding of the Cameron–Martin space into H_h is Hilbert–Schmidt, we have $P_0[H_h] = 1$. Similarly, $P_0 \circ T^{-1}[H_h] = 1$, whence $\bar{P}_0[H_h] = 1$. Thus the restriction of $\bar{P}_0$ to $\tilde{H}$ is a probability measure with support contained in the closed subspace H_h. Let $\mathcal{F}C_b^\infty(\tilde{H}')$ be the space of all smooth cylinder functions on $\tilde{H}$ based on functionals in $\tilde{H}'$. Since $\tilde{H}$ is densely and continuously embedded into $L^2(0,1)$, the space $\{g^*; g \in C_0^\infty(0,1)\}$ is a dense subspace of $\tilde{H}'$. By (5.19), we hence obtain

$$(5.20) \qquad \bar{E}_0[\partial_h G] = \bar{E}_0[Z_h \, G] \qquad \text{for all } G \in \mathcal{F}C_b^\infty(\tilde{H}'),$$

where $\bar{P}_0$ is now viewed as a probability measure on $\tilde{H}$, and Z_h denotes the element in $L^2(\tilde{H}; \bar{P}_0)$ corresponding to the element $L^0\ell$ in $L^2(E; \bar{P}_0)$. However, by a result of S. Albeverio, S. Kusuoka and M. Röckner [AlbKusRö 90, Thm. 2.5], an integration by parts identity of type (5.20) cannot hold, because $P_0[H_h] = 1$. In fact, we have the trivial disintegration

$$\int_{\tilde{H}} u(z)\, \bar{P}_0(dz) = \int_{H_h} u(x)\, \bar{P}_0(dx) = \int_{H_h}\int_{\mathbf{R}} u(x+sk)\, \rho_h(x, ds)\, \bar{P}_0(dx)$$

for every bounded measurable function u on $\tilde{H}$, where $\rho_h(x, ds) := \delta_0(ds)$ for all x. By Theorem 2.5 in [AlbKusRö 90], (5.20) now implies that $\rho_h(x \cdot)$ is an absolutely continuous function for $\bar{P}_0$-a.e. $x \in H_h$. This is obviously a contradiction, whence h has to be in $H^{1,2}(0,1)$.

Now suppose that h is in $H^{1,2}(0,1)$ but not in $H_0^{1,2}(0,1)$. Let $\tilde{h}$ be the absolutely continuous ds–version of h. Without loss of generality, we may assume $\tilde{h}(0) \neq 0$. Let $E_0 := \{\omega \in E;\ \omega(0) = 0\}$. Obviously, E_0 is a closed subspace of E, and $E = E_0 \oplus \operatorname{span}\{\tilde{h}\}$. Again, we can disintegrate the measure $\bar{P}_0$, i.e., there exist a probability measure ν_h on E_0, and a kernel $\rho_h : E_0 \times \mathcal{B}(\mathbf{R}) \to [0,1]$ such that

$$\int_E u(z)\ \bar{P}_0(dz) \;=\; \int_{E_0} \int_{\mathbf{R}} u(x + s\tilde{h})\ \rho_h(x, ds)\ \nu_h(dx)$$

for every bounded measurable function u on E. In particular,

$$\int_{E_0} \rho_h(x, \{0\})\ \nu_h(dx) \;=\; \bar{P}_0[E_0] \;=\; \frac{1}{2},$$

whence $\rho_h(x, \cdot)$ is **not** for ν_h–a.e. $x \in E_0$ absolutely continuous. By [AlbKusRö 90], this is a contradiction to the integration by parts identity

$$\int \partial_h G\ d\bar{P}_0 \;=\; \int L_0 \ell\ G\ d\bar{P}_0,$$

which holds for all $G \in \mathcal{F}C_b^\infty(E')$, as one can show easily by (5.19) and an approximation argument. Hence $\tilde{h}(0) = 0$, and, similarly, $\tilde{h}(1) = 0$, i.e., h is in $H_0^{1,2}(0,1)$. This completes the proof of (i).

(ii) There are various ways to see that the operator $(L^0,\ \mathcal{F}C_b^\infty(H_0^{1,2}(0,1)))$ is not Markov unique on $L^2(E;\ \bar{P}_0)$. The most explicit way is perhaps to show the difference between the corresponding strong and weak Sobolev space. Note that by Lemma 5.1 (i), $\mathcal{F}C_b^\infty(H_0^{1,2}(0,1))$ is contained in the domain of the closure $\bar{\mathcal{L}}$ of the operator $(\mathcal{L},\ \mathcal{F}C_b^\infty(C_0^\infty(0,1)))$ on $L^2(E;\ \bar{P}_0)$. Since L^0 is the Friedrichs extension of $\mathcal{L}$, L^0 and $\bar{\mathcal{L}}$ coincide on $\mathcal{F}C_b^\infty(H_0^{1,2}(0,1))$. Thus it is enough to show that the operator $(\mathcal{L},\ \mathcal{F}C_b^\infty(C_0^\infty(0,1)))$ is not Markov unique on $L^2(E;\ \bar{P}_0)$. The corresponding strong Sobolev space $H_0^{1,2}(E,\ L^2(0,1);\ \bar{P}_0)$ is the domain of the closure of the quadratic form

$$\mathcal{E}(F,G) \;=\; \int (\nabla F, \nabla G)_{L^2}\ d\bar{P}_0, \qquad F, G \in \mathcal{F}C_b^\infty(C_0^\infty(0,1)),$$

where ∇ is the L^2 gradient, cf. also the next subsection. Since $P_0 \leq 2\bar{P}_0$, every function in $H_0^{1,2}(E,\ L^2(0,1);\ \bar{P}_0)$ represents a P_0–class in the corresponding strong Sobolev space $H_0^{1,2}(E,\ L^2(0,1);\ P_0)$ w.r.t. P_0.

We now show that the function $W_1 : \omega \mapsto \omega(1)$ is contained in the weak Sobolev space correponding to the operator $(\mathcal{L},\ \mathcal{F}C_b^\infty(C_0^\infty(0,1)))$ on $L^2(E;\ \bar{P}_0)$, but not in the corresponding strong Sobolev space. It is easy to see that W_1 is in the weak Sobolev space corresponding to $(\mathcal{L},\ \mathcal{F}C_b^\infty(C_0^\infty(0,1)))$ w.r.t. the measure P_0, cf. Subsection 4), in particular Lemma 5.6. The same is true with P_0 replaced by $P_0 \circ T^{-1}$, and thus also with P_0 replaced by $\bar{P}_0$. In any case, the corresponding weak gradient of W_1 vanishes. If W_1 would be in $H_0^{1,2}(E,\ L^2(0,1);\ P_0)$ as well, then the strong gradient of W_1 would also vanish, i.e., W_1 would be in the kernel of the Friedrichs extension of the operator $(\mathcal{L}, \mathcal{F}C_b^\infty(C_0^\infty(0,1)))$ on $L^2(E; P_0)$. This is not possible, since the kernel consists only of constant functions, cf. the proof of Lemma 5.2. Thus W_1 is not in $H_0^{1,2}(E, L^2(0,1); P_0)$, and therefore, not in $H_0^{1,2}(E, L^2(0,1); \bar{P}_0)$ either. Hence the weak and strong Sobolev space corresponding to the operator $(\mathcal{L}, \mathcal{F}C_b^\infty(C_0^\infty(0,1)))$ on $L^2(E; \bar{P}_0)$ do not coincide, i.e., the operator is not Markov unique. This completes the proof of Lemma 5.3. ∎

3) Strong Sobolev space and Friedrichs extension of $\mathcal{L}$

Let H_{CM} denote the Cameron–Martin space. We extend the map β introduced above to the space $\{h^* ; h \in H_{\text{CM}}\}$ by defining

$$(5.21) \qquad \beta_{h^\cdot} := -\int_0^1 h'(s) \, dW_s \qquad \text{for all } h \in H_{\text{CM}}.$$

Clearly, β takes values in $L^2(E; P_0)$, and, by the Cameron–Martin theorem,

$$(5.22) \qquad E_0[\partial_h F \, G] = -E_0[F \, \partial_h G] - E_0[\beta_{h^\cdot} \, F G]$$

for all $h \in H_{\text{CM}}$ and $F, G \in \mathcal{F}C_b^\infty(E')$, cf. e.g. [MaRö 92, Ch. II, Thm. 3.11]. Let $(\mathcal{E}, \mathcal{F}C_b^\infty(C_0^\infty(0,1)))$ be the quadratic form on $L^2(E; P_0)$ corresponding to the operator $\mathcal{L}$. By (5.22),

$$\mathcal{E}(F, G) = -E_0[\mathcal{L}F \, G] = E_0[(\nabla F, \nabla G)_{L^2}]$$

for all $F, G \in \mathcal{F}C_b^\infty(C_0^\infty(0,1))$, where

$$(5.23) \qquad \nabla(f(g_1^*, \ldots, g_n^*)) = \sum_{i=1}^n g_i \frac{\partial f}{\partial x_i}(g_1^*, \ldots, g_n^*)$$

for all $n \in \mathbf{N}$, $g_1, \ldots, g_n \in C_0^\infty(0,1)$, and $f \in C_b^\infty(\mathbf{R}^n)$. Since $\mathcal{L}$ is a diffusion operator, $\mathcal{E}$ is a pre–Dirichlet form, cf. the proof of Lemma 1.10. We denote the closure again by $\mathcal{E}$, and its domain by $H_0^{1,2}(E, L^2(0,1); P_0)$, or, briefly, by $H_0^{1,2}\langle\nabla\rangle$. These notations are justified, because $H_0^{1,2}(E, L^2(0,1); P_0)$ is the domain of the closure $\bar\nabla$ of the L^2 gradient $\nabla : \mathcal{F}C_b^\infty(C_0^\infty(0,1)) \subset L^2(E; P_0) \to L^2(E \to L^2(0,1); P_0)$, which allows us to think of it as a strong Sobolev space over E equipped with tangent space $L^2(0,1)$. Not only the cylinder functions based on $C_0^\infty(0,1)$ functions are in the strong Sobolev space, but even those based on $L^2(0,1)$ functions :

Lemma 5.4 $\mathcal{F}C_b^\infty(L^2(0,1)) \subset H_0^{1,2}(E, L^2(0,1); P_0)$, and

$$\mathcal{E}(F, G) = E_0[(\nabla F, \nabla G)_{L^2}] \qquad \text{for all } F, G \in \mathcal{F}C_b^\infty(L^2(0,1)),$$

where the gradient $\nabla F \in L^2(E \to L^2(0,1); P_0)$ of a function F in $\mathcal{F}C_b^\infty(L^2(0,1))$ is also defined by (5.23).

The proof is an easy approximation argument similar to that used to prove that $\mathcal{F}C_b^\infty(H_0^{1,2}(0,1))$ is contained in the domain of $\bar{\mathcal{L}}$, cf. Lemma 5.1 (ii).

REMARK. The assertion of Lemma 5.4 also holds with P_0 replaced by an arbitrary measure P of type $P = \int P_{x,y} \, \sigma(dxdy)$, where σ is a probability measure on $\mathbf{R}^2$.

We now calculate the *Friedrichs extension* L^0 of $(\mathcal{L}, \mathcal{F}C_b^\infty(C_0^\infty(0,1)))$, i.e., the negative definite self–adjoint operator on $L^2(E; P_0)$ associated with the Dirichlet form $(\mathcal{E}, H_0^{1,2}(\nabla))$.

Lemma 5.5 *The space $\mathcal{F}C_b^\infty(H_{\mathrm{CM}})$ is an operator core for the generator L^0 of the Dirichlet form $(\mathcal{E}, H_0^{1,2}(\nabla))$ on $L^2(E; P_0)$. On functions $F \in \mathcal{F}C_b^\infty(H_{\mathrm{CM}})$, the operator L^0 is given by*

(5.24)
$$L^0 F = \Delta_{L^2} F + \langle \beta, DF \rangle \qquad P_0\text{–}a.s.,$$

where β is defined by (5.21).

REMARKS. (i) Let C_N^2 denote the space of all twice continuously differentiable functions $h : [0,1] \to \mathbf{R}$ that satisfy the boundary conditions $h(0) = 0$ and $h'(1) = 0$. We use the notation C_N^2, although the Neumann boundary condition holds only at 1 but not at 0. If we would analyze the operator on $L^2(E; P_{dx})$ instead of $L^2(E; P_0)$, then Neuman boundary conditions would appear both at 0 and 1. Integration by parts in (5.21) yields

$$L^0 F = \sum_{i,j=1}^n (h_i, h_j)_{L^2} \frac{\partial^2 f}{\partial x_i \partial x_j}(h_1^*, \ldots, h_n^*) + \sum_{i=1}^n (h_i'')^* \frac{\partial f}{\partial x_i}(h_1^*, \ldots, h_n^*)$$

P_0–a.s. for all $F = f(h_1^*, \ldots, h_n^*)$ with $n \in \mathbf{N}$, $f \in C_b^\infty(\mathbf{R}^n)$, and $h_1, \ldots h_n \in C_N^2$.

(ii) In the proof of Lemma 5.5, we give an explicit formula for the C^0 semigroup generated by L^0, cf. (5.26). It is obvious from this representation, that the *diffusion process generated by L^0 is the solution of the stochastic evolution equation (5.14)*, provided Δ denotes the operator $\frac{d^2}{ds^2}$ on $L^2(0,1; ds)$ with boundary conditions $h(0) = 0$ and $h'(1) = 0$..

PROOF OF LEMMA 5.5. Fix $F \in \mathcal{F}C_b^\infty(H_{\mathrm{CM}})$. We choose $n \in \mathbf{N}$, $f \in C_b^\infty(\mathbf{R}^n)$, and $h_1, \ldots, h_n \in H_{\mathrm{CM}}$, such that $F = f(h_1^*, \ldots, h_n^*)$. Let $G \in \mathcal{F}C_b^\infty(C_0^\infty(0,1))$. Then

$$(\nabla F, \nabla G)_{L^2} = \sum_{i=1}^n \frac{\partial f}{\partial x_i}(h_1^*, \ldots, h_n^*) \, \partial_{h_i} G.$$

By (5.22), we obtain

$$
\begin{aligned}
\mathcal{E}(F, G) &= E_0 \left[(\nabla F, \nabla G)_{L^2} \right] \\
&= -\sum_{i=1}^n E_0 \left[\partial_{h_i} \left(\frac{\partial f}{\partial x_i}(h_1^*, \ldots, h_n^*) \right) G \right]
\end{aligned}
$$

$$-\sum_{i=1}^{n} E_0 \left[\beta_{h_i^*} \frac{\partial f}{\partial x_i} (h_1^*, \ldots, h_n^*) \, G \right]$$

$$= -\sum_{i,j=1}^{n} E_0 \left[(h_i, h_j)_{L^2} \frac{\partial^2 f}{\partial x_i \partial x_j} (h_1^*, \ldots, h_n^*) \, G \right]$$

$$-\sum_{i=1}^{n} E_0 \left[\beta_{h_i^*} \frac{\partial f}{\partial x_i} (h_1^*, \ldots, h_n^*) \, G \right]$$

$$= -E_0 \left[(\Delta_{L^2} F + \langle \beta, DF \rangle) \, G \right].$$

This proves that F is in the domain of L^0, and $L^0 F$ is given by (5.24), because $\mathcal{F}C_b^\infty(C_0^\infty(0,1))$ is dense in $H_0^{1,2}(\nabla)$ w.r.t. the inner product $\mathcal{E}(\cdot, \cdot) + (\cdot, \cdot)_{L^2(E; P_0)}$.

Thus $\mathcal{F}C_b^\infty(H_{\mathrm{CM}})$ is contained in $\mathrm{Dom}(L^0)$. To show that it is a core, let Δ^N denote the operator $\frac{d^2}{ds^2}$ on $L^2(0,1)$ with boundary conditions $h(0) = 0$ and $h'(1) = 0$, i.e., Δ^N is the closure of the essentially self-adjoint operator $(\frac{d^2}{ds^2}, C_N^2)$ on $L^2(0,1)$, where C_N^2 is defined as in Remark (i) above. We fix an orthonormal basis $\{e_i; i \in \mathbf{N}\}$ of $L^2(0,1)$ consisting of the eigenfunctions of Δ^N. The corresponding eigenvalues are denoted by λ_i, $i \in \mathbf{N}$. Let $\mathcal{F}C_b^\infty(\{e_i\})$ be the space of all cylinder functions of type $f(e_1^*, \ldots, e_n^*)$ with $n \in \mathbf{N}$ and $f \in C_b^\infty(\mathbf{R}^n)$. The eigenfunctions e_i, $i \in \mathbf{N}$, are in C_N^2. In particular, $\mathcal{F}C_b^\infty(\{e_i\})$ is a subspace of $\mathcal{F}C_b^\infty(H_{\mathrm{CM}})$. For $n \in \mathbf{N}$ and $\omega \in E$, let $\Pi_n(\omega) := (e_1^*(\omega), \ldots, e_n^*(\omega))$. By Remark (i) above, we have

$$L^0 (f \circ \Pi_n) = \sum_{i=1}^{n} \left(\frac{\partial^2 f}{\partial x_i^2} \circ \Pi_n + (e_i'')^* \frac{\partial f}{\partial x_i} \circ \Pi_n \right)$$

$$(5.25) \qquad = \sum_{i=1}^{n} \left(\frac{\partial^2 f}{\partial x_i^2} \circ \Pi_n + \lambda_i e_i^* \frac{\partial f}{\partial x_i} \circ \Pi_n \right) = (\mathcal{L}^n f) \circ \Pi_n$$

P_o–a.s. for all $n \in \mathbf{N}$ and $f \in C_b^\infty(\mathbf{R}^n)$, where $\mathcal{L}^n$ denotes the operator on $C_b^\infty(\mathbf{R}^n)$ given by

$$(\mathcal{L}^n f) (x) = (\Delta f) (x) + \sum_{i=1}^{n} \lambda_i x_i \frac{\partial f}{\partial x_i} (x).$$

Note that $\mathcal{L}^n$ is the generator of the $\mathbf{R}^n$–valued Ornstein–Uhlenbeck process composed of n independent one–dimensional Ornstein–Uhlenbeck processes with parameters λ_i, $1 \le i \le n$. Let p_t^n denote the corresponding heat kernel. It is well–known and easy to show that $p_t^n f$ is in $C_b^\infty(\mathbf{R}^n)$ for every $f \in C_b^\infty(\mathbf{R}^n)$. We define linear operators T_t, $t \ge 0$, on $\mathcal{F}C_b^\infty(\{e_i\})$ by

$$(5.26) \qquad T_t (f \circ \Pi_n) = (p_t^n f) \circ \Pi_n .$$

The operators T_t, $t \ge 0$, are well–defined, and they form a semigroup. Moreover, the following claim is not difficult to verify :

Claim : The operators T_t, $t \geq 0$, are symmetric contractions on $L^2(E; P_0)$. The unique continuous extensions to all of $L^2(E; P_0)$ form a symmetric C^0 contraction semigroup, which we again denote by $(T_t)_{t \geq 0}$. The space $\mathcal{F}C_b^\infty(\{e_i\})$ is a core for the generator L of $(T_t)_{t \geq 0}$, and $L(f \circ \Pi_n) = (\mathcal{L}^n f) \circ \Pi_n$ for all $n \in \mathbf{N}$ and $f \in C_b^\infty(\mathbf{R}^n)$.

Note that, once we have proven the claim, we are done. In fact, the operator L^0 also generates a C^0 semigroup, and it coincides with L on the core $\mathcal{F}C_b^\infty(\{e_i\})$. Thus $L^0 = L$, whence $\mathcal{F}C_b^\infty(\{e_i\})$, and, therefore, $\mathcal{F}C_b^\infty(H_{\mathrm{CM}})$ is a core for L^0. It remains to prove the claim.

PROOF OF THE CLAIM. We first show the symmetry of $(T_t)_{t \geq 0}$ w.r.t. P_0. Note that P_0 is a centered Gauss measure with covariance

$$E_0 \left[g^* h^* \right] = \int_0^1 \int_0^1 (s \wedge t) \, g(s) \, h(t) \, dt \, ds,$$

$g, h \in L^2(0,1)$. The function $G(s,t) := s \wedge t$ is the Green's function of the operator Δ^N. Since the functions e_i, $i \in \mathbf{N}$, are eigenfunctions of Δ^N with eigenvalue λ_i, the marginals $P_0 \circ \Pi_n^{-1}$, $n \in \mathbf{N}$, are centered Gauss measures on $\mathbf{R}^n$ with covariance matrix C given by

$$C_{ij} = E_0 \left[e_i^* e_j^* \right] = \int_0^1 \int_0^1 e_i(s) \, G(s,t) \, e_j(t) \, ds \, dt = \lambda_i^{-1} \cdot \delta_{ij},$$

i.e.,

(5.27) $$P_0 \circ \Pi_n^{-1} = \prod_{i=1}^n N(0, \lambda_i^{-1}).$$

On the other hand, this measure is known to be the symmetrizing measure for the transition function p_t^n of the $\mathbf{R}^n$-valued Ornstein–Uhlenbeck process composed of n independent one–dimensional Ornstein–Uhlenbeck processes with parameters λ_i, $1 \leq i \leq n$. Hence

$$E_0 \left[T_t (f \circ \Pi_n) \, g \circ \Pi_n \right] = \int p_t^n f \, g \, d(P_0 \circ \Pi_n^{-1})$$

$$= \int f \, p_t^n g \, d(P_0 \circ \Pi_n^{-1}) = E_0 \left[f \circ \Pi_n \, T_t (g \circ \Pi_n) \right]$$

for all $n \in \mathbf{N}$, $f, g \in C_b^\infty(\mathbf{R}^n)$, and $t \geq 0$.

Since p_t^n is a transition function, the operators $(T_t, \mathcal{F}C_b^\infty(\{e_i\}))$ are contractions w.r.t. the $L^\infty(E; P_0)$ norm. Since they are symmetric, they are also contractions w.r.t. the $L^1(E; P_0)$ norm, and hence, by interpolation, w.r.t. the $L^2(E; P_0)$ norm. In particular, the operators can be uniquely extended to contractions defined on all of $L^2(E; P_0)$.

We have already remarked that the semigroup property for $(T_t)_{t \geq 0}$ holds on cylinder functions. By continuity, it holds for the extended operators as well. Similarly, the strong continuity is easily verified on cylinder functions, since

$T_t(f \circ \Pi_n) = (p_t^n f) \circ \Pi_n \to f \circ \Pi_n$ pointwise, and thus in $L^2(E;\ P_0)$, as $t \downarrow 0$ for every $n \in \mathbf{N}$ and $f \in C_b^\infty(\mathbf{R}^n)$. By the uniform boundedness of the operators T_t, $t \geq 0$, we see that they form a strongly continuous semigroup on $L^2(E;\ P_0)$. Finally, as $t \downarrow 0$, $\frac{1}{t}(p_t^n f - f)$ converges to $\mathcal{L}^n f$ in $L^2(\mathbf{R}^n;\ \prod_{i=1}^n N(0, \lambda_i^{-1}))$ for each $n \in \mathbf{N}$ and $f \in C_b^\infty(\mathbf{R}^n)$. Thus

$$\frac{1}{t}\ (T_t(f \circ \Pi_n) - f \circ \Pi_n)\ \longrightarrow\ (\mathcal{L}^n f) \circ \Pi_n \quad \text{as } t \downarrow 0 \ \text{ in } L^2(E;\ P_0),$$

i.e., $f \circ \Pi_n$ is in the domain of the generator L of $(T_t)_{t \geq 0}$ on $L^2(E;\ P_0)$, and $L(f \circ \Pi_N) = (\mathcal{L}^n f) \circ \Pi_n$. Hence $\mathcal{F}C_b^\infty(\{e_i\})$ is a subspace of the domain of L. Actually, it is even an operator core for L, because it is invariant under T_t for all $t \geq 0$, cf. Theorem 1.2 and 1.3 in Appendix A. This completes the proof of the claim, and hence that of Lemma 5.5 as well. ∎

4) Weak Sobolev space and maximal Markovian extension of $\mathcal{L}$

Recall from Corollary 3.1, that there exists a maximal element $\hat{L}$ among all negative–definite self–adjoint extensions L of $(\mathcal{L}, \mathcal{F}C_b^\infty(C_0^\infty(0,1)))$ such that the semigroup $(e^{tL})_{t \geq 0}$ is sub–Markovian. The operator $\hat{L}$ is the generator of the Dirichlet form $(\hat{\mathcal{E}}, W^{1,2}(\nabla))$. Here $W^{1,2}(\nabla)$ is the *weak Sobolev space* corresponding to the operator $(\mathcal{L}, \mathcal{F}C_b^\infty(C_0^\infty(0,1)))$, respectively the gradient $(\nabla, \mathcal{F}C_b^\infty(C_0^\infty(0,1)))$, i.e., it consists of those functions $F \in L^2(E;\ P_0)$ for which there exists $\hat{\nabla}F \in L^2(E \to L^2(0,1);\ P_0)$ such that

$$(5.28) \qquad E_0[F\, \partial_g G]\ =\ -E_0\Big[\big(g, \hat{\nabla}F\big)_{L^2}\ G\Big] - E_0[\beta_g \cdot F\, G]$$

for all $g \in C_0^\infty(0,1)$ and $G \in \mathcal{F}C_b^\infty(C_0^\infty(0,1))$, and, on such functions, $\hat{\mathcal{E}}$ is given by

$$\hat{\mathcal{E}}(F, \tilde{F})\ =\ E_0\left[\big(\hat{\nabla}F, \hat{\nabla}\tilde{F}\big)_{L^2}\right].$$

We will now identify the weak Sobolev space and the operator $\hat{L}$ explicitly.

For a subspace $V \subseteq L^2(0,1)$, let $\hat{\mathcal{F}}C_b^\infty(V)$ denote the space of all cylinder functions F on E of type

$$(5.29) \qquad\qquad F(\omega)\ =\ f(h_1^*(\omega), \dots, h_n^*(\omega), \omega(1))$$

for some $n \in \mathbf{N}$, $f \in C_b^\infty(\mathbf{R}^{n+1})$, and $h_1, \dots, h_n \in V$.

Lemma 5.6 $\hat{\mathcal{F}}C_b^\infty(L^2(0,1)) \subseteq W^{1,2}(\nabla)$.
For a function $F \in \hat{\mathcal{F}}C_b^\infty(L^2(0,1))$ of type (5.29) with $n \in \mathbf{N}$, $f \in C_b^\infty(\mathbf{R}^{n+1})$, and $h_1, \dots, h_n \in L^2(0,1)$, the weak gradient $\hat{\nabla}F$ is given by

$$(5.30) \qquad \big(\hat{\nabla}F\big)(\omega)\ =\ \sum_{i=1}^n h_i\, \frac{\partial f}{\partial x_i}(h_1^*(\omega), \dots, h_n^*(\omega), \omega(1))$$

for P_0-a.e. $\omega \in E$.

REMARKS. (i) Note that every cylinder function F on E of type

(5.31) $F(\omega) = f(h_1^*(\omega), \dots, h_n^*(\omega), \omega(0), \omega(1))$

for some $n \in \mathbf{N}$, $f \in C_b^\infty(\mathbf{R}^{n+2})$, and $h_1, \dots, h_n \in L^2(0,1)$, coincides P_0-a.s. with the function $\tilde{F}$ in $\hat{\mathcal{F}}C_b^\infty(L^2(0,1))$ defined by

$$\bar{F}(\omega) = f(h_1^*(\omega), \dots, h_n^*(\omega), 0, \omega(1)).$$

Hence F is in $W^{1,2}(\nabla)$, and

(5.32) $\left(\hat{\nabla}F\right)(\omega) = \sum_{i=1}^{n} h_i \frac{\partial f}{\partial x_i}(h_1^*(\omega), \dots, h_n^*(\omega), \omega(0), \omega(1))$

for P_0-a.e. ω.

(ii) Similarly, one can show that for each probability measure P of type $P = \int P_{x,y}\,\sigma(dxdy)$, where σ is a probability measure on $\mathbf{R}^2$, the cylinder functions F of type (5.31) are in the corresponding weak Sobolev space w.r.t. P, and (5.32) holds. Note, that if P is an extremal symmetrizing measure, i.e., $P = P_{x,y}$ for some $x, y \in \mathbf{R}$, then every such cylinder function coincides P-a.e. with a function in $\mathcal{F}C_b^\infty(L^2(0,1))$. Hence in this case, the corresponding P-classes are even contained in the strong Sobolev space, cf. the remark below Lemma 5.4.

PROOF OF LEMMA 5.6. Let $F \in \hat{\mathcal{F}}C_b^\infty(L^2(0,1))$ and $g \in C_0^\infty(0,1)$. By the integration by parts identity (5.22), the equation (5.28) holds for all $G \in \mathcal{F}C_b^\infty(C_0^\infty(0,1))$, provided $\hat{\nabla}F$ is defined by the right-hand side of (5.30). Note that in fact, $\partial_g F = (g, \hat{\nabla}F)$ in this case, because $g(0) = g(1) = 0$. Since $g \in C_0^\infty(0,1)$ has been chosen arbitrarily, F is in $W^{1,2}(\nabla)$ and (5.30) holds. ∎

Actually, $\hat{\mathcal{F}}C_b^\infty(L^2(0,1))$ is a form core for $(\hat{\mathcal{E}}, W^{1,2}(\nabla))$. This is a particular sequence of the following stronger result :

Lemma 5.7 *The space $\hat{\mathcal{F}}C_b^\infty(H_0^{1,2}(0,1))$ is an operator core for $\hat{L}$. For $n \in \mathbf{N}$, $f \in C_b^\infty(\mathbf{R}^{n+1})$, and $g_1, \dots, g_n \in H_0^{1,2}(0,1)$, we have*

(5.33) $\hat{L}\left(f(g_1^*, \dots, g_n^*, W_1)\right)$

$= \sum_{i,j=1}^{n} (g_i, g_j)_{L^2} \frac{\partial^2 f}{\partial x_i \partial x_j}(g_1^*, \dots, g_n^*, W_1) + \sum_{i=1}^{n} \beta_{g_i} \frac{\partial f}{\partial x_i}(g_1^*, \dots, g_n^*, W_1).$

REMARK. Again, we derive an explicit formula for the C^0 semigroup generated by $\hat{L}$, cf. (5.35). It is not difficult to see from this formula that the diffusion process generated by $\hat{L}$ is a solution of the stochastic evolution equation (5.14) with fixed end-points.

PROOF OF LEMMA 5.7. We first show that $\hat{\mathcal{F}}C_b^\infty(H_0^{1,2}(0,1))$ is contained in the domain of $\hat{L}$, and (5.33) holds. Note that for $G \in W^{1,2}(\nabla)$, the integration by parts identity

$$(5.34) \qquad E_0\left[F\,(g\,,\hat{\nabla}G)_{L^2}\right] \;=\; -E_0\,[\partial_g F\,G] \;-\; E_0\,[\beta_{g^\bullet}\,F\,G],$$

which is known to hold for $F \in \mathcal{F}C_b^\infty(C_0^\infty(0,1))$ and $g \in C_0^\infty(0,1)$, also holds for $F \in \hat{\mathcal{F}}C_b^\infty(H_0^{1,2}(0,1))$ and $g \in H_0^{1,2}(0,1)$. In fact, fix $g \in C_0^\infty(0,1)$. Since g vanishes in a neighbourhood of 1, we can find for each $f \in \hat{\mathcal{F}}C_b^\infty(H_0^{1,2}(0,1))$ a uniformly bounded sequence $(F_k)_{k\in\mathbf{N}}$ in $\mathcal{F}C_b^\infty(C_0^\infty(0,1))$ such that $F_k \to F$ P_0–a.s., and $\partial_g F_k \to \partial_g F$ in $L^2(E;P_0)$. Hence, by dominated convergence, (5.34) holds for all $g \in C_0^\infty(0,1)$ and $F \in \hat{\mathcal{F}}C_b^\infty(H_0^{1,2}(0,1))$. Moreover, since $g \mapsto \beta_{g^\bullet}$ is a continuous map from $H_0^{1,2}(0,1)$ to $L^2(E;P_0)$, the equation can be extended to all $g \in H_0^{1,2}(0,1)$.

Now fix $F \in \hat{\mathcal{F}}C_b^\infty(H_0^{1,2}(0,1))$ and $G \in W^{1,2}(\nabla)$. Suppose $F = f(g_1^*,\dots,g_n^*,W_1)$ with $n \in \mathbf{N}$, $f \in C_b^\infty(\mathbf{R}^{n+1})$, and $g_1,\dots,g_n \in H_0^{1,2}(0,1)$. By Lemma 5.6, F is also in $W^{1,2}(\nabla)$, and $\hat{\nabla}F$ is given by (5.30). Hence

$$\hat{\mathcal{E}}\,(F,G) \;=\; E_0\left[\sum_{i=1}^n \frac{\partial f}{\partial x_i}(g_1^*,\dots,g_n^*,W_1)\,\left(g_i,\hat{\nabla}G\right)_{L^2}\right]$$

$$=\; -E_0\left[\left\{\sum_{i,j=1}^n (g_i,g_j)_{L^2}\,\frac{\partial^2 f}{\partial x_i \partial x_j}(g_1^*,\dots,g_n^*,W_1)\right.\right.$$

$$\left.\left. +\; \sum_{i=1}^n \beta_{g_i^*}\,\frac{\partial f}{\partial x_i}(g_1^*,\dots,g_n^*,W_1)\right\}\,G\right].$$

Since this is true for all $G \in W^{1,2}(\nabla)$, F is in the domain of $\hat{L}$, and (5.33) holds.

We have shown that $\hat{\mathcal{F}}C_b^\infty(H_0^{1,2}(0,1))$ is a subspace of the domain of $\hat{L}$. We now prove that it is an operator core for $\hat{L}$. The proof is similar to that of the core property in Lemma 5.5, but we need some additional considerations.

Let C_D^2 denote the space of all functions in $C^2([0,1])$ that satisfy Dirichlet boundary conditions, and let Δ^D be the closure of the essentially self-adjoint operator $(\frac{d^2}{ds^2},C_D^2)$ on $L^2(0,1)$. We fix an orthonormal basis $\{e_i;\,i \in \mathbf{N}\}$ of $L^2(0,1)$ consisting of eigenfunctions of Δ^D, which are in C_D^2. The corresponding eigenvalues are denoted by λ_i, $i \in \mathbf{N}$. Let E_D denote the space consisting of all functions $\omega \in E$ that satisfy Dirichlet boundary conditions. For $n \in \mathbf{N}$ and $\omega \in E$, let $\Pi_n(\omega) := (e_1^*(\omega),\dots,e_n^*(\omega))$, and $\hat{\Pi}_n(\omega) := (e_1^*(\omega),\dots,e_n^*(\omega),\omega(1))$. We denote the space of all cylinder functions on E_D of type $f \circ \Pi_n$, $n \in \mathbf{N}$, $f \in C_b^\infty(\mathbf{R}^n)$, by $\mathcal{F}C_b^\infty(\{e_i\})$, and that of all cylinder functions on E of type $f \circ \hat{\Pi}_n$, $n \in \mathbf{N}$, $f \in C_b^\infty(\mathbf{R}^{n+1})$, by $\hat{\mathcal{F}}C_b^\infty(\{e_i\})$. For $y \in \mathbf{R}$, let $\varphi_y \in E$ denote the function $\varphi_y(s) = s \cdot y$. Below, we will show the following representation for the operator $\hat{L}$:

Claim : Let $\mathcal{L}^D$ denote the linear operator on $\mathcal{F}C_b^\infty(\{e_i\})$ defined by

$$\mathcal{L}^D (f \circ \Pi_n) = (\mathcal{L}^n f) \circ \Pi_n \quad \text{for all } n \in \mathbf{N} \text{ and } f \in C_b^\infty(\mathbf{R}^n),$$

where

$$\mathcal{L}^n f := \Delta f + \sum_{i=1}^{n} \lambda_i \, x_i \, \frac{\partial f}{\partial x_i}.$$

Then

$$\left(\hat{L}F\right)(\omega) = \mathcal{L}^D\left(F(\cdot + \varphi_{\omega(1)})\right)(\omega - \varphi_{\omega(1)})$$

for P_0–a.e. $\omega \in E$, and each $F \in \hat{\mathcal{F}}C_b^\infty(\{e_i\})$.

Hence for $y \in \mathbf{R}$, the operator $\hat{L}$ leaves the "fiber" E_y consisting of all paths from 0 to y invariant. On each fiber, it acts in the same way as $\mathcal{L}^D$ acts on $E_D = E_0$.

Once we have verified the claim, the proof of Lemma 5.7 can be completed similarly to that of Lemma 5.5 above. In fact, let p_t^n denote the heat kernel of the operator $\mathcal{L}^n$, $n \in \mathbf{N}$. We define linear operators T_t^D, $t \geq 0$, on $\mathcal{F}C_b^\infty(\{e_i\})$, and $\hat{T}_t$, $t \geq 0$, on $\hat{\mathcal{F}}C_b^\infty(\{e_i\})$ by

$$T_t^D (f \circ \Pi_n) = (p_t^n f) \circ \Pi_n$$

for all $n \in \mathbf{N}$ and $f \in C_b^\infty(\mathbf{R}^n)$, and

(5.35) $$\left(\hat{T}_t F\right)(\omega) = T_t^D\left(F(\cdot + \varphi_{\omega(1)})\right)(\omega - \varphi_{\omega(1)})$$

for all $\omega \in E$ and $F \in \hat{\mathcal{F}}C_b^\infty(\{e_i\})$.

The Brownian bridge measure $P_{0,0}$ is a centered Gauss measure on the space E^D with covariance

$$E_{0,0}\left[g^* h^*\right] = \int_0^1 \int_0^1 G(s,t) \, g(s) \, h(t) \, dt \, ds,$$

where $G(s,t) := \begin{cases} s(1-t) & \text{if } s \leq t \\ t(1-s) & \text{if } s \geq t \end{cases}$ is the Green's function of the operator Δ^D. Hence for $n \in \mathbf{N}$, $P_{0,0} \circ \Pi_n^{-1} = \prod_{i=1}^n N(0, \lambda_i^{-1})$, which is the symmetrizing measure for the transition function p_t^n. Therefore, $P_{0,0}$ is a symmetrizing measure for the operators $(T_t^D, \mathcal{F}C_b^\infty(\{e_i\}))$, $t \geq 0$. For $y \in \mathbf{R}$, the measure $P_{0,y}$ on E is the image of $P_{0,0}$ under the translation by φ_y. Hence

$$E_{0,y}\left[\hat{T}_t F \, \bar{F}\right] = \int T_t^D\left(F(\cdot + \varphi_y)\right)(\omega - \varphi_y) \, \bar{F}(\omega) \, P_{0,y}(d\omega)$$

$$= \int T_t^D\left(F(\cdot + \varphi_y)\right)(\omega) \, \tilde{F}(\omega + \varphi_y) \, P_{0,0}(d\omega)$$

for all $t \geq 0$ and $F, \tilde{F} \in \mathcal{F}C_b^\infty(\{e_i\})$, i.e., $P_{0,y}$ is a symmetrizing measure for the operators $(\hat{T}_t, \mathcal{F}C_b^\infty(\{e_i\}))$, $t \geq 0$. Since this is true for all $y \in \mathbf{R}$, the operators $\hat{T}_t$ are also **symmetric w.r.t.** P_0, because

$$P_0 = \int P_{0,y} \, p_1^{BM}(0, y) \, dy,$$

where p_t^{BM} denotes the heat kernel of Brownian motion. Now, we can conclude similarly as in the proof of Lemma 5.5 that $\hat{T}_t$ is a contraction w.r.t. the $L^2(E; P_0)$ norm for every $t \geq 0$, and the unique continuous extensions of the operators $\hat{T}_t$, $t \geq 0$, to $L^2(E; P_0)$ form a symmetric C^0 contraction semigroup. Moreover, we can show that the generator of this C^0 semigroup extends the operator $(\hat{L}, \mathcal{F}C_b^\infty(\{e_i\}))$. Since the space $\mathcal{F}C_b^\infty(\{e_i\})$ is invariant under $\hat{T}_t$ for all $t \geq 0$, it is a core for the generator. Thus $\hat{L}$ must be the generator. In particular, $\mathcal{F}C_b^\infty(\{e_i\})$, and, therefore, $\mathcal{F}C_b^\infty(H_0^{1,2}(0,1))$, is a core for $\hat{L}$. This completes the proof of Lemma 5.7. It only remains to prove the claim :

PROOF OF THE CLAIM : Fix $n \in \mathbf{N}$ and $f \in C_b^\infty(\mathbf{R}^{n+1})$, and let $F := f \circ \hat{\Pi}_n$. For $0 \leq s \leq 1$, let $W_s^D := W_s - s \cdot W_1$. By (5.33),

$$(5.36) \qquad \hat{L}F = \sum_{i=1}^n \left(\frac{\partial^2 f}{\partial x_i^2} \circ \hat{\Pi}_n + \beta_{e_i} \frac{\partial f}{\partial x_i} \circ \hat{\Pi}_n \right),$$

where

$$\begin{aligned} \beta_{h\cdot} &= -\int_0^1 h'(s) \, dW_s = -\int_0^1 h'(s) \, dW_s^D + W_1 \int_0^1 h'(s) \, ds \\ &= \int_0^1 h''(s) \, W_s^D \, ds = (h'')^*(\cdot - \varphi_{W_1}) \end{aligned}$$

P_0–a.s. for all $h \in C_D^2$. Here we have used that $h(0) = h(1) = 0$ and $W_0^D = W_1^D = 0$. Hence for $i \in \mathbf{N}$,

$$(5.37) \qquad \beta_{e_i}(\omega) = \lambda_i \, e_i^*(\omega - \varphi_{\omega(1)}) \qquad \text{for } P_0\text{–a.e. } \omega.$$

Note that for $\omega \in E$ and $\tilde{\omega} \in E_D$,

$$\begin{aligned} F(\tilde{\omega} + \varphi_{\omega(1)}) &= f(e_1^*(\tilde{\omega} + \varphi_{\omega(1)}), \ldots, e_n^*(\tilde{\omega} + \varphi_{\omega(1)}), \omega(1)) \\ &= f(e_1^*(\tilde{\omega}) + e_1^*(\varphi_{\omega(1)}), \ldots, e_n^*(\tilde{\omega}) + e_n^*(\varphi_{\omega(1)}), \omega(1)). \end{aligned}$$

Therefore,

$$\begin{aligned} & \mathcal{L}^D(F(\cdot + \varphi_{\omega(1)})) \\ &= \sum_{i=1}^n \frac{\partial^2 f}{\partial x_i^2}(e_1^* + e_1^*(\varphi_{\omega(1)}), \ldots, e_n^* + e_n^*(\varphi_{\omega(1)}), \omega(1)) \\ & \qquad + \sum_{i=1}^n \lambda_i \, e_i^* \frac{\partial f}{\partial x_i}(e_1^* + e_1^*(\varphi_{\omega(1)}), \ldots, e_n^* + e_n^*(\varphi_{\omega(1)}), \omega(1)), \end{aligned}$$

whence, by (5.36) and (5.37),

$$\mathcal{L}^D \left(F(\cdot + \varphi_{\omega(1)}) \right) (\omega - \varphi_{\omega(1)})$$

$$= \sum_{i=1}^n \frac{\partial^2 f}{\partial x_i^2} \left(e_1^*(\omega), \dots, e_n^*(\omega), \omega(1) \right)$$

$$+ \sum_{i=1}^n \lambda_i \, e_i^*(\omega - \varphi_{\omega(1)}) \, \frac{\partial f}{\partial x_i} \left(e_1^*(\omega), \dots, e_n^*(\omega), \omega(1) \right)$$

$$= (\hat{L}F)(\omega)$$

for P_0-a.e. $\omega \in E$. This proves the claim, and thus completes the proof of Lemma 5.7. ∎

c) Markov uniqueness of projective limits

In contrast to the other parts of this chapter, *we do not restrict ourselves to the flat case in this section*. The infinite dimensional state spaces of diffusion operators we are interested in can usually be viewed in some natural way as a projective limit of finite dimensional spaces. We are considering uniqueness problems for diffusion operators defined on cylinder functions over the projective limit, i.e., on functions that only depend on a finite dimensional projection. The most obvious idea to attack such problems, is to show that under certain conditions, uniqueness of the finite dimensional projections of the operator implies uniqueness of the operator defined on cylinder functions itself. The aim of this section is to derive such conditions for Markov uniqueness.

We point out, however, that the assumptions needed are restrictive, and not satisfied in many applications we are interested in. In fact, an alternative approach to L^p uniqueness presented in the next section seems to be often more fruitful. The projective limit approach is included here nevertheless, because it gives some theoretical insight, and it can be used to prove Markov uniqueness (and even L^p uniqueness) of certain "nice" infinite dimensional diffusion operators, as well as small perturbations of them.

After introducing the general framework of diffusion operators on projective limit spaces in Subsection 1), we prove our basic result on stability of Markov uniqueness w.r.t. projective limits in Subsection 2). This result is a general version of a condition for Markov uniqueness of diffusion operators with trivial geometry on Banach spaces proven in [AlbRöZha 92, Thm. 1.4]. A similar argument has already been used in [AlbHK 77]. In Subsection 3), we look at the Banach space case. Here we make our condition for stability of Markov uniqueness more precise. In particular, we show how to recover the result in [AlbRöZha 92] from our general result, and we derive a condition for Markov uniqueness of strictly elliptic diffusion operators on Wiener space. The latter result has been announced in [Eb 95].

1) Diffusion operators on projective limits

Let E be a set. Suppose that we are given a sequence of sets E_N, $N \in \mathbf{N}$, σ-algebras $\mathcal{B}_N$ on E_N, and projections $\Pi_N : E \to E_N$. Let $\mathcal{F}_N := \{\Pi_N^{-1}(A);$ $A \in \mathcal{B}_N\}$, $N \in \mathbf{N}$, be the induced σ-algebras on E_N. We assume that the sequence $(\mathcal{F}_N)_{N \in \mathbf{N}}$ is increasing. We endow E with the σ-algebra $\mathcal{F}_\infty := \bigcup_{N \in \mathbf{N}} \mathcal{F}_N$.

Now suppose, moreover, that for each N, we are given a vector space $\mathcal{A}_N$ of $\mathcal{B}_N$-measurable bounded functions on E_N, such that $\sigma(\mathcal{A}_N) = \mathcal{B}_N$, and $\phi(f_1, \ldots, f_k)$ is in $\mathcal{A}_N$ for all $k \in \mathbf{N}$, $f_1, \ldots, f_k \in \mathcal{A}_N$, and $\phi \in C^\infty(\mathbf{R}^k)$. In particular, the constant functions are in $\mathcal{A}_N$. We assume that the spaces $\{f \circ \Pi_N ; f \in \mathcal{A}_N\}$ of functions on E increase as N increases. Let

$$\mathcal{A} := \{f \circ \Pi_N ; N \in \mathbf{N}, f \in \mathcal{A}_N\} = \bigcup_{N \in \mathbf{N}} \{f \circ \Pi_N ; f \in \mathcal{A}_N\}$$

be the corresponding space of cylinder functions on E. Clearly, the functions in $\mathcal{A}$ generate the σ-algebra $\mathcal{F}_\infty$.

Let m be a probability measure on E such that $L^2(E; m)$ is separable, and $\mathcal{A}$ is dense in $L^2(E; m)$. Suppose that we are given a symmetric diffusion operator $\mathcal{L}$ with domain $\mathcal{A}$ on $L^2(E; m)$, cf. Appendix B. Let $m_N := m \circ \Pi_N^{-1}$, $N \in \mathbf{N}$, be the image measures on E_N, and let $(\mathcal{L}_N, \mathcal{A}_N)$ be the projected operators on $L^2(E_N; m_N)$ defined by

$$(\mathcal{L}_N f) \circ \Pi_N = E[\mathcal{L}(f \circ \Pi_N) | \mathcal{F}_N] \qquad m\text{-a.e.,}$$

where $E[\cdot | \cdot]$ denotes the conditional expectation w.r.t. m.

Lemma 5.8 *For every $N \in \mathbf{N}$, the operator $(\mathcal{L}_N, \mathcal{A}_N)$ is a symmetric diffusion operator on $L^2(E_N; m_N)$. The corresponding carré du champ operator $\Gamma_N :$ $\mathcal{A}_N \times \mathcal{A}_N \to L^2(E_N; m_N)$ is uniquely determined by*

$$(\Gamma_N(f, g)) \circ \Pi_N = E[\Gamma(f \circ \Pi_N, g \circ \Pi_N) | \mathcal{F}_N] \qquad m\text{-a.e.,}$$

where Γ denotes the carré du champ of $(\mathcal{L}, \mathcal{A})$.

The lemma can be verified easily.

2) Stability of Markov uniqueness w.r.t. projective limits

From now on, we make the following assumption on the carré du champ of the diffusion operator $(\mathcal{L}, \mathcal{A})$ on E :

> **Assumption :** For every $N \in \mathbf{N}$ and all $f, g \in \mathcal{A}_N$,
> $\Gamma(f \circ \Pi_N, g \circ \Pi_N)$ is $\mathcal{F}_N$-measurable.

REMARKS. (i) The assumption is in particular satisfied (w.r.t. a natural sequence of projections) for operators that have, in some generalized sense, a "constant diffusion matrix". This includes the usually considered lattice systems and more general particle systems from classical statistical mechanics. The assumption *does not hold* w.r.t. the natural projections, if $\mathcal{L}$ is the Ornstein–Uhlenbeck operator on the path or loop space over a non–flat Riemannian manifold. This is one of the reasons, why the analysis of this Ornstein–Uhlenbeck operator is extraordinarily difficult.

(ii) In principle, it is possible to relax the assumption above. However, in this case, the calculations below become much more intricate, and the resulting condition for Markov uniqueness does not seem to be very useful for applications.

Note that by the assumption and by Lemma 5.8,

$$(5.38) \quad \Gamma(f \circ \Pi_N, g \circ \Pi_N) = \Gamma_N(f, g) \circ \Pi_N \quad \forall N \in \mathbf{N}, \ f, g \in \mathcal{A}_N.$$

Let $H_0^{1,2}(d)$, $W^{1,2}(d)$ and $W_\infty^{1,2}(d)$ be the strong and weak Sobolev spaces w.r.t. the operator $(\mathcal{L}, \mathcal{A})$ on $L^2(E; m)$, and let $H_0^{1,2}(d_N)$, $W^{1,2}(d_N)$ and $W_\infty^{1,2}(d_N)$ be the corresponding Sobolev spaces w.r.t. the operators $(\mathcal{L}_N, \mathcal{A}_N)$ on $L^2(E_N; m_N)$, $N \in \mathbf{N}$, cf. Chapter 3, Section b). Here d and d_N are corresponding generalized differentials, but by Lemma 3.7, we can even define the Sobolev spaces in terms of the operators $\mathcal{L}$ and $\mathcal{L}_N$, and their carré du champs Γ and Γ_N only.

Theorem 5.1 *Suppose that $W_\infty^{1,2}(d_N) = H_0^{1,2}(d_N)$ for all $N \in \mathbf{N}$. Moreover, assume that there exists a bounded sequence of functions B_N, $N \in \mathbf{N}$, in $L^1(E; m)$, such that*

$$(5.39) \qquad \left| \sum_{i=1}^{k} (\mathcal{L}(f_i \circ \Pi_N) - E[\mathcal{L}(f_i \circ \Pi_N)|\mathcal{F}_N]) \right|^2$$
$$\leq B_N \cdot \sum_{i,j=1}^{k} \Gamma(f_i \circ \Pi_N, f_j \circ \Pi_N)$$

m-a.e. for all $N \in \mathbf{N}$, $k \in \mathbf{N}$, and $f_1, \dots, f_k \in \mathcal{A}_N$.

Then $W_\infty^{1,2}(d) = H_0^{1,2}(d)$. In particular, the operator $(\mathcal{L}, \mathcal{A})$ is Markov unique on $L^2(E; m)$.

REMARKS. (iii) In applications, E_N is usually some finite dimensional space, where it can be easily verified that $W_\infty^{1,2}(d_N)$ is a Dirichlet space. Then the assumption $W_\infty^{1,2}(d_N) = H_0^{1,2}(d_N)$ holds if and only if the operator $(\mathcal{L}_N, \mathcal{A}_N)$ is Markov unique on $L^2(E_N; m_N)$. In this sense, Theorem 5.1 shows that Markov uniqueness is stable under projective limits provided Condition (5.39) is satisfied for a bounded sequence $(B_N)_{N \in \mathbf{N}}$ in $L^1(E; m)$.

(iv) At first glance, Condition (5.39) might seem inappropriate, because on the left–hand side second order objects appear, whereas the carré du champ on the right is first order. Note however, that by the assumption made above, the second order part of $\mathcal{L}(f_i \circ \Pi_N)$ is $\mathcal{F}_N$–measurable, and thus cancels out on the left–hand side of (5.39).

(v) Because of the general situation we are considering, Condition (5.39) looks somehow obscure. We will see below, that in concrete applications it takes a much simpler looking form.

(vi) Even if Assumption (5.39) is not satisfied, Theorem 5.1 can often be used to prove that the weak differential $(\hat{d}, W_\infty^{1,2}(d))$ satisfies a **chain rule** (which implies in particular that $W_\infty^{1,2}(d)$ is a **Dirichlet space**). To see this, fix $G \in \mathcal{A}$, and let $X : \mathcal{A} \to L^2(E \, ; \, m)$ be the derivation defined by $XF = \Gamma(G, F)$. By applying Theorem 5.1 w.r.t. the diffusion operator $(\mathcal{L}_X, \mathcal{A})$, $\mathcal{L}_X F := -X^* X F$, one can prove that $\mathcal{A}$ is dense in the weak Sobolev space $W_\infty^{1,2}(X)$ introduced in Chapter 3, Section b), 2), provided $\mathcal{A}_N$ is dense in the corresponding projected weak Sobolev space $W_\infty^{1,2}(X_N)$. In fact, Assumption (5.39) is always fulfilled for the operator $\mathcal{L}_X$. As a consequence, the chain rule for the weak derivative $(\hat{X}, W_\infty^{1,2}(X))$ follows from the chain rule for $(X, \mathcal{A})$. Since the function G has been chosen arbitrarily, we also obtain a chain rule for the weak differential $(\hat{d}, W_\infty^{1,2}(d))$, cf. Chapter 3, Section b), 2).

Before proving the theorem, we give a preparatory lemma, whoch is also of independent interest. Recall the definition of weak differentiability in direction X, where X is a vector field, cf. Chapter 4, Section b), 2).

Lemma 5.9 *Let $N \in \mathbf{N}$. Fix $k \in \mathbf{N}$, $g_1, \ldots, g_k \in \mathcal{A}_N$, and $h_1, \ldots, h_k \in \mathcal{A}_N$. For $1 \le i \le k$ let $G_i := g_i \circ \Pi_N$, and $H_i := h_i \circ \Pi_N$. Let $X_N : \mathcal{A}_N \to L^2(E_N; m_N)$ and $X : \mathcal{A} \to L^2(E \, ; \, m)$ be the derivations defined by*

$$X_N f = \sum_{i=1}^k h_i \, \Gamma_N(g_i, f), \qquad and$$

$$X F = \sum_{i=1}^k H_i \, \Gamma(G_i, F).$$

Let u be a bounded function in $W^{1,2}(X)$, and let $u_N \in L^\infty(E; m)$ be such that $u_N \circ \Pi_N = E[u \,|\, \mathcal{F}_N]$ m–a.e. Then u_N is in $W^{1,2}(X_N)$, and the weak derivative $\hat{X}_N u_N$ is uniquely determined by

$$\left(\hat{X}_N u_N \right) \circ \Pi_N = E\left[\hat{X} u + u \sum_{i=1}^k H_i \left(\mathcal{L} G_i - E[\mathcal{L} G_i | \mathcal{F}_N] \right) \,\Big|\, \mathcal{F}_N \right] \quad m\text{–a.e.}$$

REMARKS.   (vii) In the tangent space notation used in Section 4 b), 2), X_N and X are the derivations corresponding to the vector fields $\sum_{i=1}^k h_i \nabla_N g_i$, $\sum_{i=1}^k H_i \nabla G_i$ respectively. Here ∇_N and ∇ denote the gradient operators on E_N and E introduced in 4 b), 2).

(viii) The lemma shows in particular, that in some sense, derivatives commute with the conditional expectaion $E[\,\cdot\,|\mathcal{F}_N]$, if $\mathcal{L}$ maps functions of type $f \circ \Pi_N$, $f \in \mathcal{A}_N$, again to $\mathcal{F}_N$–measurable functions. Otherwise, an extra term appears.

PROOF OF LEMMA 5.9. For $f \in \mathcal{A}_N$ and $F := f \circ \Pi_N$ we have

$$(5.40) \qquad \int u_N \, X_N f \, dm_N$$

$$= \sum_{i=1}^{k} \int E[u|\mathcal{F}_N] \, h_i \circ \Pi_N \, \Gamma_N(g_i, f) \circ \Pi_N \, dm$$

$$= \sum_{i=1}^{k} \int u \, H_i \, \Gamma(G_i, F) \, dm \;\; = \;\; \int u \, X F \, dm$$

$$= -\int \hat{X}u \, F \, dm \; - \; \int u \, F \left(\sum_{i=1}^{k} H_i \, \mathcal{L}G_i + \Gamma(H_i, G_i) \right) dm$$

$$= -\int E\left[\hat{X}u + u \sum_{i=1}^{k} H_i \left(\mathcal{L}G_i - E[\mathcal{L}G_i|\mathcal{F}_N] \right) \Big| \mathcal{F}_N \right] f \circ \Pi_N \, dm$$

$$\qquad - \int u_N \, f \left(\sum_{i=1}^{k} h_i \, \mathcal{L}_N g_i + \Gamma(h_i, g_i) \right) dm_N \; .$$

Here we have used that, with the notation from Section 4 b), 2),

$$\nabla^* X \;\; = \;\; -\sum_{i=1}^{N} \left(H_i \, \mathcal{L}G_i + \Gamma(H_i, G_i) \right),$$

because $X = \sum_{i=1}^{N} H_i \, \nabla G_i$. Similarly,

$$\nabla_N^* X_N \;\; = \;\; -\sum_{i=1}^{N} \left(h_i \, \mathcal{L}_N g_i + \Gamma_N(h_i, g_i) \right).$$

Hence, by (5.40), u_n is in $W^{1,2}(X_N)$, and $\hat{X}_N u_N$ has the claimed representation. ∎

PROOF OF THEOREM 5.1. We have to show that every bounded function $u \in W^{1,2}(d)$ is in $H_0^{1,2}(d)$. The idea is to verify that the conditional expectations $E[u|\mathcal{F}_N]$, $N \in \mathbb{N}$, form a bounded sequence in $H_0^{1,2}(d)$ that converges to u in $L^2(E; m)$.

Fix a bounded function $u \in W^{1,2}(d)$, and let $N \in \mathbb{N}$. Let $X_N : \mathcal{A}_N \to L^2(E_N; m_N)$ and $X : \mathcal{A} \to L^2(E; m)$ be derivations as in Lemma 5.9. By Lemma 3.5, and the remark above, u is in $W^{1,2}(X)$. Let $u_N \in L^\infty(E; m)$ be such that $u_N \circ \Pi_N = E[u|\mathcal{F}_N]$ m-a.e. Then, by Lemma 5.9, u_N is in $W^{1,2}(X_N)$, and the weak derivative $\hat{X}_N u_N$ is given as shown in the lemma. In particular, the following estimate holds :

$$(5.41) \qquad \left| \int u_N \left(\sum_{i=1}^{k} h_i \, \mathcal{L}_N g_i + \Gamma_N(h_i, g_i) \right) dm_N \right|$$

$$
\begin{aligned}
&= \left| \int u_N \, X_N^* 1 \, dm_N \right| = \left| \int \hat{X}_N u_N \, dm_N \right| \\
&= \left| \int (\hat{X}_N u_N) \circ \Pi_N \, dm \right| \\
&\leq \int |\hat{X} u| \, dm + \|u\|_{L^\infty(E;m)} \cdot \int \left| \sum_{i=1}^{k} H_i \left(\mathcal{L} G_i - E[\mathcal{L} G_i | \mathcal{F}_N] \right) \right| \, dm \\
&\leq \left(\left\| \hat{d} u \right\|_{L^2(E \to T'E;m)} + \|u\|_{L^\infty(E;m)} \cdot \|B_N\|_{L^1(E;m)}^{1/2} \right) \\
&\qquad\qquad \times \left(\int \sum_{i,j=1}^{k} H_i H_j \, \Gamma(G_i, G_j) \, dm \right)^{1/2} \\
&= C(u) \cdot \left(\int \sum_{i,j=1}^{k} h_i h_j \, \Gamma_N(g_i, g_j) \, dm_N \right)^{1/2}.
\end{aligned}
$$

Here $C(u)$ is a finite constant that is in particular independent of N, and $\hat{d}$ is the weak differential on $W^{1,2}(d)$. We have used that, by (5.39),

$$
\left| \sum_{i=1}^{k} \lambda_i \left(\mathcal{L} G_i - E[\mathcal{L} G_i | \mathcal{F}_N] \right) \right| \leq B_N^{1/2} \cdot \left(\sum_{i,j=1}^{k} \lambda_i \lambda_j \Gamma(G_i, G_j) \right)^{1/2}
$$

for all $\lambda_i \in \mathbf{R}$, $1 \leq i \leq k$, holds m–a.e.

By (5.41) and Lemma 3.7, u_N is in $W^{1,2}(d_N)$, and the $L^2(E_N \to T'E_N; m_N)$ norm of the weak differential $\hat{d}_N u_N$ is bounded by $C(u)$. Since u_N is bounded, it is also contained in $W^{1,2}_\infty(d_N)$, which coincides by assumption with $H_0^{1,2}(d_N)$. Hence there exists a sequence of functions $f_k \in A_N$, $k \in \mathbf{N}$, such that $f_k \to u_N$ in $L^2(E_N; m_N)$, and $(f_k)_{k \in \mathbf{N}}$ is Cauchy w.r.t. the $H_0^{1,2}(d_N)$ norm. For $K \in \mathbf{N}$, let $F_k := f_k \circ \Pi_N$. Then $F_k \to E[u|\mathcal{F}_N]$ in $L^2(E; m)$, and, by (5.38),

$$
\int \Gamma(F_k - F_l, F_k - F_l) \, dm = \int \Gamma_N(f_k - f_l, f_k - f_l) \, dm_N \to 0
$$

as $k, l \to \infty$. Hence $(F_k)_{k \in \mathbf{N}}$ is a Cauchy sequence in $H_0^{1,2}(d)$, whence the $L^2(E; m)$ limit $E[u|\mathcal{F}_N]$ is in $H_0^{1,2}(d)$. Moreover,

$$
\begin{aligned}
\int \Gamma \left(E[u|\mathcal{F}_N], E[u|\mathcal{F}_N] \right) dm &= \lim_{k \to \infty} \int \Gamma(F_k, F_k) \, dm \\
&= \lim_{k \to \infty} \int \Gamma_N(f_k, f_k) \, dm_N = \int \Gamma_N(u_N, u_N) \, dm \leq (C(u))^2,
\end{aligned}
$$

where Γ and Γ_N denote the unique continuous bilinear extensions of the carré du champ operators to $H_0^{1,2}(d)$, $H_0^{1,2}(d_N)$ respectively.

The considerations above show that the conditional expectations $E[u|\mathcal{F}_N]$, $N \in$

N, form a bounded sequence in the Sobolev space $H_0^{1,2}(d)$. Since $E[u|\mathcal{F}_N] \to u$ in $L^2(E\,;\,m)$, the usual arguments now imply that u is in $H_0^{1,2}(d)$ as well. Since u was an arbitrary bounded function in $W^{1,2}(d)$, we obtain $W_\infty^{1,2}(d) = H_0^{1,2}(d)$.
∎

3) Application to diffusion operators on linear spaces

To demonstrate the meaning of Condition (5.39) assumed in Theorem 5.1 above, we apply it to symmetric diffusion operators on vector spaces. Suppose E is a vector space, and H is a separable Hilbert space such that $E \cap H$ is dense in H. Moreover, we assume that we are given an orthonormal basis $\{e_k;\, k \in \mathbf{N}\}$ of H, and linear functionals $e_k^* : E \to \mathbf{R}$, $k \in \mathbf{N}$, such that $e_k^*(h) = (e_k, h)_H$ for all $h \in H$. We endow E with the σ–algebra generated by the functionals e_k^*, $k \in \mathbf{N}$. Let $K := \operatorname{span}\{e_k^*;\, k \in \mathbf{N}\}$, and let $\mathcal{A} := \mathcal{F}C_b^\infty(K)$. For a cylinder function $F \in \mathcal{A}$, $F = f(e_1^*, e_2^*, \ldots, e_N^*)$ for some $N \in \mathbf{N}$ and $f \in C_b^\infty(\mathbf{R}^N)$, let $\partial_k F$, $k \in \mathbf{N}$, denote the directional derivative in direction e_k, i.e.,

$$\partial_k F \;=\; (DF)\,(e_k) \;=\; \begin{cases} \frac{\partial f}{\partial z_k}(e_1^*, \ldots, e_N^*) & \text{if } 1 \le k \le N \\ 0 & \text{else} \end{cases}$$

We assume that we are given a probability measure m on E such that the integration by parts identities

$$(5.42) \qquad \int \partial_k F \, dm \;=\; -\int \beta_k^m \, F \, dm \qquad \text{for all } F \in \mathcal{A} \text{ and } k \in \mathbf{N}$$

hold for some functions $\beta_k^m \in L^2(E\,;\,m)$, $k \in \mathbf{N}$.

Let A be a map from E to the symmetric bounded linear operators on H. For simplicity, we also assume

$$(5.43) \qquad \varepsilon \cdot \mathrm{id}_H \;\le\; A(z) \;\le\; \varepsilon^{-1} \cdot \mathrm{id}_H \qquad \text{for all } z \in E$$

in the form sense for some $\varepsilon > 0$. This condition can be relaxed considerably, but our purpose is just to present a simple example for the considerations above. For $k, l \in \mathbf{N}$ and $z \in E$, let $A^{kl}(z) := (e_k, A(z)e_l)_H$. We assume that A^{kl} is a measurable function for all k, l.

We consider the symmetric diffusion operator $(\mathcal{L}, \mathcal{F}C_b^\infty(K))$ on $L^2(E\,;\,m)$ corresponding to the pre–Dirichlet form $(\mathcal{E}, \mathcal{F}C_b^\infty(K))$ defined by

$$\mathcal{E}(F, G) \;=\; \int (A\nabla F, \nabla G)_H \, dm \,.$$

Here $\nabla : \mathcal{F}C_b^\infty(K) \to L^2(E \to H;\, m)$ is the gradient operator, i.e., $(\nabla F, e_k)_H = \partial_k F$ for all $k \in \mathbf{N}$. To ensure that $\mathcal{F}C_b^\infty(K)$ is contained in the domain of the generator, we assume that for $k \in \mathbf{N}$, the H–valued function $z \mapsto A(z)e_k$ is contained in the domain of the adjoint ∇^* of the operator $\nabla :$ $\mathcal{F}C_b^\infty(K) \subset L^2(E\,;\,m) \to L^2(E \to H;\, m)$.

REMARK. By the integration by parts formula (5.42), the constant vector fields $z \mapsto e_k$, $k \in \mathbf{N}$, are in the domain of ∇^*, and $\nabla^* e_k = -\beta_k^m$. Hence if, for example, for every fixed $k \in \mathbf{N}$, there exists $k_0 \in \mathbf{N}$ such that A^{kl} vanishes for $l > k_0$, and A^{kl} is in the Sobolev space $H_0^{1,2}(E, H'; m)$ for all $l \leq k_0$ (i.e., in the closure of $\mathcal{F}C_b^\infty(K)$ w.r.t. the norm $F \mapsto \left(\int (F^2 + (\nabla F, \nabla F)_H) \, dm \right)^{1/2}$, then the functions Ae_k, $k \in \mathbf{N}$, are in the domain of ∇^*, and

$$\nabla^*(Ae_k) = \nabla^*\left(\sum_{l=1}^{k_0} A^{kl} e_l \right) = -\sum_{l=1}^{k_0} (A^{kl}\beta_l^m + \partial_l A^{kl}).$$

In general, $\nabla^* Ae_k$ is **informally** given as

$$ \text{``} \nabla^*(Ae_k) = -\sum_{l=1}^{\infty} (A^{kl}\beta_k^m + \partial_k A^{kl}) \text{''},$$

but of course, it is not clear in general if the sum converges.

Under the assumptions listed above, we have $\mathcal{E}(F, G) = -\int F \, \mathcal{L}G \, dm$ for all $F, G \in \mathcal{F}C_b^\infty(K)$, where

$$\mathcal{L}F = -\nabla^*(A\nabla F) = -\sum_{k=1}^{N} \nabla^*(Ae_k \cdot \partial_k F)$$

$$= \sum_{k,l=1}^{N} A_{kl} \, \partial_k \partial_l F - \sum_{k=1}^{N} \nabla^*(Ae_k) \, \partial_k F$$

whenever $F = f(e_1^*, \ldots, e_N^*)$ for some $N \in \mathbf{N}$ and $f \in C_b^\infty(\mathbf{R}^N)$. For $N \in \mathbf{N}$ let $E_N := \mathbf{R}^N$, and let $\Pi_N : E \to E_N$ be the canonical projection defined by $\Pi_N(z) = (e_1^*(z), \ldots, e_N^*(z))$. Suppose that there exists $N_0 \in \mathbf{N}$ such that A^{kl} is measurable w.r.t. the σ–algebra $\mathcal{F}_N = \sigma(e_1^*, \ldots, e_N^*)$ for all $1 \leq k, l \leq N$ and $N \geq N_0$. Then the assumption imposed at the beginning of Subsection 2) is satisfied for $N \geq N_0$, which is sufficient to apply the considerations from Subsection 2).

Lemma 5.10 *Consider the situation described, and let*

$$B_N := \varepsilon^{-1} \sum_{k=1}^{N} \left(\nabla^*(Ae_k) - E[\nabla^*(Ae_k)| \, \mathcal{F}_N] \right)^2.$$

Then for $N \geq N_0$, Condition (5.39) is satisfied.

PROOF. Fix $N \geq N_0$. Let $n \in \mathbf{N}$, $f_1, \ldots f_n \in \mathcal{A}_N$, and let $F_i := f_i \circ \Pi_N$, $1 \leq i \leq n$. Since the functions $\partial_k F_i$, $\partial_k \partial_l F_i$, and A^{kl}, $1 \leq i \leq n$, $1 \leq k, l \leq N$, are $\mathcal{F}_N$–measurable, we have

$$\mathcal{L}F_i - E[\mathcal{L}F_i| \, \mathcal{F}_N] = \sum_{k=1}^{N} \left(\nabla^*(Ae_k) - E[\nabla^*(Ae_k)| \, \mathcal{F}_N] \right) \partial_k F_i$$

for all $1 \leq i \leq n$. Hence

$$
\left| \sum_{i=1}^{n} (\mathcal{L}F_i - E[\mathcal{L}F_i| \mathcal{F}_N]) \right|^2
$$

$$
= \left| \sum_{k=1}^{N} (\nabla^*(Ae_k) - E[\nabla^*(Ae_k)| \mathcal{F}_N]) \, \partial_k (\sum_{i=1}^{n} F_i) \right|^2
$$

$$
\leq \sum_{k=1}^{N} (\nabla^*(Ae_k) - E[\nabla^*(Ae_k)| \mathcal{F}_N])^2 \cdot \sum_{k=1}^{N} \left(\partial_k (\sum_{i=1}^{n} F_i) \right)^2
$$

$$
= \varepsilon B_N \cdot \left| \sum_{i=1}^{n} \nabla F_i \right|_H^2 \leq B_N \cdot \left(\sum_{i=1}^{n} \nabla F_i, A \sum_{j=1}^{n} \nabla F_j \right)_H
$$

$$
= B_N \cdot \sum_{i,j=1}^{N} (\nabla F_i, A \nabla F_j)_H \quad \blacksquare
$$

By Theorem 5.1 and Lemma 5.10, *Markov uniqueness of the operator*
$(\mathcal{L}, \mathcal{F}C_b^\infty(K))$ *holds, provided it holds for the finite dimensional projections, and the functions B_N, $N \geq N_0$, defined in Lemma 5.10 form a bounded sequence in $L^1(E\,;m)$.* We finally look at two classes of applications for this Markov uniqueness criterion : Diffusion operators with trivial geometry (i.e., $A(z)$ is the identity on H for every z), and elliptic differential operators on Wiener space.

Application 1 (*Flat case*)

Suppose that $A(z) = \mathrm{id}_H$ for every $z \in E$. Then $\nabla^*(Ae_k) = \nabla^* e_k = -\beta_k^m$ for all $k \in \mathbf{N}$, cf. the remark above. The operator $(\mathcal{L}, \mathcal{F}C_b^\infty(K))$ is hence given by

$$
(5.44) \qquad \mathcal{L}F = \sum_{k=1}^{N} \left(\partial_k^2 F + \beta_k^m \partial_k F \right)
$$

whenever $F = f(e_1^*, \ldots, e_N^*)$ for some $N \in \mathbf{N}$ and $f \in C_b^\infty(\mathbf{R}^N)$. We obtain the following criterion for Markov uniqueness, which has first been proven in [AlbRöZha 92, Thm. 1.4], generalizing previous results from [Ta 87] and [AlbHK 77] :

Corollary 5.1 *Suppose that*

$$
\sup_{N \in \mathbf{N}} \sum_{k=1}^{N} \int (\beta_k^m - E[\beta_k^m | \mathcal{F}_N])^2 \, dm < \infty.
$$

Then the operator $(\mathcal{L}, \mathcal{F}C_b^\infty(K))$ defined by (5.44) is Markov unique on $L^2(E\,;m)$

REMARK.   In Theorem 5.2 below, we will show that under slightly stronger conditions, the operator $(\mathcal{L}, \mathcal{F}C_b^\infty(K))$ is even $L^p(E\,;m)$ unique for small p.

PROOF OF THE COROLLARY.   By Lemma 5.10, the assumption guarantees that Condition (5.39) holds for a bounded sequence $B_N \in L^1(E\,;m)$, $N \in \mathbf{N}$. Moreover, the finite dimensional projections $(\mathcal{L}_N, C_b^\infty(\mathbf{R}^N))$ of the operator $(\mathcal{L}, \mathcal{F}C_b^\infty(K))$ are given by

$$ - \int \mathcal{L}_N f\, g\, dm_N \;=\; \int \nabla f \cdot \nabla g\, dm_N \,, $$

$m_N = m \circ \Pi_N^{-1}$. It is known that the integration by parts formula (5.42) implies that $m_N = \varphi^2\, dx$ for some function $\varphi \in H^{1,2}(\mathbf{R}^n;\, dx)$, cf. [BogRö 95, Thm. 3.1]. Hence by Theorem 3.3, the spaces $W^{1,2}$ and $H_0^{1,2}$ corresponding to the operator $(\mathcal{L}_N, C_b^\infty(\mathbf{R}^N))$ on $L^2(\mathbf{R}^N;\, m_N)$ coincide. Now Theorem 5.1 implies Markov uniqueness of $(\mathcal{L}, \mathcal{F}C_b^\infty(K))$. ∎

Application 2 (*Elliptic operators on Wiener space*)

Suppose that (E, H, m) is a Wiener space, e.g., $E = \{\omega \in C([0,1] \to \mathbf{R}^d);\; \omega(0) = 0\}$, H is the Cameron Martin subspace of E, and m is the law of Brownian motion in $\mathbf{R}^d$ starting at 0. It is well-known that in this case, the weak and strong Sobolev spaces $H_0^{1,2}(E, H';\, m)$ and $W^{1,2}(E, H';\, m)$ coincide. This follows for example from our results above, but it can also be verified easily in many other ways. Following a usual convention, we denote the resulting unique Sobolev space by $\mathcal{D}^{1,2}$. The operator ∇ is now the usual *Malliavin gradient*, and the operator ∇^* is the *Skorokhod integral*, which we also denote by δ.

Now suppose, we are given an operator–valued function A on E as above. Let $\{e_k;\, k \in \mathbf{N}\}$ be an orthonormal basis of H as above, and let P_N, $N \in \mathbf{N}$, denote the orthogonal projections in H onto the span of $e_1, \dots, e_N$. Since (E, H, m) is a Wiener space, the divergence of the constant vector fields $z \mapsto e_k$, $k \in \mathbf{N}$, is given by $\delta e_k = e_k^*$. We assume that the coefficients A^{kl}, $k, l \in \mathbf{N}$, of the map A are in $\mathcal{D}^{1,2}$. Hence the functions $z \mapsto P_N A(z)e_k$, $k, N \in \mathbf{N}$, are in the domain of δ, and

$$ \delta\,(P_N A e_k) \;=\; \sum_{l=1}^{N} \delta\,(A^{kl} e_l) \;=\; \sum_{l=1}^{N} (A^{kl} e_l^* - \partial_l A^{kl}). $$

Corollary 5.2 *Suppose that the function $z \mapsto A(z)e_k$ is contained in the domain of δ for every $k \in \mathbf{N}$. Moreover, assume that there exists $N_0 \in \mathbf{N}$ such that A^{kl} is $\mathcal{F}_N$–measurable for all $N \geq N_0$ and $1 \leq k, l \leq N$. If*

$$ \sup_{N \geq N_0} \sum_{k=1}^{N} \int \big|\delta\,((1 - P_N)A(\,\cdot\,)e_k)\big|^2\, dm \;<\; \infty, $$

then the operator $(-\delta(AD\,\cdot\,), \mathcal{F}C_b^\infty(K))$ is Markov unique on $L^2(E\,;m)$.

PROOF. Fix $N \geq N_0$. By the assumption, $A^{kl} = a^{kl} \circ \Pi_N$ for some measurable functions a^{kl} on $\mathbf{R}^N$ whenever $1 \leq k, l \leq N$. Since the functions A^{kl} are in $\mathcal{D}^{1,2}$, the functions a^{kl} are in $H^{1,2}(\mathbf{R}^N; m_N)$, where $m_N = m \circ \Pi_N^{-1}$ is the standard normal distribution on $\mathbf{R}^N$. Moreover, $\partial_l A^{kl} = \frac{\partial a^{kl}}{\partial x_l} \circ \Pi_N$ m-a.s. Thus $\delta(P_N A e_k)$ is $\mathcal{F}_N$-measurable for all $1 \leq k \leq N$, whence

$$\int \left(\delta(A e_k) - E[\delta(A e_k) | \mathcal{F}_N] \right)^2 dm \leq \int \left(\delta((1 - P_N) A e_k) \right)^2 dm$$

for all $1 \leq k \leq N$. The assertion now follows by Theorem 5.1 and Lemma 5.10.
∎

d) Approximative approaches to uniqueness and existence in L^p

In this section, we prove existence and uniqueness for C^0 semigroups generated by diffusion operators defined on cylinder functions in $L^p(E; m)$, where E is an infinite dimensional space, and m is a probability measure on E, as described in Section a). Similar results in situations with non-trivial geometry will be proven in forthcoming work. We point out that even the existence problem for non-symmetric infinite dimensional diffusion operators is much harder than the corresponding finite dimensional problem, and can not so easily be decoupled from the uniqueness problem. The reason is, that an assumption saying that a non-symmetric operator is, in some sense, a "small perturbation" of a symmetric operator is much more restrictive in infinite dimensions than in finite dimensions. Therefore, we will emphasize the existence part of our results stronger than we did in the finite dimensional case.

Our uniqueness and existence proofs are based on the approximative criterion, cf. Corollary 1.5 in Appendix A, and on suitable gradient estimates for finite-dimensional resolvents. This technique has the advantage that we by the way obtain finite dimensional approximations of the resolvent corresponding to the infinite dimensional diffusion operator. The disadvantage of the use of approximative criteria in finite dimensions is that they usually require *global* integrability conditions on the operator coefficients, see e.g. [LiSem 92]. In the infinite dimensionial applications we are interested in, however, the coefficients of the finite dimensional projections typically satisfy global integrability conditions.

REMARK. All the proofs and approximations below can be carried out similarly with the resolvents replaced by the semigroups, and the solutions of elliptic equations replaced by solutions of the corresponding parabolic equations. This is sometimes more convenient, since the semigroup usually has better smoothing properties. In our situation, however, the resolvent approach seems to be slightly easier and more clear.

To state our results, we consider the framework described in Section a) above. We fix E, m, K, the "tangent space" H, $p \in [1, \infty)$, and a linear map $\beta : K \to L^p(E; m)$ as introduced there. We assume that the integration by parts identity (IP) holds for some map $\beta^m : K \to L^p(E; m)$, and that $b := \beta - \beta^m$ satisfies the divergence bound (A 1).

Let $(\mathcal{L}, \mathcal{A})$ be the diffusion operator on $L^p(E; m)$ defined by (5.1) respectively (5.9). We assume that we are given finite sets Λ_N, $N \in \mathbf{N}$, and finite-dimensional projections $\Pi_N : E \to \mathbf{R}^{\Lambda_N}$, such that the assumptions listed in a), 8) hold. As in a), 8), we introduce the vector fields $\beta_N \in L^p(E \to \mathbf{R}^{\Lambda_N}; m)$ defined by

$$\beta_N := \left(\beta_{\Pi_N^i} \right)_{i \in \Lambda_N}, \qquad N \in \mathbf{N}.$$

We now state the key assumption for our first uniqueness and existence result:

(A 2) p is in $[1, 2]$. There exist vector fields $\gamma_N \in C_b^\infty(\mathbf{R}^{\Lambda_N} \to \mathbf{R}^{\Lambda_N})$, $N \in \mathbf{N}$, such that

$$(5.45) \qquad \lim_{N \to \infty} \| \, |\beta_N - \gamma_N \circ \Pi_N| \, \|_{L^{2p/(2-p)}(E; m)} = 0.$$

Here, and in the sequel, we set $2p/(2-p) := \infty$ if $p = 2$, and use $| \cdot |$ to denote the Euclidean norm on $\mathbf{R}^{\Lambda_N}$, $N \in \mathbf{N}$.

REMARKS. (i) For $p < 2$, Assumption (A 2) is satisfied if $|\beta_N|$ is in $L^{2p/(2-p)}(E; m)$, and

$$(5.46) \qquad \lim_{N \to \infty} \| \, |\beta_N - E^m[\beta_N \mid \Pi_N]| \, \|_{L^{2p/(2-p)}(E; m)} = 0,$$

where $E^m[\, \cdot \mid \Pi_N]$ denotes the conditional expectation given Π_N w.r.t. the probability measure m. In fact, in this case

$$(5.47) \qquad E^m[\beta_N \mid \Pi_N] = \tilde{\beta}_N \circ \Pi_N \qquad m\text{-a.e.}$$

for some vector field $\tilde{\beta}_N \in L^{2p/(2-p)}(\mathbf{R}^{\Lambda_N} \to \mathbf{R}^{\Lambda_N}; m \circ \Pi_N^{-1})$, whence there exist smooth vector fields $\gamma_N \in C_b^\infty(\mathbf{R}^{\Lambda_N} \to \mathbf{R}^{\Lambda_N})$, $N \in \mathbf{N}$, such that

$$\int | E^m[\beta_N \mid \Pi_N] - \gamma_N \circ \Pi_N |^{2p/(2-p)} \, dm$$

$$= \int \left| \tilde{\beta}_N - \gamma_N \right|^{2p/(2-p)} d(m \circ \Pi_N^{-1}) \leq \frac{1}{N}.$$

By (5.46), this implies (5.45).

For $p = 1$, Condition (5.46) is even equivalent to (A 2), because the conditional expectation w.r.t. Π_N is the best $\sigma(\Pi_N)$-measurable approximation in $L^2(E; m)$.

For $p = 2$, Assumption (A 2) is satisfied provided (5.47) holds for some bounded *continuous* functions $\tilde{\beta}_N$, $N \in \mathbf{N}$, and (5.46) holds with $2p/(2-p) = \infty$.

(ii) Condition (5.46) with $p = 1$, which is equivalent to (A 2) in this case, has already been used in [AlbHK 77] to prove that the cylinder functions are dense in weak Sobolev spaces. This has been applied by several authors to show Markov uniqueness for symmetric infinite dimensional Dirichlet operators, cf. [Ta 87], [AlbRöZha 92], [So 92], cf. also Section c) above. To obtain Markov uniqueness, it is already enough to assume that the supremum of the norms in (5.46) is finite. However, I do not know a previous result where the condition (5.46) has been used to show L^P uniqueness for some $p \in [1, 2]$.

Let γ_N, $N \in \mathbf{N}$, be smooth vector fields as in (A 2). Fix $N \in \mathbf{N}$. Since γ_N is in $C_b^\infty(\mathbf{R}^{\Lambda_N} \to \mathbf{R}^{\Lambda_N})$, it can be shown by classical elliptic regularity theory, that the PDE

$$(5.48) \qquad (\lambda - \Delta - \gamma_N \cdot \nabla)\left(R_\lambda^N f\right)(x) = f(x), \qquad x \in \mathbf{R}^{\Lambda_N},$$

has a unique solution $x \mapsto \left(R_\lambda^N f\right)(x)$ in $C_b^\infty(\mathbf{R}^{\Lambda_N})$ for every function $f \in C_b^\infty(\mathbf{R}^{\Lambda_N})$, and all $\lambda > 0$, cf. e.g. [Kry 96, Thm. 4.3.2 and Cor 4.3.4].

We now state the first main result of this section:

Theorem 5.2 *Suppose (A 1) and (A 2) hold. Then the closure $\bar{\mathcal{L}}$ on $L^P(E ; m)$ of the operator $(\mathcal{L}, \mathcal{A})$ given by (5.1) generates a C^0 semigroup.*
Moreover, suppose that γ_N, $N \in \mathbf{N}$, are vector fields in $C_b^\infty(\mathbf{R}^{\Lambda_N} \to \mathbf{R}^{\Lambda_N})$ such that (5.45) holds. Then for R_λ^N as defined above, we have

$$(5.49) \qquad \left(R_\lambda^N f\right) \circ \Pi_N \xrightarrow{N \uparrow \infty} (\lambda - \bar{\mathcal{L}})^{-1} F \qquad in \ L^P(E ; m)$$

for all $\lambda > \alpha/p$, every bounded function F on E, and any uniformly bounded functions $f_N \in C_b^\infty(\mathbf{R}^{\Lambda_N})$, $N \in \mathbf{N}$, such that $f_N \circ \Pi_N \to F$ in $L^P(E ; m)$. Here α is the divergence bound of b, cf. Assumption (A 1).

The proof of Theorem 5.2 will be given below. Before, we will state our second uniqueness and existence result. Here, (A 2) will be replaced by an alternative assumption, which is more appropriate for many applications.

For $N \in \mathbf{N}$, let K_N denote the subspace of K generated by the projection Π_N, i.e.,

$$K_N := \operatorname{span}\left\{\Pi_N^i ; i \in \Lambda_N\right\} \subseteq K,$$

cf. also Subsection a), 8) above. We assume that we are given a second inner product $(\cdot, \cdot)_+$ on $\bigcup_{N \in \mathbf{N}} K_N$. In applications, the corresponding norm $(\cdot, \cdot)_+^{1/2}$ typically dominates the norm $(\cdot, \cdot)_H^{1/2}$ from above. For $N \in \mathbf{N}$, we can identify K_N and $\mathbf{R}^{\Lambda_N}$ via the isomorphism

$$\Phi_N : \mathbf{R}^{\Lambda_N} \to K_N, \qquad \Phi_N(v) = \sum_{i \in \Lambda_N} v^i \Pi_N^i.$$

By the assumptions on Π_N, the pull-back of the inner product $(\cdot, \cdot)_{H'}$ w.r.t. Φ_N is the Euclidean inner product on $\mathbf{R}^{\Lambda_N}$, cf. Subsection a), 8) above. We denote the pull–back to $\mathbf{R}^{\Lambda_N}$ of the inner product $(\cdot,\cdot)_+$ by $(\cdot,\cdot)_+$ as well, i.e.,

$$(v,w)_+ := (\Phi_N(v), \Phi_N(w))_+ \qquad \text{for all } v, w \in \mathbf{R}^{\Lambda_N}.$$

Moreover, we denote the dual inner products on $\mathbf{R}^{\Lambda_N}$, $N \in \mathbf{N}$, by $(\cdot,\cdot)_-$, i.e., the matrix $\left((e_i, e_j)_- \right)_{i,j \in \Lambda_N}$ is the inverse of the matrix $\left((e_i, e_j)_+ \right)_{i,j \in \Lambda_N}$, where e_i is the i–th Euclidean unit vector in $\mathbf{R}^{\Lambda_N}$. Hence

$$|v \cdot w| \leq |v|_- \cdot |w|_+ \qquad \text{for all } v, w \in \mathbf{R}^{\Lambda_N},$$

where the dot denotes the Euclidean inner product, $|\cdot|_- := (\cdot,\cdot)_-^{1/2}$, and $|\cdot|_+ := (\cdot,\cdot)_+^{1/2}$.

We now state the key assumption for our second uniqueness result :

(A 2′) p is in $[1,\infty)$. There exists an inner product $(\cdot,\cdot)_+$ on $\bigcup_{N \in \mathbf{N}} K_N$, and vector fields $\gamma_N \in C_b^\infty(\mathbf{R}^{\Lambda_N} \to \mathbf{R}^{\Lambda_N})$, $N \in \mathbf{N}$, such that

$$(5.50) \qquad \lim_{N \to \infty} \left\| \, |\beta_N - \gamma_N \circ \Pi_N|_- \right\|_{L^p(E;m)} = 0, \qquad \text{and}$$

$$(5.51) \qquad (v, \partial_v \gamma_N(x))_- \leq \kappa \cdot (v,v)_-$$

for all $N \in \mathbf{N}$, and $x, v \in \mathbf{R}^{\Lambda_N}$, where κ is a finite constant.

Here $(\cdot,\cdot)_-$ is the inner product on $\mathbf{R}^{\Lambda_N}$, $N \in \mathbf{N}$, defined as described above, and $\partial_v \gamma_N$ denotes the directional derivative.

REMARK.   (iii) An assumption similar to (A 2′) has been applied in [AlbKoRö 95] to prove essential self–adjointness of infinite dimensional Dirichlet operators. However, Albeverio, Kondratiev and Röckner do not, in general, use finite dimensional approximations of the operator, but they approximate β by Hilbert–space valued drifts. For the applications known so far, finite dimensional approximations seem to be sufficient.

(iv) Let κ be a finite constant, and let $N \in \mathbf{N}$ and $x \in \mathbf{R}^{\Lambda_N}$. Then Condition (5.51) holds for all $v \in \mathbf{R}^{\Lambda_N}$ if and only if

$$\left(v, (\partial \gamma_N(x))^{\mathrm{tr}} v \right)_+ \leq \kappa \cdot (v,v)_+ \quad \text{for all } v \in \mathbf{R}^{\Lambda_N},$$

where $(\partial \gamma_N(x))^{\mathrm{tr}}$ denotes the matrix $\left(\frac{\partial \gamma_N^j}{\partial x_i}(x) \right)_{i,j \in \Lambda_N}$.

We now state the second main result of this section:

Theorem 5.3 *Suppose (A 1) and (A 2′) hold. Then the closure $\bar{\mathcal{L}}$ on $L^p(E;m)$ of the operator $(\mathcal{L}, A)$ given by (5.1) generates a C^0 semigroup.*
Moreover, suppose that γ_N, $N \in \mathbf{N}$, are vector fields in $C_b^\infty(\mathbf{R}^{\Lambda_N} \to \mathbf{R}^{\Lambda_N})$ such that (5.50) and (5.51) hold for some finite constant κ. Then the resolvent approximation (5.49) holds, provided $\lambda > \max(\kappa, \alpha/p)$, $F = f \circ \Pi_M$ for some $M \in \mathbf{N}$ and $f \in C_b^\infty(\mathbf{R}^{\Lambda_M})$, and for $N \geq M$, f_N is the unique function on $\mathbf{R}^{\Lambda_N}$ satisfying $f_N \circ \Pi_N = F$.

The proof of Theorem 5.3 will be given below.

If one is just interested in **existence** of a C^0 semigroup on $L^p(E; m)$, $1 \leq p < \infty$, which is generated by an extension of $(\mathcal{L}, \mathcal{A})$, then it is enough to assume that (A 2) or (A 2′) holds for $p = 1$:

Corollary 5.3 *Suppose (A 1) holds, and (A 2) or (A 2′) holds with $p = 1$. Then the closure of the operator $(\mathcal{L}, \mathcal{A})$ on $L^1(E; m)$ generates a <u>sub-Markovian</u> C^0 semigroup $(T_t)_{t \geq 0}$. The restriction of $(T_t)_{t \geq 0}$ to $L^p(E; m)$ is a C^0 semigroup on $L^p(E; m)$ for each $p \in [1, \infty)$. The generator of this semigroup again extends $(\mathcal{L}, \mathcal{A})$, provided the range of this operator is in $L^p(E; m)$ (i.e., $\beta_\ell \in L^p(E; m)$ for all $\ell \in K$).*

The assertion follows from Theorem 5.2 respectively Theorem 5.3, and Lemma 1.9 and 1.11 in Appendix B. The corollary demonstrates once more, that in contrast to uniqueness, existence of C^0 semigroups generated by diffusion operators on L^p does not depend so much on the value of p.

We now start with preparations for the proofs of Theorem 5.2 and 5.3. Suppose that (A 2) or (A 2′) holds, and fix vector fields γ_N, $N \in \mathbb{N}$, as in (A 2) resp. (A 2′). For $N \in \mathbb{N}$, let m_N denote the probability measure $m \circ \Pi_N^{-1}$ on $\mathbb{R}^{\Lambda_N}$. The basic ingredients in the proofs of Theorem 5.2 and Theorem 5.3 are the following gradient estimates for the resolvents R_λ^N :

Lemma 5.11 *For every $\lambda > 0$, there exists a finite constant C_λ, such that*

$$\int \left| \nabla R_\lambda^N f \right|^2 dm_N \leq C_\lambda \cdot \left(1 + \int \left| \beta_N - \gamma_N \circ \Pi_N \right|^2 dm \right) \cdot \sup_{x \in \mathbb{R}^{\Lambda_N}} |f(x)|^2$$

for all $N \in \mathbb{N}$ and $f \in C_b^\infty(\mathbb{R}^{\Lambda_N})$.

Lemma 5.12 *Suppose that (A 2′) holds, and let κ be the finite constant in (5.51). Then*

$$\sup_{x \in \mathbb{R}^{\Lambda_N}} \left| (\nabla R_\lambda^N f)(x) \right|_+ \leq \frac{1}{\lambda - \kappa} \cdot \sup_{x \in \mathbb{R}^{\Lambda_N}} \left| (\nabla f)(x) \right|_+$$

for all $N \in \mathbb{N}$, $\lambda > \kappa$, and $f \in C_b^\infty(\mathbb{R}^{\Lambda_N})$.

REMARKS. (iv) L^p gradient estimates for resolvents and semigroups similar to Lemma 5.11 have been proven by V. Liskevič, Y. Semenov, and E. Tuv even for $p > 2$, cf. [LiSem 92], [LiTuv 93], and [Li 94]. However, there is an essential difference between these results and Lemma 5.11: The estimates given by the above authors depend *both* on the L^{2p} norms of γ_N, and those of the logarithmic derivative of the measure m_N, whereas the estimate in Lemma 5.11 only depends on the L^2 norm of the *difference* of β_N and $\gamma_N \circ \Pi_N$. This is crucial for infinite

dimensional applications. To obtain the estimate in Lemma 5.11, we use the
divergence bound (A 1) in an essential way.

(v) Lemma 5.12 is essentially a special case of the to experts well–known
fact, that pointwise estimates for the gradients of resolvents and semigroups
generated by diffusion operators hold under a condition on the operator Γ_2 as
introduced by D. Bakry, cf. e.g. [Ba 90]. The proof is a generalization of the
classical Bochner technique, which implies (in particular) gradient bounds on
Riemannian manifolds provided the Ricci curvature is bounded from below, cf.
e.g. [GalHuLaf 90].

PROOF OF LEMMA 5.11.   Fix $\lambda > 0$, $N \in \mathbf{N}$, and $f \in C_b^\infty(\mathbf{R}^{\Lambda_N})$. Let
$g := R_\lambda^N f$. Let $\tilde\beta_N \in L^p(\mathbf{R}^{\Lambda_N} \to \mathbf{R}^{\Lambda_N}; m_N)$ such that

$$E^m[\beta_N \mid \Pi_N] = \tilde\beta_N \circ \Pi_N \qquad m\text{–a.e.}$$

By (5.8) and (5.13), we have

$$\int |\nabla g|^2 \, dm_N = \int |\nabla g|^2 \circ \Pi_N \, dm$$
$$= -\int \mathcal{L}(g \circ \Pi_N) \, g \circ \Pi_N \, dm + \int \langle b, D(g \circ \Pi_N) \rangle \, g \circ \Pi_N \, dm$$
$$= -\int \left(\Delta g + \tilde\beta_N \cdot \nabla g\right) g \, dm_N + \frac{1}{2} \int \langle b, D(g^2 \circ \Pi_N) \rangle \, dm$$
$$\leq \int (f - \lambda g) \, g \, dm_N + \int \left(\gamma_N - \tilde\beta_N\right) \cdot \nabla g \, g \, dm_N + \frac{\alpha}{2} \int g^2 \circ \Pi_N \, dm.$$

Here we have used the divergence bound (A 1) and the equation
$-\Delta g - \gamma_N \cdot \nabla g = f - \lambda g$ in the last step. Now note, that

$$\lambda \sup_{x \in \mathbf{R}^{\Lambda_N}} |g(x)| = \sup_{x \in \mathbf{R}^{\Lambda_N}} |(\lambda R_\lambda^N f)(x)| \leq \sup_{x \in \mathbf{R}^{\Lambda_N}} |f(x)|.$$

In fact, it is well–known that the resolvent $(R_\lambda^N)_{\lambda>0}$ is sub–Markov, i.e.,
λR_λ^N is sub–Markov for all $\lambda > 0$. Thus we obtain

$$\int |\nabla g|^2 \, dm_N \leq \left(\frac{1}{\lambda} + \frac{\alpha}{2\lambda^2}\right) \cdot \sup |f|^2$$
$$+ \left(\int \left|\gamma_N - \tilde\beta_N\right|^2 \, dm_N\right)^{1/2} \cdot \left(\int |\nabla g|^2 \, dm_N\right)^{1/2} \cdot \frac{1}{\lambda} \sup |f|,$$

whence

$$\int |\nabla g|^2 \, dm_N \leq C_\lambda \cdot \left(1 + \int \left|\gamma_N - \tilde\beta_N\right|^2 \, dm_N\right) \cdot \sup |f|^2$$

for some constant C_λ which only depends on λ. This proves the lemma by
definition of $\tilde\beta_N$. ∎

PROOF OF LEMMA 5.12. The method of proof is in the spirit of the Bochner technique from analysis on manifolds, cf. e.g. [GalHuLaf 90, 4.15]. Fix $N \in \mathbf{N}$. Let $\mathcal{L}^N := \Delta + \gamma_N \cdot \nabla$, and let $\vec{\mathcal{L}}^N$ be the differential operator on $C^\infty(\mathbf{R}^{\Lambda_N} \to \mathbf{R}^{\Lambda_N})$ defined by

$$\vec{\mathcal{L}}^N v = \Delta v + (\gamma_N \cdot \nabla) v + (\partial \gamma_N)^{\mathrm{tr}} v,$$

where $\Delta v := (\Delta v^i)_{i \in \Lambda_N}$, and $(\gamma_N \cdot \nabla) v := (\gamma_N \cdot \nabla v^i)_{i \in \Lambda_N}$. One immediately verifies the following **intertwining formula** :

(5.52) $$\nabla \mathcal{L}^N g = \vec{\mathcal{L}}^N \nabla g \qquad \text{for all } g \in C^\infty(\mathbf{R}^{\Lambda_N}).$$

Moreover, for any $v \in C^\infty(\mathbf{R}^{\Lambda_N} \to \mathbf{R}^{\Lambda_N})$ we have

$$
\begin{aligned}
\frac{1}{2} \mathcal{L}^N |v|_+^2 &= \frac{1}{2} \Delta (v,v)_+ + \frac{1}{2} \gamma_N \cdot \nabla (v,v)_+ \\
&\geq (\Delta v, v)_+ + ((\gamma_N \cdot \nabla) v, v)_+ \\
&= \left(\vec{\mathcal{L}}^N v, v \right)_+ - \left((\partial \gamma_N)^{\mathrm{tr}} v, v \right)_+ \\
&\geq \left(\vec{\mathcal{L}}^N v, v \right)_+ - \kappa \cdot (v,v)_+ ,
\end{aligned}
$$
(5.53)

where the last inequality holds by (5.51) and Remark (iv).
Now fix $\lambda > \kappa$. Let $f \in C_0^\infty(\mathbf{R}^{\Lambda_N})$, and let $g := R_\lambda^N f$. Then, by (5.52) and (5.53), we obtain :

$$
\begin{aligned}
\frac{1}{2} \mathcal{L}^N |\nabla g|_+^2 &\geq \left(\vec{\mathcal{L}}^N \nabla g, \nabla g \right)_+ - \kappa \cdot (\nabla g, \nabla g)_+ \\
&= (\nabla (\mathcal{L}^N - \kappa) g, \nabla g)_+ \\
&= (\lambda - \kappa) \cdot (\nabla g, \nabla g)_+ - (\nabla f, \nabla g)_+ \\
&\geq (\lambda - \kappa) \cdot |\nabla g|_+^2 - |\nabla f|_+ \cdot |\nabla g|_+ .
\end{aligned}
$$
(5.54)

Since f has compact suppport, $\nabla g(x)$ converges to 0 as $|x| \to \infty$. This well-known fact can, for example, be deduced from the representation

$$g(x) = \int_0^\infty e^{-\lambda t} \int_{\mathbf{R}^{\Lambda_N}} p_t^N(x,y) f(y) \, dy \, dt,$$

where p_t^N denotes the heat kernel of the operator $\mathcal{L}^N$, and the standard estimate

(5.55) $$|\nabla_x p_t^N(x,y)| \leq C_1 \cdot t^{-\frac{1+d}{2}} e^{-C_2 \frac{|x-y|^2}{t}}$$

for all $t \geq 0$, and $x, y \in \mathbf{R}^{\Lambda_N}$, where C_1 and C_2 are finite constants, and d is the dimension of $\mathbf{R}^{\Lambda_N}$, cf. e.g. [Fri 64, (6.13)]. Hence there exists $x_0 \in \mathbf{R}^{\Lambda_N}$ such that $|\nabla g(x_0)|_+^2 = \sup |\nabla g|_+^2$. Evaluation of (5.54) at x_0 yields

$$0 \geq \frac{1}{2} \left(\mathcal{L}^N |\nabla g|_+^2 \right)(x_0) \geq (\lambda - \kappa) |\nabla g(x_0)|_+^2 - |\nabla f(x_0)|_+ \cdot |\nabla g(x_0)|_+ ,$$

whence

$$\sup_{x \in \mathbf{R}^{\Lambda_N}} |\nabla g(x)|_+ = |\nabla g(x_0)|_+ \leq \frac{1}{\lambda - \kappa} \cdot |\nabla f(x_0)|_+ .$$

Thus the assertion of Lemma 5.12 holds if f has compact support.

If f is an arbitrary function in $C_b^\infty(\mathbf{R}^{\Lambda_N})$, then we can find a sequence $(f_n)_{n \in \mathbf{N}}$ of functions in $C_0^\infty(\mathbf{R}^{\Lambda_N})$ such that $f_n = f$ on the ball of radius n around 0, $\sup_x |f_n(x)| \leq \sup_x |f(x)|$, and $\sup_x |\nabla f_n(x)| \leq n^{-1} + \sup_x |\nabla f(x)|$ for all n. By (5.55), we can apply Lebesgue's theorem to show that

$$\begin{aligned} \left| (\nabla R_\lambda^N f)(x) \right|_+ &= \lim_{n \to \infty} \left| (\nabla R_\lambda^N f_n)(x) \right|_+ \\ &\leq \frac{1}{\lambda - \kappa} \liminf_{n \to \infty} \sup_y |\nabla f_n(y)|_+ \leq \frac{1}{\lambda - \kappa} \sup_y |\nabla f(y)|_+ \end{aligned}$$

for all $x \in \mathbf{R}^{\Lambda_N}$. Hence the assertion holds in this case as well. ∎

PROOF OF THEOREM 5.2. We fix a bounded function $F : E \to \mathbf{R}$, and uniformly bounded functions $f_N \in C_b^\infty(\mathbf{R}^{\Lambda_N})$, $N \in \mathbf{N}$, such that $f_N \circ \Pi_N \to F$ in $L^p(E; m)$. Note that such functions f_N exist for every bounded function F, since the maps Π_N, $N \in \mathbf{N}$, generate the σ-algebra on E. Fix $\lambda > \alpha/p$. Let $\mathcal{L}^N := \Delta + \gamma_N \cdot \nabla$, and $g_N := R_\lambda^N f_N$, $N \in \mathbf{N}$. Then $g_N \circ \Pi_N$ is in $\mathcal{F}C_b^\infty(K)$. We will show :

Claim: $(\lambda - \mathcal{L})(g_N \circ \Pi_N) \longrightarrow F$ in $L^p(E; m)$ as $N \uparrow \infty$.

The claim implies that the equation $\lambda v - \mathcal{L} v = F$ is approximately solvable in $L^p(E; m)$, cf. Appendix A. Now recall, that the operator $(\mathcal{L} - \frac{\alpha}{p}, \mathcal{F}C_b^\infty(K))$ is dissipative on $L^p(E; m)$, and $\lambda > \alpha/p$. Since the equation $\lambda v - \mathcal{L} v = F$ is approximately solvable for each bounded function F on E, we obtain that the operator $\bar{\mathcal{L}} - \frac{\alpha}{p}$ generates a C^0 contraction semigroup on $L^p(E; m)$, cf. Corollary 1.5 in Appendix A. Here $\bar{\mathcal{L}}$ denotes the closure of $(\mathcal{L}, \mathcal{A})$ on $L^p(E; m)$. Thus $\bar{\mathcal{L}}$ generates a C^0 semigroup, and the inverse $(\lambda - \bar{\mathcal{L}})^{-1}$ exists for $\lambda > \alpha/p$, and is a bounded operator on $L^p(E; m)$. The claim now implies that

$$g_N \circ \Pi_N = (\lambda - \bar{\mathcal{L}})^{-1}(\lambda - \mathcal{L})(g_N \circ \Pi_N) \to (\lambda - \bar{\mathcal{L}})^{-1} F$$

as $N \to \infty$ for any functions F, f_N, and $g_N = R_\lambda^N f_N$ as above, i.e., (5.49) holds. This completes the proof of Theorem 5.2. It remains to prove the claim :

PROOF OF THE CLAIM. Obviously, it suffices to show

$$\lim_{N \to \infty} \| f_N \circ \Pi_N - (\lambda - \mathcal{L})(g_N \circ \Pi_N) \|_{L^p(E; m)} = 0.$$

For $N \in \mathbf{N}$, we have $f_n \circ \Pi_N = ((\lambda - \mathcal{L}^N) g_N) \circ \Pi_N$, whence

(5.56) $$\int |f_N \circ \Pi_N - (\lambda - \mathcal{L})(g_N \circ \Pi_N)|^p \, dm$$

$$= \int \left| \mathcal{L}(g_N \circ \Pi_N) - (\mathcal{L}^N g_N) \circ \Pi_N \right|^p \, dm$$

$$= \int \left| (\beta_N - \gamma_N \circ \Pi_N) \cdot ((\nabla g_N) \circ \Pi_N) \right|^p \, dm$$

$$\leq \left(\int |\beta_N - \gamma_N \circ \Pi_N|^{\frac{2p}{2-p}} \, dm \right)^{\frac{2-p}{2}} \cdot \left(\int_{\mathbf{R}^{\Lambda_N}} |\nabla g_N|^2 \, dm_N \right)^{\frac{p}{2}}$$

$$\leq \left(\int |\beta_N - \gamma_N \circ \Pi_N|^{\frac{2p}{2-p}} \, dm \right)^{\frac{2-p}{2}} \cdot C_\lambda^{\frac{p}{2}}$$

$$\times \left(1 + \int |\beta_N - \gamma_N \circ \Pi_N|^2 \, dm \right)^{\frac{p}{2}} \cdot \sup_{x \in \mathbf{R}^{\Lambda_N}} |f_N(x)|^p.$$

Here we have applied Lemma 5.11 to obtain the last estimate. By (5.45), and since the measure m is finite, and the functions f_N, $N \in \mathbf{N}$, are uniformly bounded, the right hand side converges to 0 as $N \to \infty$. This proves the claim. ∎

PROOF OF THEOREM 5.3. The proof is similar to that of Theorem 5.2, only that we now use the gradient estimate in Lemma 5.12 instead of that in Lemma 5.11. Fix vector fields γ_N, $N \in \mathbf{N}$, as in the assertion of Theorem 5.3, and let R_λ^N be as defined above. Moreover, we fix $M \in \mathbf{N}$, and functions f, F, and f_N, $N \geq M$, as in the assertion of Theorem 5.3. Let $\lambda > \max(\kappa, \alpha/p)$, and let $g_N := R_\lambda^N f_N$, $N \geq M$. We show that $(\lambda - \mathcal{L})(g_N \circ \Pi_N)$ converges to F in $L^p(E; m)$ as $N \to \infty$. This implies the assertion in the same way as in Theorem 5.2, because $\mathcal{F}C_b^\infty \left(\bigcup_{N \in \mathbf{N}} K_N \right)$ is dense in $L^p(E; m)$.

For $N \geq M$ we have $F = f \circ \Pi_M = f_N \circ \Pi_N$. In particular, $f_N = f \circ \varphi$ for some linear map $\varphi : \mathbf{R}^{\Lambda_N} \to \mathbf{R}^{\Lambda_M}$, whence f_N is in $C_b^\infty(\mathbf{R}^{\Lambda_N})$. A similar estimation as in (5.56) yields

$$\int \left| F - (\lambda - \mathcal{L})(g_N \circ \Pi_N) \right|^p \, dm$$

$$= \int \left| (\beta_N - \gamma_N \circ \Pi_N) \cdot ((\nabla g_N) \circ \Pi_N) \right|^p \, dm$$

(5.57) $$\leq \int |\beta_N - \gamma_N \circ \Pi_N|_-^p \, dm \cdot \sup_{x \in \mathbf{R}^{\Lambda_N}} |\nabla g_N(x)|_+^p$$

$$\leq \frac{1}{\lambda - \kappa} \int |\beta_N - \gamma_N \circ \Pi_N|_-^p \, dm \cdot \sup_{x \in \mathbf{R}^{\Lambda_N}} |\nabla f_N(x)|_+^p.$$

For $N \geq M$, $x \in \mathbf{R}^{\Lambda_N}$, and $\omega \in E$ such that $\Pi_N(\omega) = x$, we have

$$(\nabla f_N(x), \nabla f_N(x))_+ = (DF(\omega), DF(\omega))_+$$
$$= (\nabla f(\Pi_M(\omega)), \nabla f(\Pi_M(\omega)))_+,$$

whence

$$\sup_{x \in \mathbf{R}^{\Lambda_N}} |\nabla f_N(x)|_+ \leq \sup_{x \in \mathbf{R}^{\Lambda_N}} |\nabla f(x)|_+ \qquad \text{for all } N \geq M.$$

Thus, by (5.50), the right hand side of (5.57) converges to 0 as $N \to \infty$. ∎

e)   Applications to lattice systems in classical statistical mechanics

We now demonstrate the advantages and disadvantages of the different approaches to uniqueness in infinite dimensions by applying them to lattice systems from classical statistical mechanics. With the exception of some refinements, the applications presented here are not new. We consider them nevertheless, because they are well–suited to compare the different approaches.

Let Λ be a countable set, e.g., $\Lambda = \mathbf{Z}^d$ for some $d \in \mathbf{N}$. Suppose the state space E is $\mathbf{R}^\Lambda$, or some linear subspace of $\mathbf{R}^\Lambda$. Let $e_i^* : E \to \mathbf{R}$, $e_i^*(x) = x_i$, be the coordinate maps. We assume that E is endowed with the σ–algebra generated by the functionals e_i^*, $i \in \Lambda$. Let $\mathcal{F}C_b^\infty$ denote the space of all cylinder functions $F : E \to \mathbf{R}$ of type

$$ F = f(e_{i_1}^*, e_{i_2}^*, \dots, e_{i_n}^*) $$

for some $n \in \mathbf{N}$, $i_1, \dots i_n \in \Lambda$, and $f \in C_b^\infty(\mathbf{R}^n)$. For F as above and $j \in \Lambda$ let

$$ (\partial_j F)(x) := \begin{cases} \frac{\partial f}{\partial x_k}(e_{i_1}^*, \dots, e_{i_n}^*) & \text{if } j = i_k \text{ for some } 1 \le k \le n \\ 0 & \text{else} \end{cases} $$

Now assume that we are given functions $\beta_i : E \to \mathbf{R}$, $i \in \Lambda$. These functions describe the interaction of the lattice system considered. Informally, "$\beta_i = -\frac{\partial U}{\partial x_i}$", where U is the total potential energy of the system. If, for example, the system is described by one–particle potentials $V_i : \mathbf{R} \to \mathbf{R}$, $i \in \Lambda$, and two–particle interaction potentials $W_{ij} : \mathbf{R}^2 \to \mathbf{R}$, $i, j \in \Lambda$, $i \ne j$, $W_{ij} = W_{ji}$, then U is informally given as

$$ \text{"} U(x) = \sum_i V_i(x_i) + \frac{1}{2} \sum_{i \ne j} W_{ij}(x_i, x_j) \text{"}, $$

whence

$$ \beta_i(x) = -V_i'(x_i) - \sum_{j \in \Lambda \setminus \{i\}} \frac{\partial W_{ij}}{\partial x_i}(x_i, x_j), \quad i \in \Lambda. $$

The expression for β_i is rigorous if, for example, for each fixed i, W_{ij} vanishes for all but finitely many j, i.e., the interaction of each particle has finite range.

Definition 5.1 *We call a probability measure m on E a **Gibbs measure** (with **exponential bound**) w.r.t. $(\beta_i)_{i \in \Lambda}$, if and only if*

(i) $\displaystyle \sup_{i \in \Lambda} \int e^{|x_i|}\, m(dx) < \infty.$

(ii) β_i is m–integrable for every $i \in \Lambda$. The integration by parts identities

$$\int \frac{\partial f}{\partial x_{i_k}}(x_{i_1}, \ldots, x_{i_n})\, m\,(dx) = -\int \beta_{i_k}(x)\, f(x_{i_1}, \ldots, x_{i_n})\, m\,(dx)$$

hold for all $n \in \mathbf{N}$, $i_1, \ldots, i_n \in \Lambda$, $f \in C_b^\infty(\mathbf{R}^n)$, and $1 \le k \le n$.

This definition of Gibbs measures does essentially coincide with the usual definition of Gibbs measures w.r.t. the specification determined by the corresponding interaction potentials, cf. e.g. [AlbKoRö 97c, Prop. 5.9].

From now on, we fix a Gibbs measure m w.r.t. $(\beta_i)_{i \in \Lambda}$. We consider the symmetric diffusion operator $(\mathcal{L}, \mathcal{F}C_b^\infty)$ on $L^2(E\,; m)$ corresponding to the pre–Dirichlet form

$$\mathcal{E}(F, G) = \int \sum_{i \in \Lambda} \partial_i F\, \partial_i G\; dm\,, \quad F, G \in \mathcal{F}C_b^\infty,$$

i.e.,

(5.58)
$$\mathcal{L}F = \sum_{i \in \Lambda}(\partial_i^2 F + \beta_i\, \partial_i F)\,.$$

The diffusion operator describes the stochastic dynamics of the corresponding lattice systems. Markov uniqueness of the operator $(\mathcal{L}, \mathcal{F}C_b^\infty)$ hence implies uniqueness of a reversible dynamics with initial distribution m, and $L^1(E\,; m)$ uniqueness implies uniqueness of a stationary dynamics with initial distribution m. We will now discuss in which cases the different methods developed above can be applied to prove Markov uniqueness and $L^p(E\,; m)$ uniqueness.

The projective approach

By a slight modification of Corollary 5.1, the operator $(\mathcal{L}, \mathcal{A})$ is Markov unique on $L^2(E\,; m)$ if there exists an increasing sequence Λ_N of finite subsets of Λ such that $\Lambda = \bigcup_{N \in \mathbf{N}} \Lambda_N$, and

(5.59)
$$\sup_{n \in \mathbf{N}} \int \sum_{i \in \Lambda_N} (\beta_i - E[\beta_i | \mathcal{F}_{\Lambda_N}])^2\; dm \;<\; \infty.$$

Here $\mathcal{F}_{\Lambda_N}$ denotes the σ–algebra generated by the functionals e_i^*, $i \in \Lambda_N$. If, moreover,

(5.60)
$$\lim_{N \to \infty} \int \left(\sum_{i \in \Lambda_N} (\beta_i - E[\beta_i | \mathcal{F}_{\Lambda_N}])^2 \right)^{p/(2-p)} dm = 0$$

for some $p \in [1, 2)$, then, by Theorem 5.2, $(\mathcal{L}, \mathcal{F}C_b^\infty)$ is even $L^p(E\,; m)$ unique, and the resolvents of the finite dimensional approximations converge to the resolvent of the operator closure in $L^p(E\,; m)$.

Condition (5.59) is very useful for one–dimensional lattice systems, i.e., if $\Lambda = \mathbf{Z}^1$. For $\Lambda = \mathbf{Z}^d$, $d \ge 2$, however, the condition is usually not satisfied for the physically relevant models. In fact, suppose that we are given a distance function d on Λ such that the d–balls are finite. We assume that there exists a finite radius r_0 such that there is no interaction between particles at positions $i, j \in \Lambda$ with $d(i,j) > r_0$. In terms of β this means that for each $i \in \Lambda$, β_i is $\mathcal{F}_{B(i,r_0)}$–measurable, where $B(i,r) := \{j \in \Lambda;\, d(i,j) \leq r\}$. For a set $\Lambda_0 \subset \Lambda$ let $|\Lambda_0|$ denote the number of points in Λ_0, and let $\partial_{r_0}\Lambda_0$ be the r_0–boundary defined by

$$\partial_{r_0}\Lambda_0 := \{i \in \Lambda_0;\, \exists j \in \Lambda \setminus \Lambda_0,\, d(i,j) \leq r_0\}.$$

Theorem 5.4 *Consider the situation just described, and fix $i_0 \in \Lambda$. Suppose that there exists a function $C : [0,\infty) \to (0,\infty)$ such that*

$$\int \beta_i^2 \, dm \leq C(d(i,i_0)) \qquad \text{for all } i \in \Lambda.$$

If

$$\limsup_{r\to\infty} C(r - r_0) \cdot |\partial_{r_0} B(i_0,r)| \; < \; \infty,$$

then Condition (5.59) is satisfied, i.e., the operator $(\mathcal{L}, \mathcal{F}C_b^\infty)$ is Markov unique on $L^2(E;m)$.

PROOF. For $N \in \mathbf{N}$, $N \geq r_0$, let $\Lambda_N := B(i_0, N)$. Note that for $i \in \Lambda_N \setminus \partial_{r_0}\Lambda_N$, β_i is $\mathcal{F}_{\Lambda_N}$–measurable. Hence

$$\int \sum_{i\in\Lambda_N} \left(\beta_i - E[\beta_i|\mathcal{F}_{\Lambda_N}]\right)^2 dm$$

$$\leq \sum_{i\in\partial_{r_0}\Lambda_N} \int \beta_i^2 \, dm \; \leq \; |\partial_{r_0}\Lambda_N| \cdot C(N - r_0).$$

The assumed condition thus implies (5.59), whence Markov uniqueness holds by Corollary 5.1. ∎

EXAMPLE. Suppose $\Lambda = \mathbf{Z}^d$ for some $d \in \mathbf{N}$, and let $d(i,j) := |i - j|$. Then for every fixed $i_0 \in \Lambda$, and $r_0 \geq 1$, $|\partial_{r_0} B(i_0,r)| \sim r^{d-1}$ as $r \to \infty$. Hence the projective approach yields Markov uniqueness provided

$$\limsup_{|i|\to\infty} |i|^{d-1} \int \beta_i^2 \, dm \; < \; \infty .$$

In particular, *for $d = 1$, Markov uniqueness holds if the interactions are bounded* in the sense that $\sup_{i\in\Lambda} \int \beta_i^2 \, dm < \infty$. This includes the physically relevant *translation invariant case. For $d \geq 2$, however, the projective approach only yields Markov uniqueness if the interactions decay rapidly enough as $|i| \to \infty$.* The translation invariant case is not included.

The considerations above show that the projective approach is only of very limited use to prove Markov uniqueness for lattice systems from classical statistical mechanics. In particular, it is not applicable to the most interesting models, which have a phase transition.

The second approximative approach

In Section d) above, we have described an alternative approximative approach to L^p uniqueness of infinite dimensional diffusion operators. Here the operator is approximated by finite dimensional operators that are not necessarily projections of the infinite dimensional diffusion operator, cf. Theorem 5.3. It has already been pointed out in several articles by S. Albeverio, Y. Kondratiev and M. Röckner that this approach can be used to prove essential self–adjointness even in situations with phase transitions, where the projective approach fails. At this place, we briefly recall their application to discrete $P(\phi)_d$–models, which serves as an illustration on how to apply Theorem 5.3. For other applications, we refer to [AlbKoRö 95, 95a, 95b, 97c], [PaYoo 97], and the references given in these articles.

In the discrete $P(\phi)_d$–models, $\Lambda = \mathbf{Z}^d$, the one–particle potentials are given by $V_i(x_i) = P(x_i)$ for all $i \in \Lambda$, where

$$P(q) \ = \ a_{2m} q^{2m} + \ldots + a_1 q + a_0 \,, \quad a_{2m} > 0, \quad m \geq 2,$$

is a polynomial of even degree, and the two–particle interactions W_{ij}, $i, j \in \Lambda$, are quadratic nearest neighbour interactions, i.e.,

$$W_{ij}\,(x_i, x_j) \ = \ \begin{cases} (x_i - x_j)^2 & \text{if } |i - j| = 1 \\ 0 & \text{else} \end{cases} .$$

Hence for $i \in \mathbf{Z}^d$,

$$\beta_i\,(x) \ = \ -P'(x_i) \ - \ 2 \sum_{j \in \mathbf{Z}^d, \, |j-i|=1} (x_i - x_j)\,, \quad x \in \mathbf{R}^{\mathbf{Z}^d}.$$

Let m be a Gibbs measure on $\mathbf{R}^{\mathbf{Z}^d}$ w.r.t. $(\beta_i)_{i \in \mathbf{Z}^d}$ in the sense defined above. It is well–known that if P is replaced by $\varepsilon \cdot P$ with $\varepsilon > 0$ sufficiently small, then m is unique, whereas for large ε, there can be a phase transition, cf. the references in [AlbKoRö 95, Sect. 4]. The following result generalizes [AlbKoRö 97c, Thm. 5.13] to the case $p \neq 2$:

Theorem 5.5 *For every (exponentially bounded) Gibbs measure m w.r.t. $(\beta_i)_{i \in \mathbf{Z}^d}$, and for every $p \in [1, \infty)$, the operator $(\mathcal{L}, FC_b^\infty)$ defined by (5.58) is $L^p(E\,;m)$ unique.*

For generalizations to other one–particle and interaction potentials, cf. [AlbKoRö 95, Sect. 4], and the other references mentioned above. It is shown in [AlbKoRö 97c], [AlbKoRö 95] respectively, how Theorem 5.5 for the case $p = 2$ can be deduced from a result similar to Theorem 5.3. In essentially the same way, we can apply Theorem 5.3 to obtain Theorem 5.5 :

PROOF OF THEOREM 5.5. Let $(\Lambda_N)_{N \in \mathbf{N}}$ be a increasing sequence of finite subsets of $\mathbf{Z}^d$ such that $\mathbf{Z}^d = \bigcup \Lambda_N$. Let $\gamma_N : \mathbf{R}^{\Lambda_N} \to \mathbf{R}^{\Lambda_N}$, $N \in \mathbf{N}$, be the smooth vector fields given by

$$\gamma_N^i(x) \;=\; -P'(x_i) \,-\, 2 \sum_{j \in \Lambda_N,\, |i-j|=1} (\,x_i - x_j\,), \quad x \in \mathbf{R}^{\Lambda_N}.$$

Let $\mathbf{R}_0^{\mathbf{Z}^d}$ be the set of all $v \in \mathbf{R}^{\mathbf{Z}^d}$ such that v_i vanishes for alll except finitely many i, and let $(\,\cdot\,,\,\cdot\,)_+$ be the inner product on $\mathbf{R}_0^{\mathbf{Z}^d}$ defined by

$$(\,v\,,\,w\,)_+ \;=\; \sum_{i \in \mathbf{R}^{\mathbf{Z}^d}} v_i\, w_i \;(1 + |i|)^s\;,$$

where s is a sufficiently large integer. Then it can be shown similarly to the arguments in [AlbKoRö 95, Sect. 4.1], that Assumption (A 2′) from Section d) is satisfied, except that γ_N is not in $C_b^\infty(\mathbf{R}^{\Lambda_N} \to \mathbf{R}^{\Lambda_N})$. Note that the assumption on the uniqueness of the Gibbs measure made in [AlbKoRö 95] is not really needed. In fact, it has been pointed out in [AlbKoRö 97c, Proof of Theorem 5.13] that it is enough to assume exponential boundedness of m. It is not difficult to replace the vector fields γ_N, $N \in \mathbf{N}$, by slightly modified vector fields $\tilde{\gamma}_N \in C_b^\infty(\mathbf{R}^{\Lambda_N} \to \mathbf{R}^{\Lambda_N})$ such that (A 2′) still holds. Now Theorem 5.3 implies the assertion. ■

f)   Stability of L^p uniqueness under H–valued perturbations

Several authors have proven essential self–adjointness respectively L^1 uniqueness for operators on Wiener space of type $\mathcal{L}^{(0)} + (B^H, \nabla \cdot)_H$, where H is the Cameron–Martin space, and $\mathcal{L}^{(0)}$ is the Ornstein–Uhlenbeck operator or some more general "nice" diffusion operator, and B^H is a measurable vector field with values in H. For example, I. Shigekawa [Shi 95] showed essential self–adjointness on $L^2(E; \varphi^2 \cdot \mu)$ for operators as above with $B^H = (\nabla\varphi)/\varphi$, where φ is a sufficiently smooth function. In particular, φ is assumed to be in the Sobolev space $W^{2,p}$ for all finite p. W. Stannat [St 96] proved that non–symmetric perturbations of type above of an essentially self–adjoint operator $\mathcal{L}^{(0)}$ are still L^1 unique if $\big|B^H\big|_H$ is in L^2. By combining both results, one hence obtains L^1 uniqueness for a general class of non–symmetric diffusion operators on Wiener space.

   The aim of this section is to demonstrate that this combined result and more general perturbation results can be directly deduced from Theorem 5.2. More precisely, consider the framework described in Section a) above. We show that the condition (A 2) from Section d) is stable under appropriate H–valued perturbations of the drift β, cf. Subsection 1). As a consequence, we prove L^p uniquenes for H–valued (not necessarily symmetric) first–order perturbations of diffusion operators with linear drift, cf. Subsection 2). In Subsection 3), we look at our results in the symmetric case. In particular, we prove L^p uniqueness for infinite dimensional generalized Schrödinger operators. Applications to perturbations of the Ornstein–Uhlenbeck operator on Wiener space, to perturbations of random motions of strings, and to uniqueness problems in quantum field theory will be considered in Section g).

1) The general perturbation result

We fix E, m, K, H and β as introduced in Section a). Moreover, we fix a strictly increasing sequence $(i_N)_{N \in \mathbf{N}}$ of positive integers, and an orthonormal basis $\{e_i \; ; \; i \in \mathbf{N}\}$ of the Hilbert space H such that the functionals $e_i^* \in H'$, $e_i^*(h) = (e_i, h)_H$, are in K for all $i \in \mathbf{N}$. We define projections $\Pi_N : E \to \mathbf{R}^{i_N}$, $N \in \mathbf{N}$, by

$$\Pi_N(\omega) = \left(e_1^*(\omega), \ldots, e_{i_N}^*(\omega) \right),$$

cf. also Section a), 8).

Lemma 5.13 *Let $p \in [1, 2)$. Suppose there exists a decomposition*

$$(5.61) \qquad \beta_\ell = \beta_\ell^{(0)} + \left(j(\ell), B^H \right)_H \qquad \text{for all } \ell \in K,$$

where $\beta^{(0)} : K \to L^p(E; m)$ is a linear map satisfying (A 2), and B^H is a vector field in $L^{2p/(2-p)}(E \to H; m)$. Then β satisfies (A 2) as well.

Note that under the assumptions of the lemma, the operator $(\mathcal{L}, \mathcal{A})$ defined by (5.1) is given by

$$\mathcal{L} F = \mathcal{L}^{(0)} F + \left(B^H, \nabla F \right)_H,$$

where $\mathcal{L}^{(0)}$ is the operator with drift $\beta^{(0)}$, and ∇ is the H-gradient.

REMARK. For $p = 2$, the assertion of the lemma holds as well, provided we assume additionally, that

$$E^m[B_N^H | \Pi_N] = \bar{B}_N^H \circ \Pi_N \qquad m\text{-a.e.}$$

for some bounded continuous vector fields $\bar{B}_N^H : \mathbf{R}^{i_N} \to \mathbf{R}^{i_N}$, $N \in \mathbf{N}$. Here B_N^H denotes the vector field in $L^{2p/(2-p)}(E \to \mathbf{R}^{i_N} \; ; \; m)$ given by

$$(5.62) \qquad B_N^H = \left((e_1, B^H)_H, \ldots, (e_{i_N}, B^H)_H \right).$$

PROOF OF THE LEMMA. For $N \in \mathbf{N}$, let $m_N := m \circ \Pi_N^{-1}$. There exists a vector field $\bar{B}_N^H \in L^{2p/(2-p)}(\mathbf{R}^{i_N} \to \mathbf{R}^{i_N} \; ; \; m_N)$ such that

$$E^m[B_N^H | \Pi_N] = \bar{B}_N^H \circ \Pi_N \qquad m\text{-a.e.},$$

where B_N^H is given by (5.62). Since $C_b^\infty(\mathbf{R}^{i_N} \to \mathbf{R}^{i_N})$ is dense in $L^{2p/(2-p)}(\mathbf{R}^{i_N} \to \mathbf{R}^{i_N} \; ; \; m_N)$ for all N, we can find smooth vector fields

$\gamma_N^{(1)} \in C_b^\infty(\mathbf{R}^{i_N} \to \mathbf{R}^{i_N})$, $N \in \mathbf{N}$, such that

$$\int \left| E^m[B_N^H|\Pi_N] - \gamma_N^{(1)} \circ \Pi_N \right|^{2p/(2-p)} dm$$

$$(5.63) \qquad = \int_{\mathbf{R}^{i_N}} \left| \tilde{B}_N^H - \gamma_N^{(1)} \right|^{2p/(2-p)} dm_N \leq N^{-1}.$$

Let $\gamma_N := \gamma_N^{(0)} + \gamma_N^{(1)}$, where $\gamma_N^{(0)} \in C_b^\infty(\mathbf{R}^{i_N} \to \mathbf{R}^{i_N})$, $N \in \mathbf{N}$, are vector fields satisfying (5.45) with β replaced by $\beta^{(0)}$. We have

$$|\beta_N - \gamma_N \circ \Pi_N| \leq \left| \beta_N^{(0)} - \gamma_N^{(0)} \circ \Pi_N \right| + \left| B_N^H - E^m[B_N^H|\Pi_N] \right|$$

$$(5.64) \qquad\qquad\qquad + \left| E^m[B_N^H|\Pi_N] - \gamma_N^{(1)} \circ \Pi_N \right|$$

for all $N \in \mathbf{N}$. By (5.45) and (5.63), the $L^{2p/(2-p)}(E; m)$ norms of the first and the last term on the right hand side converge to 0 as $N \to \infty$. Moreover, for m-a.e. $\omega \in E$, $B_N^H(\omega) - E^m[B_N^H|\Pi_N](\omega)$ is the orthogonal projection in H of $B^H(\omega) - E^m[B^H|\Pi_N](\omega)$ onto $\mathbf{R}^{i_N} \cong \mathrm{span}\{e_1, \ldots, e_{i_N}\}$, whence

$$\left| B_N^H - E^m[B_N^H|\Pi_N] \right| \leq \left| B^H - E^m[B^H|\Pi_N] \right|_H \qquad m\text{-a.e.}$$

Since, by assumption, B^H is in $L^{2p/(2-p)}(E \to H; m)$, the $L^{2p/(2-p)}(E; m)$ norm of the right hand side converges to 0 as $N \to \infty$. Thus, by (5.64),

$$\lim_{N \to \infty} \| |\beta_N - \gamma_N \circ \Pi_N| \|_{L^{2p/(2-p)}(E;m)} = 0.$$

Since γ_N is in $C_b^\infty(\mathbf{R}^{i_N} \to \mathbf{R}^{i_N})$ for any $N \in \mathbf{N}$, this proves the assertion. ∎

2) Perturbations of operators with linear drift

We now apply the perturbation lemma to the case where the drift β has a decomposition into a linear part and a small perturbation. We fix E, H, and m as before. To simplify the notation, the functional $j^{-1}(g)$ in H' corresponding to an element $g \in H$ will in the sequel be briefly denoted by g^*, i.e.,

$$g^*(h) := (g, h)_H \qquad \text{for all } h \in H.$$

Informally, we consider a linear operator $\mathcal{L}$ on $L^p(E; m)$ with drift β given by

$$\text{``} \quad \beta_{g^*}(\omega) = (g, A^*\omega + B^H(\omega))_H \quad \text{''}$$

for a (not neccessarily bounded) linear operator A on H, and a vector field B^H as above, where A^* denotes the adjoint operator of A on H. The operator $\mathcal{L}$ is hence informally given by

$$\text{``} \quad (\mathcal{L}F)(\omega) = (\Delta_H F)(\omega) + (A^*\omega + B^H(\omega), (\nabla F)(\omega))_H \quad \text{''}.$$

To make the expressions for β and $\mathcal{L}$ rigorous, suppose V is a dense subspace of H. We assume that the functionals g^*, $g \in V$, are continuous on $E \cap H$ w.r.t. the norm on E. Hence they can be uniquely extended to continuous linear functionals on E, which we also denote by g^*. To get into the framework used in Subsection 1), we set

$$K := \{ g^* ; g \in V \}.$$

The operator $\mathcal{L}$ will be defined on functions in $\mathcal{F}C_b^\infty(K)$, but in the sequel we prefer to use the notation $\mathcal{F}C_b^\infty(V)$ for this space, as we did in Section b).

Suppose A is a linear operator on H with domain V. We assume that for h in the range of A, the functional h^* is in $E' \cap H'$. In other words : The image $\bar{A} := j^{-1} \circ A \circ j$ of the operator A under the Riesz isometry j^{-1} maps functionals in K to functionals in $E' \cap H'$. We refer to the examples in Section g), if the reader is confused by the variety of different spaces.

Let B^H be a vector field in $L^1(E \to H ; m)$. We assume

$$(5.65) \qquad \beta_{g^*}(\omega) = (Ag)^*(\omega) + (g, B^H(\omega))_H$$

for all $g \in V$ and $\omega \in E$. The operator $(\mathcal{L}, \mathcal{F}C_b^\infty(V))$ is hence given by

$$(5.66) \qquad \mathcal{L}F = \mathcal{L}^A F + (B^H, \nabla F)_H ,$$

where

$$\mathcal{L}^A \left(f(g_1^*, \ldots, g_n^*) \right)$$
$$= \sum_{\lambda,\mu=1}^{n} (g_\lambda, g_\mu)_H \frac{\partial^2 f}{\partial x_\lambda \partial x_\mu}(g_1^*, \ldots, g_n^*) + \sum_{\mu=1}^{n} (Ag_\mu)^* \frac{\partial f}{\partial x_\mu}(g_1^*, \ldots, g_n^*)$$

for all $n \in \mathbf{N}$, $g_1, \ldots, g_n \in V$, and $f \in C_b^\infty(\mathbf{R}^n)$.

Assumption (A 2) can be most easily verified if the operator A has discrete spectrum. If A is non–symmetric, a spectral resolution can only exist if we complexify. Hence let $H_{\mathbf{C}}$ be the complexification of H, i.e., $H_{\mathbf{C}} = H \times H$ with addition given by $[g_1, h_1] + [g_2, h_2] = [g_1 + g_2, h_1 + h_2]$, scalar multiplication given by $(x + iy)[g, h] = [xg - yh, xh + yg]$, and inner product given by $([g_1, h_1], [g_2, h_2])_{H_{\mathbf{C}}} := (g_1, g_2)_H + i(g_1, h_2)_H - i(h_1, g_2)_H + (h_1, h_2)_H$, and let $A_{\mathbf{C}}$ denote the complexification of the operator A, i.e., the domain of $A_{\mathbf{C}}$ is $V_{\mathbf{C}} := V \times V \subseteq H_{\mathbf{C}}$, and

$$A_{\mathbf{C}}([g, h]) = [Ag, Ah] \qquad \text{for } g, h \in V.$$

Assumptions on A and m, under which we can easily show that β satisfies (A 2), are :

(DS) (i) There exists a basis $\{\psi_i ; i \in \mathbf{N}\}$ of the complex Hilbert space $H_{\mathbf{C}}$ consisting of eigenfunctions of the operator $(A_{\mathbf{C}}, V_{\mathbf{C}})$.

(ii) $\int |\ell(\omega)|^{2p/(2-p)} m(d\omega) < \infty \qquad$ for all $\ell \in E' \cap H'$.

However, many operators one is interested in in applications do not have discrete spectrum. If we impose an additional condition on the measure m, then we can also verify (A 2) under a weaker assumption on A :

(CS) (i) There exist a strictly increasing sequence $(i_N)_{N \in \mathbf{N}}$ of positive integers, reals λ_{ik}, $1 \leq i, k \leq \infty$, and a complete orthonormal system $\{e_i; i \in \mathbf{N}\}$ of H, such that

- e_i is in V for all $i \in \mathbf{N}$.
- $\lambda_{ik} = 0$ whenever $i \leq i_N < k$ for some $N \in \mathbf{N}$.
- $\sum_{i=1}^{\infty} |Ae_i - \sum_{k=1}^{\infty} \lambda_{ik} e_k|_H < \infty$.

Note that by the assumption on λ_{ik}, only finitely many summands of $\sum_{k=1}^{\infty} \lambda_{ik} e_k$ do not vanish.

(ii) There exists a finite constant C such that

$$\int |g^*(\omega)|^{2p/(2-p)} \, m(d\omega) \leq C \cdot |g|_H^{2p/(2-p)}$$

holds for all $g \in H$ such that g^* is in $E' \cap H'$.

REMARKS.   (i) Suppose the operator (A, V) is **essentially self–adjoint** on H. Then (CS) (i) is satisfied. In fact, in this case there exists an orthogonal spectral resolution of the closure $\bar{A}$ over $\mathbf{R}$, from which one can construct easily an orthonormal basis $\{e_i; i \in \mathbf{N}\}$ of approximate eigenfunctions of A, i.e., e_i is in V for all $N \in \mathbf{N}$, and

$$\sum_{i=1}^{\infty} |Ae_i - \lambda_i e_i| < \infty$$

for some sequence $\lambda_i \in \mathbf{R}$, $i \in \mathbf{N}$. Hence (CS) (i) holds with $i_N = N$ and $\lambda_{ik} = \lambda_i \delta_{ik}$.

(ii) Moreover, Assumption (CS) (i) is satisfied whenever Assumtion (DS) holds, cf. the proof of Theorem 5.6 below.

(iii) Because of the orthogonality assumption in (CS) (i), I am not sure whether (CS) (i) can be verified for every (non–symmetric) closable operator (A, V) such that the closure admits a (non–discrete) spectral resolution over $\mathbf{C}$. However, it should be possible to verify (CS) (i) for a large class of non–symmetric operators with non–discrete spectrum as well.

Theorem 5.6 *Let $p \in [1, 2)$. Suppose (A, V) is a densely defined linear operator on H as above, and B^H is a vector field in $L^{2p/(2-p)}(E \to H; m)$. Assume that (CS) or (DS) holds, and that the map $b := \beta - \beta^m$ defined by (5.65) and (5.4) satisfies the divergence bound (A 1).*
Then the closure on $L^p(E; m)$ of the operator $(\mathcal{L}, \mathcal{F}C_b^{\infty}(V))$ given by (5.66) is the generator of a C^0 semigroup.

REMARK. Again, a similar result for $p = 2$ can be proven under an additional continuity assumption on the finite–dimensional projections of the vector field B^H, cf. the remark below Lemma 5.13.

We first prove Theorem 5.6 provided Assumption (DS) holds. The proof under Assumption (CS) will be given below.

PROOF OF THEOREM 5.6 UNDER ASSUMPTION (DS). We fix an eigenbasis $\{\psi_N; N \in \mathbf{N}\}$ of the operator $(A_{\mathbf{C}}, V_{\mathbf{C}})$. For $N \in \mathbf{N}$, let E_N be the *real* vector space spanned by the real and imaginary part of ψ_N. Then $E_N \subseteq V$ and $A(E_N) \subseteq E_N$ for all N. We can find a strictly increasing sequence of positive integers $(i_N)_{N \in \mathbf{N}}$, and a complete *orthonormal* system $\{e_i; i \in \mathbf{N}\}$ of H, such that $\{e_1, e_2, \ldots, e_{i_N}\}$ is a basis of $\bigoplus_{M=1}^{N} E_M$ for each $N \in \mathbf{N}$. In particular, the span of $\{e_1, e_2, \ldots, e_{i_N}\}$ is invariant under A for all N, whence

$$A e_i = \sum_{k=1}^{\infty} a_{ik} e_k \qquad \text{for all } i \in \mathbf{N},$$

where a_{ik}, $1 \le i, k < \infty$, are reals such that $a_{ik} = 0$ whenever $i \le i_N < k$ for some $N \in \mathbf{N}$. Let $\Pi_N : E \to \mathbf{R}^{i_N}$, $N \in \mathbf{N}$, be the projections given by

$$\Pi_N(\omega) = \left(e_1^*(\omega), \ldots, e_{i_N}^*(\omega) \right) \qquad \text{for all } \omega.$$

Recall that $K = \{g^*; g \in V\}$. We define $\beta^{(0)} : K \to L^p(E; m)$ by

$$\beta_{g^*}^{(0)}(\omega) := (Ag)^*(\omega) \qquad \text{for all } \omega \in E \text{ and } g \in V.$$

Note that, by (DS) (ii), $\beta_{g^*}^{(0)}$ is indeed in $L^p(E; m)$ for every $g \in V$. For $N \in \mathbf{N}$ and $1 \le i \le i_N$, we have

$$\beta_{e_i^*}^{(0)} = (Ae_i)^* = \sum_{k=1}^{\infty} a_{ik} e_k^* = \sum_{k=1}^{i_N} a_{ik} e_k^*.$$

Let $\beta_N^{(0)}$ be the projected vector field in $L^p(E \to \mathbf{R}^{i_N}; m)$ defined as above, i.e.,

$$\beta_N^{(0)}(\omega) = \left(\beta_{e_1^*}^{(0)}(\omega), \ldots, \beta_{e_{i_N}^*}^{(0)}(\omega) \right) \qquad \text{for all } \omega.$$

Then $\beta_N^{(0)} = \tilde{\beta}_N^{(0)} \circ \Pi_N$, where $\tilde{\beta}_N^{(0)}$ is the smooth vector field on $\mathbf{R}^{i_N}$ given by

$$\tilde{\beta}_N^{(0)}(x) = \left(\sum_{k=1}^{i_N} a_{ik} x_k \mid 1 \le i \le i_N \right).$$

Let $r := 2p/(2-p)$, and $m_N := m \circ \Pi_N^{-1}$, $N \in \mathbf{N}$. By Assumption (DS) (ii), $\tilde{\beta}_N^{(0)}$ is in $L^r(\mathbf{R}^{i_N} \to \mathbf{R}^{i_N}; m_N)$. Thus we can find a vector field $\gamma_N^{(0)} \in C_b^{\infty}(\mathbf{R}^{i_N} \to \mathbf{R}^{i_N})$ such that

$$\int_{\mathbf{R}^{i_N}} \left| \tilde{\beta}_N^{(0)} - \gamma_N^{(0)} \right|^r dm_N \le N^{-1}.$$

We obtain

$$\lim_{N\to\infty} \int \left| \beta_N^{(0)} - \gamma_N^{(0)} \circ \Pi_N \right|^r dm = \lim_{N\to\infty} \int \left| \tilde{\beta}_N^{(0)} - \gamma_N^{(0)} \right|^r dm_N = 0.$$

Thus, Assumption (A 2) is satisfied for $\beta^{(0)}$, whence, by (5.65) and Lemma 5.13, it is satisfied for β as well. Now, Theorem 5.2 implies that the closure of $(\mathcal{L}, \mathcal{F}C_b^\infty(V))$ on $L^p(E\,;\,m)$ generates a C^0 semigroup. ∎

PROOF OF THEOREM 5.6 UNDER ASSUMPTION (CS). We fix a sequence $(i_N)_{N\in\mathbf{N}}$, reals λ_{ik}, $1 \le i, k < \infty$, and a complete orthonormal system $\{e_i;\, i \in \mathbf{N}\}$ of H as in (CS) (i). Let $\Pi_N : E \to \mathbf{R}^{i_N}$, $N \in \mathbf{N}$, be the corresponding projections, i.e., $\Pi_N(\omega) = (e_1^*(\omega), \dots, e_{i_N}^*(\omega))$ for all ω. We define $\beta^{(0)} : K \to L^p(E\,;\,m)$ and $\beta_N^{(0)} \in L^p(E \to \mathbf{R}^{i_N}\,;\,m)$, $N \in \mathbf{N}$, as in the proof under Assumption (DS) above. Supose $\tilde{\gamma}_N^{(0)}$, $N \in \mathbf{N}$, are smooth vector fields on $\mathbf{R}^{i_N}$ given by

$$\tilde{\gamma}_N^{(0)}(x) = \left(\sum_{k=1}^{i_N} \mu_{ik}^{(N)} x_k \mid 1 \le i \le i_N \right)$$

for some reals $\mu_{ik}^{(N)}$, $N \in \mathbf{N}$, $1 \le i, k \le i_N$. Let $r := 2p/(2 - p)$, and $m_N := m \circ \Pi_N^{-1}$, $N \in \mathbf{N}$. By (CS) (ii), $\tilde{\gamma}_N^{(0)}$ is in $L^r(\mathbf{R}^{i_N}\,;\,m_N)$ for each $N \in \mathbf{N}$. Moreover, since $\beta_{e_i^*}^{(0)} = (Ae_i)^*$ for all $i \in \mathbf{N}$, we obtain

(5.67)
$$\left\| \left| \beta_N^{(0)} - \tilde{\gamma}_N^{(0)} \circ \Pi_N \right| \right\|_{L^r(E;m)}$$
$$\le \sum_{i=1}^{i_N} \left\| (Ae_i)^* - \sum_{k=1}^{i_N} \mu_{ik}^{(N)} e_k^* \right\|_{L^r(E;m)}$$
$$\le C^{1/r} \cdot \sum_{i=1}^{i_N} \left| Ae_i - \sum_{k=1}^{i_N} \mu_{ik}^{(N)} e_k \right|_H .$$

Suppose we can find reals $\mu_{ik}^{(N)}$, such that the right–hand side converges to 0 as $N \to \infty$. Then Condition (5.45) in Assumption (A 2) holds with β_N replaced by $\beta_N^{(0)}$, and γ_N replaced by $\tilde{\gamma}_N^{(0)}$. The vector fields $\tilde{\gamma}_N^{(0)}$ are not in $C_b^\infty(\mathbf{R}^{i_N} \to \mathbf{R}^{i_N})$, since they are unbounded, but as in the proof under Assumption (DS) above, we can find vector fields $\gamma_N^{(0)} \in C_b^\infty(\mathbf{R}^{i_N} \to \mathbf{R}^{i_N})$, $N \in \mathbf{N}$, such that

$$\lim_{N\to\infty} \left\| \left| \beta_N^{(0)} - \gamma_N^{(0)} \circ \Pi_N \right| \right\|_{L^r(E;m)} = 0$$

as well. Therefore, Assumption (A 2) is satisfied for $\beta^{(0)}$, which implies the claimed assertion by Lemma 5.13 and Theorem 5.2.

It remains to find reals $\mu_{ik}^{(N)}$, $N \in \mathbf{N}$, $1 \le i, k \le i_N$, such that the right–hand side of (5.67) converges to 0. Since $\{e_k; \, k \in \mathbf{N}\}$ is an orthonormal basis of H, there exist $a_{ik} \in \mathbf{R}$, $1 \leq i, k < \infty$, such that $\sum_{k=1}^{\infty} a_{ik}^2 < \infty$, and

$$A e_i = \sum_{k=1}^{\infty} a_{ik} \, e_k \qquad \text{for all } i.$$

In particular, $\sum_{k=1}^{i_N} a_{ik} \, e_k$ converges to $A e_i$ as $N \to \infty$ for each i. Hence there exists an increasing sequence of positive integers $(\alpha_N)_{N \in \mathbf{N}}$ converging to infinity, such that $\alpha_N \leq i_N$ for all N, and

$$(5.68) \qquad \lim_{N \to \infty} \sum_{i=1}^{\alpha_N} \Big| A e_i - \sum_{k=1}^{i_N} a_{ik} \, e_k \Big|_H = 0.$$

For $N \in \mathbf{N}$ and $1 \leq i, k \leq i_N$, we define

$$\mu_{ik}^{(N)} := \begin{cases} a_{ik} & \text{if } i \leq \alpha_N. \\ \lambda_{ik} & \text{else.} \end{cases}$$

Since $\lambda_{ik} = 0$ whenever $i \leq i_N < k$ for some $N \in \mathbf{N}$, we obtain

$$\sum_{i=1}^{i_N} \Big| A e_i - \sum_{k=1}^{i_N} \mu_{ik}^{(N)} e_k \Big|_H$$

$$\leq \sum_{i=1}^{\alpha_N} \Big| A e_i - \sum_{k=1}^{i_N} a_{ik} e_k \Big|_H + \sum_{i=\alpha_N+1}^{\infty} \Big| A e_i - \sum_{k=1}^{i_N} \lambda_{ik} e_k \Big|_H$$

By (5.68) and (CS) (i), the right–hand side converges to 0 as $N \to \infty$, whence the $\mu_{ik}^{(N)}$ have the desired property. This completes the proof of Theorem 5.6. ∎

3) Infinite dimensional generalized Schrödinger operators

We finally look at the results from Subsection 2) in the special case where m is a symmetrizing measure for the perturbed operator $(\mathcal{L}, \mathcal{F}C_b^{\infty}(V))$. This case includes in particular infinite dimensional generalized Schrödinger operators. The results from this subsection will be applied in the next section to prove L^p uniqueness for Euclidean quantum fields with polynomial interaction in finite volume.

We consider the framework from Subsection 2), but we assume that A is a *positive definite self-adjoint invertible linear operator* defined on a subspace of H that contains V, and that the range of A, i.e., the domain of the inverse A^{-1}, also contains V. In applications typically one of the operators A and A^{-1} is bounded, and can thus be defined on all of H, whereas the other operator is defined on a dense subspace containing V.

Suppose that there exists a **mean zero Gaussian measure** μ on E with **covariance operator** A^{-1}, i.e., the functionals g^*, $g \in V$, are centered Gaussian random variables w.r.t. μ, and

$$\int_E g^*(\omega) \, h^*(\omega) \, \mu(d\omega) = (g, A^{-1}h)_H$$

for all $g, h \in V$.

REMARK. It is well understood, under which conditions on the state space E there exists a Gaussian measure as above, cf. e.g. [Kuo 75] or [dPZa 92].

The measure μ satisfies the integration by parts formula

$$\int \partial_g F \, d\mu \;=\; -\int (Ag)^* \, F \, d\mu \quad \text{for all } F \in \mathcal{F}C_b^\infty(V) \text{ and } g \in V.$$

In particular, the operator $\mathcal{L}^A$ introduced in Subsection 2) is symmetric w.r.t. μ, and the corresponding pre–Dirichlet form $(\mathcal{E}^A, \mathcal{F}C_b^\infty(V))$ on $L^2(E; \mu)$ is given by

$$\mathcal{E}^A(F, G) \;=\; \int (\nabla F, \nabla G)_H \, d\mu \, .$$

Now fix $1 \leq p < 2$. We assume that (DS) or (CS) holds with $m := \mu$. Then the *unperturbed* operator $(\mathcal{L}^A, \mathcal{F}C_b^\infty(V))$ is $L^p(E; m)$ unique.

To study perturbations of the operator $\mathcal{L}^A$, we first introduce Gaussian Sobolev spaces on E. For $1 \leq r < \infty$, let $W^{1,r}(E, H'; \mu)$ denote the weak Sobolev space consisting of all functions $u \in L^r(E; \mu)$ for which there exists $\hat{\nabla} u \in L^r(E \to H; \mu)$ such that

$$\int (\hat{\nabla} u, g) \, F \, d\mu \;=\; -\int u \, \partial_g F \, d\mu \,-\, \int (Ag)^* \, u \, F \, d\mu$$

for all $F \in \mathcal{F}C_b^\infty(V)$ and $g \in V$. For $r = 2$, this is precisely the weak Sobolev space corresponding to the operator $(\mathcal{L}^A, \mathcal{F}C_b^\infty(V))$, which we have introduced in Chapter 3, Section a), 3), above to study Markov uniqueness.

REMARKS.   (i) We point out that under very weak assumptions on the operator A, $\mathcal{F}C_b^\infty(V)$ is dense in $W^{1,r}(E, H'; \mu)$ w.r.t. the norm $\|u\|_{1,r} = \left(\int (|u|^r + |\hat{\nabla} u|_H^r) \, dm\right)^{1/r}$. Hence the reader may think of $(\hat{\nabla}, W^{1,r}(E, H'; \mu))$ as the L^r closure of ∇, if he prefers this point of view. For the following considerations, however, the density of the cylinder functions is not needed.

(ii) In Lemma 3.3, we have given an explicit characterization of elements in $W^{1,2}(E, H'; \mu)$. It can be shown similarly, that an analogue characterization holds for elements in $W^{1,r}(E, H'; \mu)$ if $r \neq 2$. In particular, it follows immediately from this characterization, that for every $r \geq 1$, and every positive function $u \in W^{1,r}(E, H'; \mu)$, the function u^r is in $W^{1,1}(E, H'; \mu)$, and $\hat{\nabla} u^r = r \cdot u^{r-1} \hat{\nabla} u$.

Now let $r \geq 2$, and fix a function $\varphi \in W^{1,r}(E, H'; \mu)$ such that $\varphi > 0$ μ-a.e. Let $\rho := \varphi^r$. By Remark (ii), ρ is in $W^{1,1}(E, H'; \mu)$. We want to consider the diffusion operator $(\mathcal{L}, \mathcal{F}C_b^\infty(V))$ defined by

$$(5.69) \quad \mathcal{L}F \;=\; \mathcal{L}^A F + \left(\frac{\hat{\nabla} \rho}{\rho}, \nabla F \right)_H \;=\; \mathcal{L}^A F + r \left(\frac{\hat{\nabla} \varphi}{\varphi}, \nabla F \right)_H .$$

Note that for $F \in \mathcal{F}C_b^\infty(V)$, $\mathcal{L}F$ is in $L^r(E; \rho \cdot \mu)$, and thus in particular in $L^2(E; \rho \cdot \mu)$. Moreover, by Remark (ii) above, one easily shows that $(\mathcal{L}, \mathcal{F}C_b^\infty(V))$ is symmetric on $L^2(E; \rho \cdot \mu)$, and the corresponding pre–Dirichlet form $(\mathcal{E}, \mathcal{F}C_b^\infty(V))$ is given by

$$\mathcal{E}(F, G) = \int (\nabla F, \nabla G) \, \rho \, d\mu \, .$$

Application of Theorem 5.6 yields the following result :

Corollary 5.4 *Let ρ be a function in $W^{1,1}(E, H'; \mu)$ such that $\rho > 0$ μ-a.e. Fix $p \in [1, 2)$, and let $r := 2p/(2 - p)$. Suppose that Assumption (DS) or (CS) is satisfied with $m := \rho \cdot \mu$. If $\rho^{1/r}$ is in $W^{1,r}(E, H'; \mu)$, then the operator $(\mathcal{L}, \mathcal{F}C_b^\infty(V))$ defined by (5.69) is $L^p(E; m)$ unique.*

REMARK. In [Eb 93, Satz 35], Markov uniqueness for operators of type (5.69) has been shown, provided $\rho^{1/2}$ is in $W^{1,2}(E, H'; \mu)$. The corollary shows that even $L^1(E; \rho \cdot \mu)$ uniqueness holds in this case.

PROOF OF THE COROLLARY. All the assumptions in Theorem 5.6 are satisfied. In particular, $\rho^{1/r} \in W^{1,r}(E, H'; \mu)$ implies

$$\frac{\hat{\nabla}\rho}{\rho} = r \cdot \frac{\hat{\nabla}\rho^{1/r}}{\rho^{1/r}} \in L^r(E \to H; \rho \cdot \mu) \, .$$

Hence the corollary follows from Theorem 5.6. ∎

g) Applications to perturbed operators

We finally look at some applications of the perturbations results from Section f). In Subsection 1), we prove L^p uniqueness for the Dirichlet operator of the $P(\phi)_2$ quantum field in finite volume. Perturbations of Ornstein–Uhlenbeck operators on Wiener space are studied in Subsection 2). In Subsection 3), we look at the generator of the Brownian string in a velocity field.

1) Finite volume quantum fields

The Dirichlet operators of the Euclidean quantum fields with polynomial interaction in *finite volume* studied by G. Jona–Lasinio and P. K. Mitter [JLMit 85] can be viewed as small perturbations of the corresponding free field operators. We will now show that by applying our results from Section f), 3), we obtain L^p uniqueness of the perturbed operators for $1 \leq p < 2$. This improves a result by M. Röckner and T. S. Zhang [RöZha 92], who have shown Markov uniqueness. The case $p = 2$ requires additional considerations, which are carried out in a forthcoming article by V. Liskevich and M. Röckner. We point out that the most interesting problem in this direction, that is uniqueness of non–trivial quantum fields in *infinite volume*, is still open, despite attempts by various mathematicians.

The framework for the finite volume $P(\phi)_2$ quantum fields is described for example in [RöZha 92, Sect. 7]. We briefly recall the framework here, see [RöZha 92] for details. Let Λ be a finite open rectangle in $\mathbf{R}^2$. The Neumann Laplacian on Λ is denoted by Δ_N, i.e., Δ_N is the generator of the quadratic form $(u,v) \mapsto \int_\Lambda \nabla u \cdot \nabla v \, dx$ with domain $H^{1,2}(\Lambda; dx)$ on $L^2(\Lambda; dx)$. For $\alpha \geq 0$ let H_α denote the domain of $(1 - \Delta_N)^{\alpha/2}$ with inner product

$$(u,v)_{H_\alpha} = \left((1 - \Delta_N)^{\alpha/2} u, \, (1 - \Delta_N)^{\alpha/2} v \right)_{L^2(\Lambda; dx)}.$$

Here $(1 - \Delta_N)^{\alpha/2}$ is defined via the spectral theorem. Moreover, let $H_{-\alpha}$ denote the dual of H_α. By identifying $L^2(\Lambda; dx)$ with its dual, we have

$$H_\alpha \subseteq H_0 = L^2(\Lambda; dx) \subseteq H_{-\alpha}.$$

Let $E := H_{-\delta}$ for some $\delta > 0$. The (time–zero) **free Euclidean field on Λ** is the mean zero Gaussian measure μ on E such that

$$\int \langle \omega, g \rangle^2 \, \mu(d\omega) = \int_\Lambda g \, (1 - \Delta_N)^{-1} g \, dx \qquad \text{for all } g \in H_\delta,$$

where $\langle \cdot, \cdot \rangle$ denotes the dualisation between $H_{-\delta}$ and H_δ. The existence of μ follows from the Gross–Minlos–Sazonov theorem, cf. [RöZha 92, Sect. 7]. Now fix a complete orthonormal system $\{e_i; i \in \mathbf{N}\}$ consisting of eigenfunctions of Δ_N, and let $V := \mathrm{span}\,\{e_i; i \in \mathbf{N}\}$. The corresponding eigenvalues of $1 - \Delta_N$ are denoted by λ_i. Note that $V \subset H_\alpha$ for all $\alpha \in \mathbf{R}$. The *free field Dirichlet form* on $L^2(E; \mu)$ is given by

$$\mathcal{E}^{(0)}(F, G) = \int \left(\nabla^{(\alpha)} F, \nabla^{(\alpha)} G \right)_{H_\alpha} d\mu, \qquad F, G \in \mathcal{F}C_b^\infty(V),$$

where $\alpha > 0$ is a fixed constant, and $\nabla^{(\alpha)}$ denotes the H_α gradient. For $F, G \in \mathcal{F}C_b^\infty(V)$, we have

$$\mathcal{E}^{(0)}(F, G) = -\int \mathcal{L}^{(0)} F \, G \, d\mu,$$

where

(5.70) $$\mathcal{L}^{(0)} F = \sum_{i=1}^k \frac{\partial^2 f}{\partial x_i^2}(e_1^*, \ldots, e_k^*) + \beta_i \frac{\partial f}{\partial x_i}(e_1^*, \ldots, e_k^*)$$

with

$$\beta_i(\omega) := \langle \omega, (1 - \Delta_N) e_i \rangle = \lambda_i \langle \omega, e_i \rangle, \quad i \in \mathbf{N},$$

whenever $F = f(e_1^*, \ldots, e_k^*)$ for some $k \in \mathbf{N}$ and $f \in C_b^\infty(\mathbf{R}^k)$. Here e_i^* denotes the unique continuous extension of the functional $(e_i, \cdot)_{H_\alpha}$ to E, i.e., $e_i^*(\omega) = \langle \omega, (1 - \Delta_N)^\alpha e_i \rangle = \lambda_i^\alpha \langle \omega, e_i \rangle$.

The unique self–adjoint extension of the operator $(\mathcal{L}^{(0)}, \mathcal{F}C_b^\infty(V))$ is the generator of the dynamics of the free field. The corresponding diffusion process is called the **space–time free field with Neumann boundary conditions in Λ**.

Now fix $n \in \mathbf{N}$ and $a_i \in \mathbf{R}$, $0 \le i \le 2n$, such that $a_{2n} > 0$. Let

$$V(z) := \sum_{i=0}^{2n} a_i : z^i : (\chi_\Lambda), \qquad z \in E.$$

Here χ_Λ denotes the indicator function of Λ, and $: z^i :$ is the regularized product defined w.r.t. the operator $1 - \Delta_N$. See for example [RöZha 92, Section 7] for a detailed description of several ways to define $: z^i :$. The measure $e^{-V} \cdot \mu$ is called the (**time–zero**) $P(\phi)_2$ **quantum field with Neumann boundary conditions in the volume Λ**. The corresponding pre–Dirichlet form on $L^2(E; e^{-V}\mu)$ is given by

(5.71) $$\mathcal{E}^V(F, G) = \int \left(\nabla^{(\alpha)}F, \nabla^{(\alpha)}G\right)_{H_\alpha} e^{-V} d\mu.$$

Fix $r \ge 2$, and let $\varphi := e^{-V/r}$. By the proof of Theorem 7.5 in [RöZha 92], φ is in the Sobolev space $W^{1,r}(E, H'_\alpha; \mu)$ defined as in Section f), 3) above. In particular, e^{-V} is in $W^{1,1}(E, H'_\alpha; \mu)$, and for $F, G \in \mathcal{F}C_b^\infty(V)$,

$$\mathcal{E}^V(F, G) = -\int \mathcal{L}^V F\, G\, e^{-V} d\mu,$$

where

$$\mathcal{L}^V F = \mathcal{L}^{(0)} F + e^V \left(\nabla^{(\alpha)} e^{-V}, \nabla^{(\alpha)} F\right)_{H_\alpha}$$

(5.72) $$= \mathcal{L}^{(0)} F + r \cdot \left(\frac{\nabla^{(\alpha)}\varphi}{\varphi}, \nabla^{(\alpha)} F\right)_{H_\alpha},$$

cf. (5.69). The range of the operator $(\mathcal{L}^V, \mathcal{F}C_b^\infty(V))$ is contained in the space $L^r(E; e^{-V} \cdot \mu)$ for every $r \in [1, \infty)$. A diffusion process generated by a self–adjoint extension of $(\mathcal{L}^V, \mathcal{F}C_b^\infty(V))$ on $L^2(E; e^{-V} \cdot \mu)$ is called a **space–time $P(\phi)_2$ quantum field in Λ**. M. Röckner and T. S. Zhang have shown that the operator $(\mathcal{L}^V, \mathcal{F}C_b^\infty(V))$ is Markov unique on $L^2(E; e^{-V} \cdot \mu)$, i.e., the space–time $P(\phi)_2$ quantum field with Neumann boundary conditions in Λ is uniquely determined. By our considerations from Section f), we obtain the following stronger result :

Corollary 5.5 *For every $1 \le p < 2$, the operator $(\mathcal{L}^V, \mathcal{F}C_b^\infty(V))$ is $L^p(E; e^{-V} \cdot \mu)$ unique.*

PROOF. Fix $1 \le p < 2$, and let $r := 2p/(2-p)$. Since $e^{-V/r}$ is in $W^{1,r}(E, H'_\alpha; \mu)$, the L^p uniqueness holds by Corollary 5.4. ∎

2) Perturbations of the Ornstein–Uhlenbeck operator on Wiener space

Fix $n \in \mathbf{N}$. As in Example (iii) in Section a), 6), let E be the space of all continuous paths $\omega : [0,1] \to \mathbf{R}^n$ starting at 0, and let H be the Cameron–Martin subspace of E. Let μ denote Wiener maesure on E, and let

$$K := \operatorname{span}\{\Pi_s^i\,;\, s \in (0,1],\, 1 \le i \le n\},$$

where Π_s^i is the functional in E' defined by $\Pi_s^i(\omega) = \omega^i(s)$. The embedding $j : E' \to H$ is uniquely determined by

$$\left(j(\Pi_s^i), h\right)_H = \Pi_s^i(h) = h^i(s)$$

for all $s \in (0,1]$, $1 \le i \le n$, and $h \in H$, whence $j(\Pi_s^i)(t) = (s \wedge t) \cdot e^{(i)}$, where $e^{(i)}$ denotes the i-th unit vector in $\mathbf{R}^n$. Therfore, the H–gradient ∇F of a cylinder function $F \in \mathcal{F}C_b^\infty(K)$, $F(\omega) = f(\omega(s_1), \dots, \omega(s_k))$ for some $k \in \mathbf{N}$, $s_1, \dots s_k \in (0,1]$, and $f \in C_b^\infty(\mathbf{R}^{n \cdot k})$, is given by

$$(\nabla F)(\omega) = j((DF)(\omega)) = \sum_{i=1}^k (\nabla_i f)(\omega(s_1), \dots, \omega(s_k)) \cdot (s_j \wedge \bullet)$$

for all $\omega \in E$, where ∇_i denotes the gradient of the function f on $(\mathbf{R}^n)^k$ w.r.t. the i-th component. ∇ is precisely the Malliavin gradient on E.

As remarked in Section a), 6) (iii) above, μ satisfies an integration by parts identity with $\beta_\ell^\mu = \ell$ m–a.e. for all $\ell \in E'$, whence the operator $\mathcal{L}^{\mathrm{OU}} := \Delta + \langle \beta^\mu, D\cdot\rangle$ $(= -\nabla^*\nabla)$ is the Ornstein–Uhlenbeck operator on the Wiener space E.

REMARK. Explicitly, $(\mathcal{L}^{\mathrm{OU}}, \mathcal{F}C_b^\infty(K))$ is given by

$$\left(\mathcal{L}^{\mathrm{OU}}F\right)(\omega) = \sum_{i,k=1}^m (s_i \wedge s_k)\,(\operatorname{tr} \nabla_i \nabla_k f)\,(\omega(s_1), \dots, \omega(s_m))$$

$$- \sum_{k=1}^m \omega(s_k) \cdot (\nabla_k f)\,(\omega(s_1), \dots, \omega(s_m)),$$

whenever $F(\omega) = f(\omega(s_1), \dots, \omega(s_m))$ for some $m \in \mathbf{N}$, $s_1, \dots, s_m \in [0,1]$, and $f \in C_b^\infty(\mathbf{R}^{n \cdot m})$.

Now suppose $(B_s^H)_{0 \le s \le 1}$ is a measurable stochastic process defined on E. We assume that $s \mapsto B_s^H(\omega)$ is in the Cameron–Martin space for all ω. Then $(B_s^H)_{0 \le s \le 1}$ induces a measurable vector field $B^H : E \to H$. We consider the operator

(5.73) $$\mathcal{L}F = \mathcal{L}^{\mathrm{OU}}F + (B^H, \nabla F)_H .$$

$\mathcal{L}$ is the generator of the diffusion process solving the infinite dimensional stochastic differential equation

$$dX_t = dW_t - X_t \, dt + B^H(X_t) \, dt,$$

where W_t is a cylindrical Brownian motion over the Cameron–Martin space H.

Suppose we are given an invariant, or, more generally, a sub–invariant measure for this operator. The existence and regularity of invariant measures for operators of this type is well–investigated, cf. [Shi 87] and [BogRö 95]. In particular, it is known that an invariant measure exists, if, for example, $\left|B^H\right|_H$ is bounded, and that every invariant measure m is absolutely continuous w.r.t. μ, and the square–root of the density is in the Gaussian Sobolev space $W^{1,2}(E, H'; \mu)$, provided B^H is in $L^2(E \to H; \mu)$.

We are now in the framework described in Section f), 2) above, where A is the identity operator on $j(K) \subset H$. Clearly, Assumption (DS) (i) is satisfied. By Theorem 5.6, we obtain :

Corollary 5.6 *Let $p \in [1, 2)$. Let $(\mathcal{L}, \mathcal{F}C_b^\infty(K))$ be the operator given by (5.73), where $\left(B_s^H\right)_{0 \le s \le 1}$ is a measurable stochastic process defined on E. Suppose m is a sub–invariant measure for $(\mathcal{L}, \mathcal{F}C_b^\infty(K))$ such that*

$$\int |\omega(s)|^{2p/(2-p)} \, m(d\omega) < \infty \qquad \text{for all } s \in [0, 1],$$

and

$$\int \left(\int_0^1 \left| \frac{d}{ds} B_s^H(\omega) \right|^2 ds \right)^{p/(2-p)} m(d\omega) < \infty.$$

Then the closure of $(\mathcal{L}, \mathcal{F}C_b^\infty(K))$ on $L^p(E; m)$ generates a C^0 semigroup.

REMARK. The condition on B^H in the corollary guarantees that the law of the diffusion process generated by closure of the operator $\mathcal{L}$ is absolutely continuous w.r.t. the law of the E–valued Ornstein–Uhlenbeck process, cf. [Eb 93, I.3 and I.4]. The absolute continuity can be viewed as the probabilistic counterpart to the analytic notion "small perturbation" as used above.

3) The Brownian string in a velocity field

Let $n \in \mathbf{N}$, and let v be a measurable vector field on $\mathbf{R}^n$. In this subsection, we show how the results from Section f) can be applied to prove L^p uniqueness for the diffusion operator describing the random motion of an elastic string in the velocity field v, provided the boundary values are fixed. Informally (and also formally if the meaning of all objects is made precise), the Brownian string in the velocity field v is a diffusion process with state space $C([0, 1] \to \mathbf{R}^n)$ solving the infinite dimensional stochastic differential equation

$$dX_t = dW_t + \left(\frac{d^2}{ds^2} X_t + v \circ X_t \right) dt$$

for a cylindrical Brownian motion $(W_t)_{t \geq 0}$ on $L^2([0,1] \to \mathbf{R}^n; ds)$, respectively the corresponding stochastic partial differential equation

$$\frac{\partial X}{\partial t}(t, s) = \frac{\partial W}{\partial t}(t, s) + \frac{\partial^2 X}{\partial s^2}(t, s) + v(X(t, s)),$$

where $\frac{\partial W}{\partial t}$ is space–time white noise, and $\frac{d^2}{ds^2}$ denotes the one–dimensional Laplacian with, for example, Dirichlet boundary conditions.

Let $E := C([0,1] \to \mathbf{R}^n)$, $H := L^2([0,1] \to \mathbf{R}^n; ds)$, and $V := \{g \in C^2([0,1] \to \mathbf{R}^n);\ g(0) = g(1) = 0\}$. The diffusion operator $(\mathcal{L}, \mathcal{F}C_b^\infty(V))$ on E corresponding to the Brownian string in the velocity field v with Dirichlet boundary conditions is given by (5.66) with $A := \frac{d^2}{ds^2}$, and the non–linear drift perturbation $B^H : E \to H$ defined by $B^H(\omega) = (B_s^H(\omega))_{0 \leq s \leq 1}$ with

$$B_s^H(\omega) = v(\omega(s)), \quad \omega \in E, \quad 0 \leq s \leq 1.$$

Now assume that we are given a probability measure m on E which satisfies an integration by parts identity of type (IP) (cf. Section a), 5)), and which is *sub–invariant* for the operator $(\mathcal{L} - \alpha, \mathcal{F}C_b^\infty(V))$ for some $\alpha \geq 0$.

REMARKS. (i) If v vanishes, then the law $P_{0,0}$ of the Brownian bridge on E is a symmetrizing (and hence invariant) measure for $(\mathcal{L}, \mathcal{F}C_b^\infty(V))$, cf. the proof of Lemma 5.1.

(ii) If $v = \nabla\psi$ for some function $\psi \in C_b^1(\mathbf{R}^n)$, then $m := e^V \cdot P_{0,0}$, $V(\omega) := \int_0^1 \psi(\omega(s))\, ds$, is a symmetrizing measure for $(\mathcal{L}, \mathcal{F}C_b^\infty(V))$. The same is true if ψ is only in $H_{\text{loc}}^{1,2}(\mathbf{R}^n; dx)$, and ψ and $\nabla\psi$ satisfy appropriate integrability assumptions.

(iii) In general, the existence of an invariant measure for $(\mathcal{L}, \mathcal{F}C_b^\infty(V))$ is known if $|B^H|_H$ is bounded, cf. [Shi 87] and [BogRöZha 97]. Obviously, this is the case whenever v is a bounded vector field on $\mathbf{R}^n$.

By applying Theorem 5.6, we obtain :

Corollary 5.7 *Let $(\mathcal{L}, \mathcal{F}C_b^\infty(V))$ and m be as described above. Fix $1 \leq p < 2$. Suppose that*

$$(5.74) \qquad \int \int_0^1 |\omega(s)|^{2p/(2-p)}\, ds\ m(d\omega) < \infty, \qquad and$$

$$(5.75) \qquad \int \int_0^1 |v(\omega(s))|^{2p/(2-p)}\, ds\ m(d\omega) < \infty.$$

Then the closure of $(\mathcal{L}, \mathcal{F}C_b^\infty(V))$ on $L^p(E; m)$ generates a C^0 semigroup.

PROOF. By (5.74), and since the operator $\frac{d^2}{ds^2}$ with Dirichlet boundary conditions on $(0,1)$ has discrete spectrum and smooth eigenfunctions, Assumption (DS) in Section f) is satisfied. By (5.75), B^H is in $L^{2p/(2-p)}(E \to H; m)$.

Moreover, since m is sub–invariant for $(\mathcal{L} - \alpha, \mathcal{F}C_b^\infty(V))$ for some $\alpha \geq 0$, the divergence bound assumed in Theorem 5.6 holds. Hence the assertion follows from Theorem 5.6. ∎

REMARK. Note that in contrast to the counterexample given in Section b), we do now consider the operator $\mathcal{L}$ on cylinder functions based on functions in V instead of $C_0^\infty(0,1)$. This is in fact essential for the proof of the corollary, because the eigenfunctions of the operator $\frac{d^2}{ds^2}$ with Dirichlet boundary conditions are not contained in $C_0^\infty(0,1)$, whence Assumption (DS) from Section f) does not hold if V is replaced by $C_0^\infty(0,1)$.

References

[Ac 94] E. ACOSTA. On the essential self-adjointness of Dirichlet operators on group-valued path space, *Proc. AMS* **122**, 581–590, (1994).

[Ai 93] S. AIDA. On the Ornstein–Uhlenbeck operators on Wiener–Riemannian manifolds, *J. Funct. Anal.* **116**, 83–110, (1993).

[Alb 97] S. ALBEVERIO. Some applications of infinite–dimensional analysis in mathematical physics, *Helv. Phys. Acta* **70**, 479–506, (1997).

[AlbBoRö 97] S. ALBEVERIO, V. BOGACHEV, M. RÖCKNER. On uniqueness of invariant measures for finite and infinite dimensional diffusions, *Comm. Pure Appl. Math.* **52**, 325–362, (1999).

[AlbHK 77] S. ALBEVERIO, R. HOEGH–KROHN. Dirichlet forms and diffusion processes on rigged Hilbert spaces, *Z. Wahrsch. verw. Geb.* **40**, 1–57, (1977).

[AlbKoRö 95] S. ALBEVERIO, Y. KONDRATIEV, M. RÖCKNER. Dirichlet operators via stochastic analysis, *J. Funct. Anal.* **128**, 102–138, (1995).

[AlbKoRö 95a] S. ALBEVERIO, Y. KONDRATIEV, M. RÖCKNER. A remark on stochastic dynamics on the infinite dimensional torus, in: Seminar on Stochastic Analysis, Random fields and Applications (Ascona 1993), Birkhäuser, Basel, (1995).

[AlbKoRö 95b] S. ALBEVERIO, Y. KONDRATIEV, M. RÖCKNER. Uniqueness of stochastic dynamics for continuous spin systems on a lattice, *J. Funct. Anal.* **133**, 10–20, (1995).

[AlbKoRö 97] S. ALBEVERIO, Y. KONDRATIEV, M. RÖCKNER. Differential geometry of Poisson spaces, *C. R. Acad. Sci. Paris* t. **323**, 1129–1134, (1997).

[AlbKoRö 97a] S. ALBEVERIO, Y. KONDRATIEV, M. RÖCKNER. Analysis and geometry on configuration spaces, *J. Funct. Anal.* **154**, 444–500, (1998).

[AlbKoRö 97b] S. ALBEVERIO, Y. KONDRATIEV, M. RÖCKNER. Analysis and geometry on configuration spaces : The Gibbsian case, *J. Funct. Anal.* **157**, 242–291, (1998).

[AlbKoRö 97c] S. ALBEVERIO, Y. KONDRATIEV, M. RÖCKNER. Ergodicity of L^2 semigroups and extremality of Gibbs states, *J. Funct. Anal.* **144**, 394–423, (1997).

[AlbKusRö 90] S. ALBEVERIO, S. KUSUOKA, M. RÖCKNER. On partial integration in infinite–dimensional space and applications to Dirichlet forms,

J. London Math. Soc. **42**, 122–136, (1990).

[AlbRö 90] S. ALBEVERIO, M. RÖCKNER. Classical Dirichlet forms on topological vector spaces – Closability and a Cameron–Martin formula, *J. Funct. Anal.* **88**, 395–436, (1990).

[AlbRö 91] S. ALBEVERIO, M. RÖCKNER. Stochastic differential equations in infinite dimensions: solutions via Dirichlet forms, *Probab. Th. Rel. Fields* **89**, 347–386, (1991).

[AlbRö 95] S. ALBEVERIO, M. RÖCKNER. Dirichlet form methods for uniqueness of martingale problems and applications, in: Stochastic analysis, Proc. Sympos. Pure Math., AMS, Providence, 513–528, (1995).

[AlbRöZha 92] S. ALBEVERIO, M. RÖCKNER, T. S. ZHANG. Markov uniqueness for a class of infinite dimensional Dirichlet operators, in: Stochastic processes and optimal control (Friedrichsroda 1992), Gordon and Breach, Montreux, 1–26, (1995).

[Alt 85] H. W. ALT. *Lineare Funktionalanalysis.* Springer, Berlin etc. (1985).

[Ar 86] W. ARENDT. The abstract Cauchy problem, special semigroups and perturbation, in: One–parameter semigroups of positive operators. Edited by R. Nagel. Springer, Berlin etc. (1986).

[Ba 90] D. BAKRY. Ricci curvature and dimension for diffusion semigroups, in: Stochastic processes and their applications. Ed.: S. Albeverio et al., Proc. Bielefeld 1985, Kluwer (1990).

[Be 86] Y. M. BEREZANSKY. *Self–adjoint operators in spaces of functions of infinitely many variables.* Tranlat. Amer. Math. Soc. **63**, Providence (1986).

[BirMos 95] M. BIROLI, U. MOSCO. A Saint–Venant principle for Dirichlet forms on discontinuous media, *Ann. Mat. Pura Appl.* **169**, 125–181, (1995).

[BoKryRö 96] V. BOGACHEV, N. V. KRYLOV, M. RÖCKNER. Elliptic regularity and essential self–adjointness of Dirichlet operators, *Ann. Scuola Norm. Sup. Pisa* **24**, 451–461, (1997).

[BoRö 95] V. BOGACHEV, M. RÖCKNER. Regularity of invariant measures on finite and infinite dimensional spaces and applications, *J. Funct. Anal.* **133**, 168–223, (1995).

[BoRöZha 97] V. BOGACHEV, M. RÖCKNER, T. S. ZHANG. Existence of invariant measures for diffusions with singular drifts, Preprint (1997).

[BoSa 96] A. N. BORODIN, P. SALMINEN. *Handbook of Brownian motion — Facts and formulae,* Birkhäuser, Basel (1996).

[BouHi 91] N. BOULEAU, F. HIRSCH *Dirichlet forms and analysis on Wiener space.* De Gruyter, Berlin, New York (1991).

[CarPar 90] E. A. CARLEN, E. PARDOUX. Differential calculus and integration by parts on Poisson space, in: Stochastics, algebra and analysis in classical and quantum dynamics, Ed.: S. Albeverio et al., Kluwer (1990).

[CatFra 97] P. CATTIAUX, M. FRADON. Entropy, reversible diffusion processes, and Markov uniqueness, *J. Funct. Anal.* **134**, 243–272, (1997).

[CrMal 94] A.-B. CRUZEIRO, P. MALLIAVIN. Répère mobile et géométrie riemannienne sur les espaces des chemins, *C. R. Acad. Sci. Paris* t. **319**, 859–

864, (1994).

[CrMal 96] A.-B. CRUZEIRO, P. MALLIAVIN. Renormalized differential geometry on path space : Structural equation, curvature, *J. Funct. Anal.* **139**, 119–181, (1996).

[dPZa 92] G. DA PRATO, J. ZABCZYK. *Stochastic equations in infinite dimensions.* Cambridge University Press (1992).

[Dav 85] E. B. DAVIES. L^1 properties of second order elliptic differential operators, *Bull. London Math. Soc.* **17**, 417–436, (1985).

[Dav 89] E. B. DAVIES. *Heat kernels and spectral theory.* Cambridge University Press (1989).

[Dix 69] J. DIXMIER. *Les algèbres d'opérateurs dans l'espace hilbertien.* Gauthier-Villars, Paris (1969).

[Dri 92] B. DRIVER. A Cameron–Martin type quasi–invariance theorem for Brownian motion on a compact Riemmanian manifolds, *J. Funct. Anal.* **110**, 272–376, (1992).

[Dri 95] B. DRIVER. A primer on Riemannian geometry and stochastic analysis on path spaces, Preprint (1995).

[DriRö 92] B. DRIVER, M. RÖCKNER. Construction of diffusions on path and loop spaces of compact Riemmanian manifolds, *C. R. Acad. Sci. Paris* t. **315**, 603–608, (1992).

[Dy 91] E. B. DYNKIN. Branching particle systems and superprocesses, *Annals Probab.* **19**, 1157–1194, (1991).

[Eb 93] A. EBERLE. Absolutstetigkeit zweier unendlichdimensionaler Diffusionen und Anwendungen. Diplomarbeit Universität Bonn (1993).

[Eb 95] A. EBERLE. Weak Sobolev spaces and Markov uniqueness of operators, *C.R. Acad. Sci. Paris* **320**, Série I, 1249–1254 (1995).

[Eb 96] A. EBERLE. Girsanov–type transformations of local Dirichlet forms: An analytic approach, *Osaka J. Math.* **33**, 497–531 (1996).

[Eb 97] A. EBERLE. Diffusions on path and loop spaces: Existence, finite dimensional approximation and Hölder continuity, *Probab. Th. Rel. Fields* **109**, 77–99 (1997).

[Eb 99] A. EBERLE. L^p uniqueness of non–symmetric diffusion operators with singular drift coefficients : I. The finite–dimensional case, submitted to *J. Funct. Anal.*.

[Eb 99a] A. EBERLE. In preparation.

[EthKur 86] S. N. ETHIER, T. G. KURTZ. *Markov processes. Characterization and convergence.* Wiley, New York etc. (1986).

[Fe 51] W. FELLER. The parabolic differential equations and the associated semi–groups of transformations, *Ann. Math.* **55**, 468–519, (1951).

[FrSeSC 95] B. FRANCHI, R. SERAPIONI, F. SERRA CASSANO. Champs de vecteurs, théorème d'approximation de Meyers–Serrin et phénomène de Lavrentev pour des fonctionelles dégénérées, *C. R. Acad. Sci. Paris* t. **320**, Série I, 695–698, (1995).

[Fr 77] J. FREHSE. Essential self–adjointness of singular elliptic operators, *Bol. Soc. Bras. Mat.* **82**, 87–107, (1977).

[Fri 64] A. FRIEDMAN. *Partial differential equations of parabolic type.* Prentice–Hall, Eaglewood Cliffs, N.J. (1964).

[Fu 87] M. FUKUSHIMA. Energy forms and diffusion processes, in: Mathematics and Physics, Vol. 1, 65–97, World Scientific Publishing, Singapore (1987).

[FuOshTa 94] M. FUKUSHIMA, Y. OSHIMA, M. TAKEDA. *Dirichlet forms and symmetric Markov processes*, de Gruyter, Berlin, New York (1994).

[Fun 82] T. FUNAKI. Random motion of strings and stochastic differential equations on the space $C([0,1], \mathbf{R}^d)$, in: Stochastic Analysis, Proc. of the Taniguchi Symp., Katata 1982. North–Holland, Amsterdam, New York (1982).

[Fun 83] T. FUNAKI. Random motion of strings and related stochastic evolution equations, *Nagoya Math. J.* **89**, 129–193, (1983).

[Ga 51] M. P. GAFFNEY. The harmonic operator for exterior differential forms, *Proc. Nat. Acad. Sci. USA* **37**, 48–50, (1951).

[GalHuLaf 90] S. GALLOT, D. HULIN, J. LAFONTAINE. *Riemannian geometry.* Second edition. Springer, Berlin etc. (1990).

[GiTru 83] D. GILBARG, N. S. TRUDINGER. *Elliptic partial differential equations of second order.* Second edition. Springer, Berlin etc. (1983).

[ItMKe 65] K. ITÔ, H. P. MC KEAN. *Diffusion processes and their sample paths.* Springer, Berlin etc. (1965).

[JLMit 85] P. JONA–LASINIO, P. K. MITTER. On the stochastic quantization of field theory, *Comm. Math. Phys.* **101**, 409–436, (1985).

[JöRel 76] K. JÖRGENS, F. RELLICH. *Eigenwerttheorie gewöhnlicher Differentialgleichungen.* Springer, Berlin etc. (1976).

[KawTa 95] T. KAWABATA, M. TAKEDA. On uniqueness problem for local Dirichlet forms, *Osaka J. Math.* **33**, 881–893, (1996).

[KoTsy 93] Y. KONDRATIEV, T. TSYCALENKO. Infinite–dimensional Dirichlet operators, I. Essential self–adjointness and associated elliptic equations, *Potential Anal.* **2**, 1–21, (1993).

[Kry 96] N. V. KRYLOV. *Lectures on elliptic and parabolic equations in Hölder spaces.* AMS Graduate Studies in Mathematics, Vol. 12 (1996).

[Kuo 75] H. KUO. *Gaussian measures in Banach spaces.* LNM 463, Springer, Berlin etc. (1975).

[Li 94] V. A. LISKEVICH. Smoothness estimates and uniqueness for the Dirichlet operators, in: Math. Results in Quantum Mechanics, Internat. conference in Blossin (Germany), Birkhäuser, Basel etc. (1994).

[LiSem 92] V. A. LISKEVICH, Y. A. SEMENOV. Dirichlet operators: A priori estimates and the uniqueness problem, *J. Funct. Anal.* **109**, 199–213, (1992).

[LiSem 96] V. A. LISKEVICH, Y. A. SEMENOV. Some problems on Markov semigroups, in: Schrödinger operators, Markov semigroups, wavelet analysis, operator algebras. Math. Top. 11, Akademie–Verlag, Berlin (1996).

[LiTuv 93] V. A. LISKEVICH, E. W. TUV. A priori estimates for second order elliptic equations and their applications, *Israel J. Math.* **81**, 257–263, (1993).

[MaRö 92] Z. M. MA, M. RÖCKNER. *Introduction to the theory of (non–symmetric) Dirichlet forms.* Springer, Berlin etc. (1992).

[Mal 97] P. MALLIAVIN. *Stochastic analysis*. Springer, Berlin etc. (1997).

[Man 68] P. MANDL. *Analytic treatment of one-dimensional Markov processes*. Springer, Berlin etc. (1968).

[Maz 85] V. G. MAZ'JA. *Sobolev spaces*. Springer, Berlin etc. (1985).

[MeySer 64] N. G. MEYERS, J. SERRIN. $H = W$, *Proc. Nat. Acad. Sci. USA*, Vol. 51, 1055–1056 (1964).

[OkVCa 96] O. OKITALOSHIMA, J. A. VAN CASTEREN. On the uniqueness of the martingale problem, *International J. Math.*, Vol. 7, No. 6 (1996).

[OvRö 97] L. OVERBECK, M. RÖCKNER. Geometric aspects of finite- and infinite-dimensional Fleming–Viot processes, *Random Oper. Stochastic Equations* 5, 35–58, (1997).

[OvRöSchm 95] L. OVERBECK, M. RÖCKNER, B. SCHMULAND. An analytic approach to Fleming–Viot processes with interactive selection, *Ann. Probab.* 23, 1–36, (1995).

[PaYoo 97] Y. M. PARK, H. J. YOO. Dirichlet operators on loop spaces: Essential self-adjointness and log–Sobolev inequality, *J. Math. Phys.* 38, 3321–3346, (1997).

[Pa 85] A. PAZY. *Semigroups of linear operators and applications to partial differential equations*. Springer, Berlin etc. (1985).

[ReSi 75] M. REED, B. SIMON. *Methods of modern mathematical physics*, *I–IV*. Academic Press, San Diego etc. (1975).

[RöSchm 95] M. RÖCKNER, B. SCHMULAND. Quasi-regular Dirichlet forms: examples and counterexamples, *Canadian J. Math.* 47, 165–200, (1995).

[RöZha 92] M. RÖCKNER, T. S. ZHANG. Uniqueness of generalized Schrödinger operators and applications, *J. Funct. Anal.* 105, 187–231, (1992).

[RöZha 94] M. RÖCKNER, T. S. ZHANG. Uniqueness of generalized Schrödinger operators — Part II, *J. Funct. Anal.* 119, 455–467, (1994).

[Schi 91] A. SCHIED. Zur Konstruktion masswertiger Verzweigungsprozesse, Diplomarbeit Universität Bonn (1991).

[Schi 97] A. SCHIED. Geometric aspects of Fleming–Viot and Dawson–Watanabe processes, *Ann. Probab.* 25, 1160–1179, (1997).

[Schm 92] B. SCHMULAND. Dirichlet forms with polynomial domain, *Math. Japonica* 37, 1015–1024, (1992).

[Shi 87] I. SHIGEKAWA. Existence of invariant measures of diffusions on an abstract Wiener space, *Osaka J. Math.* 24, 37–59, (1987).

[Shi 95] I. SHIGEKAWA. An example of regular (r,p)-capacity and essential self-adjointness of a diffusion operator in infinite dimensions, *J. Math. Kyoto Univ.* 35, 639–651, (1995).

[Sil 74] M. SILVERSTEIN. *Symmetric Markov processes*. LNM 426, Springer, Berlin etc. (1974).

[So 94] S. SONG. A study on markovian maximality, change of probability and regularity, *Potential Anal.* 3, 391–422, (1994).

[St 96] W. STANNAT. First order perturbations of Dirichlet operators: Existence and uniqueness, *J. Funct. Anal.* 141, 216–248, (1996).

[St 97] W. STANNAT. (Nonsymmetric) Dirichlet operators on L^1: existence, uniqueness and associated Markov processes, *Ann. Scuola Norm. Sup. Pisa* **28**, 99–140, (1999).

[StrVar 79] D. W. STROOCK, S. R. S. VARADHAN. *Multidimensional diffusion processes*. Springer, Berlin etc. (1979).

[Stu 95] K.-T. STURM. On the geometry defined by Dirichlet forms, in: Seminar on stochastic analysis, random fields and applications (Ascona 1993), Birkhäuser, Basel, 231–242 (1995).

[Ta 87] M. TAKEDA. On the uniqueness of the Markovian self–adjoint extension, in: Stochastic processes – Mathematics and Physics, LNM 1250, Springer, Berlin etc., 319–325 (1987).

[Ta 91] M. TAKEDA. On the maximum Markovian extensions of one–dimensional diffusion operators, in: Gaussian random fields, Nagoya 1990, World Sci. Publishing, River Edge, N.J., 374–383 (1991).

[Ta 92] M. TAKEDA. The maximum Markovian extensions of generalized Schrödinger operators, *J. Math. Soc. Japan* **44**, 113–130, (1992).

[Ta 96] M. TAKEDA. Two classes of extensions for generalized Schrödinger operators, *Pot. Anal.* **5**, 1–13, (1996).

[Wei 87] J. WEIDMANN. *Spectral theory of ordinary differential operators.* LNM 1258, Springer, Berlin etc. (1987).

[Wie 85] N. WIELENS. The essential self–adjointness of generalized Schrödinger operators, *J. Funct. Anal.* **65**, 98–115, (1985).

[Wu 97] L. M. WU. Uniqueness of Schrödinger operators restricted in a domain, *J. Funct. Anal.* **153**, 276–319, (1998)

[Yo 80] K. YOSIDA. *Functional analysis.* Sixth edition. Springer, Berlin etc. (1980).

Index

adjoint criterion 32
approximative criterion 33, 225

Bochner technique 230
Brownian string 197, 251

Cameron–Martin space 194, 250
Carré du champ 35, 149, 162
C^0 semigroup 14
conformally flat 92
conservativity
    of semigroups 134
    of diffusion processes 178
co–tangent space (generalized) 90,
    149, 188

degenerate operator 65, 118, 129
differential
    generalized 148
    strong 105
    weak 91, 93, 105
diffusion operator 1, 35, 147
    infinite dimensional 187
    on Wiener space 224
    representation theorem 162
direct integral 104
Dirichlet form 22, 90
Dirichlet operator 37, 114
dissipative 31, 195
divergence form representation 192
divergence operator (generalized) 160

energy image density 106
energy measure 116
entrance boundary, entrance point
    171
ergodicity 136, 178

essentially self–adjoint 21, 61
existence of C^0 semigroups 30, 229
exit boundary, exit point 171
extremality of symmetrizing measures
    141

Feller classification 170
Fleming–Viot process 156
form core 20
free field 248
Friedrichs extension 37, 114, 207

Gaussian measure 245
generalized Schroedinger operator 54,
    59, 69, 95, 125, 245
Gibbs measure 142, 234
gradient
    gradient estimate for resolvent
      229
    gradient estimate for semigroup
      229
    gradient operator (generalized)
      108
    H–gradient 191
    Malliavin gradient 191

integration by parts identity 93, 101,
    142, 191
intertwining formula 231
intrinsic metric 132
invariant measure 12
irreducibility 138

Kato type inequality 81
Krein extension 115
Kunita–Watanabe inequality 116

L^p entrance boundary 181

L^p uniqueness 19, 41, 179, 225, 238
L^2 derivation 163
L^2 differential 104
L^2 metric 195
lattice system 146, 193, 234
limit point case 45,67
Lipschitz continuous coefficients 55
localization 82
loop space 153
Lumer–Phillips theorem 31

Markov uniqueness 17, 89
    basic criterion for 115
    strong 15
martingale problem 10
maximal Dirichlet extension 112, 210
maximality problem 115
Mazja criterion 130
measurable field of Hilbert spaces 148
measure–valued diffusion 155
Meyers–Serrin type theorem 125,126,128

natural boundary, natural point 171
non–uniqueness 44, 145, 179, 197

one-form 105
operator core 18
Ornstein–Uhlenbeck operator
    flat 194, 250
    non–flat 153

$P(\phi)_d$ model (discrete) 237
$P(\phi)_2$ quantum field 249
particle system 179
    interacting 156
path space 153
perturbation
    H–valued 238, 247
    small 45, 54, 193, 238
phase transition 145, 237
Poisson space 155, 157
projective limit 215, 235

quantum field 247

random motion of strings 195, 197, 251

regular boundary, regular point 171
regularity of ODE 46, 85
resolvent approximation 227

Schroedinger operator (generalized)
    54, 59, 69, 95, 125, 245
Silverstein
    extension 23, 135
    uniqueness 23, 132, 135
singularity 66
singularity set 94, 119
Sobolev space
    fractional 85
    strong 105, 206
    weak 91, 105, 210
    weighted 92, 124
stationary distribution 18, 134
strong uniqueness 19
Sturm–Liouville operator 42, 170
sub–invariant measure 36
sub–Markov 13
super Brownian motion 155
symmetric operator 28
symmetrizing probability measure 142, 199

weak derivative 91, 93 ,98, 109
Wiener space 194, 250
    abstract 200

Vol. 1626: J. Azéma, M. Emery, M. Yor (Eds.), Séminaire de Probabilités XXX. VIII, 382 pages. 1996.

Vol. 1627: C. Graham, Th. G. Kurtz, S. Méléard, Ph. E. Protter, M. Pulvirenti, D. Talay, Probabilistic Models for Nonlinear Partial Differential Equations. Montecatini Terme, 1995. Editors: D. Talay, L. Tubaro. X, 301 pages. 1996.

Vol. 1628: P.-H. Zieschang, An Algebraic Approach to Association Schemes. XII, 189 pages. 1996.

Vol. 1629: J. D. Moore, Lectures on Seiberg-Witten Invariants. VII, 105 pages. 1996.

Vol. 1630: D. Neuenschwander, Probabilities on the Heisenberg Group: Limit Theorems and Brownian Motion. VIII, 139 pages. 1996.

Vol. 1631: K. Nishioka, Mahler Functions and Transcendence. VIII, 185 pages. 1996.

Vol. 1632: A. Kushkuley, Z. Balanov, Geometric Methods in Degree Theory for Equivariant Maps. VII, 136 pages. 1996.

Vol. 1633: H. Aikawa, M. Essén, Potential Theory – Selected Topics. IX, 200 pages. 1996.

Vol. 1634: J. Xu, Flat Covers of Modules. IX, 161 pages. 1996.

Vol. 1635: E. Hebey, Sobolev Spaces on Riemannian Manifolds. X, 116 pages. 1996.

Vol. 1636: M. A. Marshall, Spaces of Orderings and Abstract Real Spectra. VI, 190 pages. 1996.

Vol. 1637: B. Hunt, The Geometry of some special Arithmetic Quotients. XIII, 332 pages. 1996.

Vol. 1638: P. Vanhaecke, Integrable Systems in the realm of Algebraic Geometry. VIII, 218 pages. 1996.

Vol. 1639: K. Dekimpe, Almost-Bieberbach Groups: Affine and Polynomial Structures. X, 259 pages. 1996.

Vol. 1640: G. Boillat, C. M. Dafermos, P. D. Lax, T. P. Liu, Recent Mathematical Methods in Nonlinear Wave Propagation. Montecatini Terme, 1994. Editor: T. Ruggeri. VII, 142 pages. 1996.

Vol. 1641: P. Abramenko, Twin Buildings and Applications to S-Arithmetic Groups. IX, 123 pages. 1996.

Vol. 1642: M. Puschnigg, Asymptotic Cyclic Cohomology. XXII, 138 pages. 1996.

Vol. 1643: J. Richter-Gebert, Realization Spaces of Polytopes. XI, 187 pages. 1996.

Vol. 1644: A. Adler, S. Ramanan, Moduli of Abelian Varieties. VI, 196 pages. 1996.

Vol. 1645: H. W. Broer, G. B. Huitema, M. B. Sevryuk, Quasi-Periodic Motions in Families of Dynamical Systems. XI, 195 pages. 1996.

Vol. 1646: J.-P. Demailly, T. Peternell, G. Tian, A. N. Tyurin, Transcendental Methods in Algebraic Geometry. Cetraro, 1994. Editors: F. Catanese, C. Ciliberto. VII, 257 pages. 1996.

Vol. 1647: D. Dias, P. Le Barz, Configuration Spaces over Hilbert Schemes and Applications. VII, 143 pages. 1996.

Vol. 1648: R. Dobrushin, P. Groeneboom, M. Ledoux, Lectures on Probability Theory and Statistics. Editor: P. Bernard. VIII, 300 pages. 1996.

Vol. 1649: S. Kumar, G. Laumon, U. Stuhler, Vector Bundles on Curves – New Directions. Cetraro, 1995. Editor: M. S. Narasimhan. VII, 193 pages. 1997.

Vol. 1650: J. Wildeshaus, Realizations of Polylogarithms. XI, 343 pages. 1997.

Vol. 1651: M. Drmota, R. F. Tichy, Sequences, Discrepancies and Applications. XIII, 503 pages. 1997.

Vol. 1652: S. Todorcevic, Topics in Topology. VIII, 153 pages. 1997.

Vol. 1653: R. Benedetti, C. Petronio, Branched Standard Spines of 3-manifolds. VIII, 132 pages. 1997.

Vol. 1654: R. W. Ghrist, P. J. Holmes, M. C. Sullivan, Knots and Links in Three-Dimensional Flows. X, 208 pages. 1997.

Vol. 1655: J. Azéma, M. Emery, M. Yor (Eds.), Séminaire de Probabilités XXXI. VIII, 329 pages. 1997.

Vol. 1656: B. Biais, T. Björk, J. Cvitanic, N. El Karoui, E. Jouini, J. C. Rochet, Financial Mathematics. Bressanone, 1996. Editor: W. J. Runggaldier. VII, 316 pages. 1997.

Vol. 1657: H. Reimann, The semi-simple zeta function of quaternionic Shimura varieties. IX, 143 pages. 1997.

Vol. 1658: A. Pumarino, J. A. Rodriguez, Coexistence and Persistence of Strange Attractors. VIII, 195 pages. 1997.

Vol. 1659: V. Kozlov, V. Maz'ya, Theory of a Higher-Order Sturm-Liouville Equation. XI, 140 pages. 1997.

Vol. 1660: M. Bardi, M. G. Crandall, L. C. Evans, H. M. Soner, P. E. Souganidis, Viscosity Solutions and Applications. Montecatini Terme, 1995. Editors: I. Capuzzo Dolcetta, P. L. Lions. IX, 259 pages. 1997.

Vol. 1661: A. Tralle, J. Oprea, Symplectic Manifolds with no Kähler Structure. VIII, 207 pages. 1997.

Vol. 1662: J. W. Rutter, Spaces of Homotopy Self-Equivalences – A Survey. IX, 170 pages. 1997.

Vol. 1663: Y. E. Karpeshina; Perturbation Theory for the Schrödinger Operator with a Periodic Potential. VII, 352 pages. 1997.

Vol. 1664: M. Väth, Ideal Spaces. V, 146 pages. 1997.

Vol. 1665: E. Giné, G. R. Grimmett, L. Saloff-Coste, Lectures on Probability Theory and Statistics 1996. Editor: P. Bernard. X, 424 pages. 1997.

Vol. 1666: M. van der Put, M. F. Singer, Galois Theory of Difference Equations. VII, 179 pages. 1997.

Vol. 1667: J. M. F. Castillo, M. González, Three-space Problems in Banach Space Theory. XII, 267 pages. 1997.

Vol. 1668: D. B. Dix, Large-Time Behavior of Solutions of Linear Dispersive Equations. XIV, 203 pages. 1997.

Vol. 1669: U. Kaiser, Link Theory in Manifolds. XIV, 167 pages. 1997.

Vol. 1670: J. W. Neuberger, Sobolev Gradients and Differential Equations. VIII, 150 pages. 1997.

Vol. 1671: S. Bouc, Green Functors and G-sets. VII, 342 pages. 1997.

Vol. 1672: S. Mandal, Projective Modules and Complete Intersections. VIII, 114 pages. 1997.

Vol. 1673: F. D. Grosshans, Algebraic Homogeneous Spaces and Invariant Theory. VI, 148 pages. 1997.

Vol. 1674: G. Klaas, C. R. Leedham-Green, W. Plesken, Linear Pro-p-Groups of Finite Width. VIII, 115 pages. 1997.

Vol. 1675: J. E. Yukich, Probability Theory of Classical Euclidean Optimization Problems. X, 152 pages. 1998.

Vol. 1676: P. Cembranos, J. Mendoza, Banach Spaces of Vector-Valued Functions. VIII, 118 pages. 1997.

Vol. 1677: N. Proskurin, Cubic Metaplectic Forms and Theta Functions. VIII, 196 pages. 1998.

Vol. 1678: O. Krupková, The Geometry of Ordinary Variational Equations. X, 251 pages. 1997.

Vol. 1679: K.-G. Grosse-Erdmann, The Blocking Technique. Weighted Mean Operators and Hardy's Inequality. IX, 114 pages. 1998.

Vol. 1680: K.-Z. Li, F. Oort, Moduli of Supersingular Abelian Varieties. V, 116 pages. 1998.

Vol. 1681: G. J. Wirsching, The Dynamical System Generated by the 3n+1 Function. VII, 158 pages. 1998.

Vol. 1682: H.-D. Alber, Materials with Memory. X, 166 pages. 1998.

Vol. 1683: A. Pomp, The Boundary-Domain Integral Method for Elliptic Systems. XVI, 163 pages. 1998.

Vol. 1684: C. A. Berenstein, P. F. Ebenfelt, S. G. Gindikin, S. Helgason, A. E. Tumanov, Integral Geometry, Radon Transforms and Complex Analysis. Firenze, 1996. Editors: E. Casadio Tarabusi, M. A. Picardello, G. Zampieri. VII, 160 pages. 1998.

Vol. 1685: S. König, A. Zimmermann, Derived Equivalences for Group Rings. X, 146 pages. 1998.

Vol. 1686: J. Azéma, M. Émery, M. Ledoux, M. Yor (Eds.). Séminaire de Probabilités XXXII. VI, 440 pages. 1998.

Vol. 1687: F. Bornemann, Homogenization in Time of Singularly Perturbed Mechanical Systems. XII, 156 pages. 1998.

Vol. 1688: S. Assing, W. Schmidt, Continuous Strong Markov Processes in Dimension One. XII, 137 page. 1998.

Vol. 1689: W. Fulton, P. Pragacz, Schubert Varieties and Degeneracy Loci. XI, 148 pages. 1998.

Vol. 1690: M. T. Barlow, D. Nualart, Lectures on Probability Theory and Statistics. Editor: P. Bernard. VIII, 237 pages. 1998.

Vol. 1691: R. Bezrukavnikov, M. Finkelberg, V. Schechtman, Factorizable Sheaves and Quantum Groups. X, 282 pages. 1998.

Vol. 1692: T. M. W. Eyre, Quantum Stochastic Calculus and Representations of Lie Superalgebras. IX, 138 pages. 1998.

Vol. 1694: A. Braides, Approximation of Free-Discontinuity Problems. XI, 149 pages. 1998.

Vol. 1695: D. J. Hartfiel, Markov Set-Chains. VIII, 131 pages. 1998.

Vol. 1696: E. Bouscaren (Ed.): Model Theory and Algebraic Geometry. XV, 211 pages. 1998.

Vol. 1697: B. Cockburn, C. Johnson, C.-W. Shu, E. Tadmor, Advanced Numerical Approximation of Nonlinear Hyperbolic Equations. Cetraro, Italy, 1997. Editor: A. Quarteroni. VII, 390 pages. 1998.

Vol. 1698: M. Bhattacharjee, D. Macpherson, R. G. Möller, P. Neumann, Notes on Infinite Permutation Groups. XI, 202 pages. 1998.

Vol. 1699: A. Inoue,Tomita-Takesaki Theory in Algebras of Unbounded Operators. VIII, 241 pages. 1998.

Vol. 1700: W. A. Woyczyński, Burgers-KPZ Turbulence,XI, 318 pages. 1998.

Vol. 1701: Ti-Jun Xiao, J. Liang, The Cauchy Problem of Higher Order Abstract Differential Equations. XII, 302 pages. 1998.

Vol. 1702: J. Ma, J. Yong, Forward-Backward Stochastic Differential Equations and Their Applications. XIII, 270 pages. 1999.

Vol. 1703: R. M. Dudley, R. Norvaiša, Differentiability of Six Operators on Nonsmooth Functions and p-Variation. VIII, 272 pages. 1999.

Vol. 1704: H. Tamanoi, Elliptic Genera and Vertex Operator Super-Algebras. VI, 390 pages. 1999.

Vol. 1705: I. Nikolaev, E. Zhuzhoma, Flows in 2-dimensional Manifolds. XIX, 294 pages. 1999.

Vol. 1706: S. Yu. Pilyugin, Shadowing in Dynamical Systems. XVII, 271 pages. 1999.

Vol. 1707: R. Pytlak, Numerical Methods for Optical Control Problems with State Constraints. XV, 215 pages. 1999.

Vol. 1708: K. Zuo, Representations of Fundamental Groups of Algebraic Varieties. VII, 139 pages. 1999.

Vol. 1709: J. Azéma, M. Émery, M. Ledoux, M. Yor (Eds.). Séminaire de Probabilités XXXIII. VIII, 418 pages. 1999.

Vol. 1710: M. Koecher, The Minnesota Notes on Jordan Algebras and Their Applications. IX, 173 pages. 1999.

Vol. 1711: W. Ricker, Operator Algebras Generated by Commuting Projections: A Vector Measure Approach. XVII, 159 pages. 1999.

Vol. 1712: N. Schwartz, J. J. Madden, Semi-algebraic Function Rings and Reflectors of Partially Ordered Rings. XI, 279 pages. 1999.

Vol. 1713: F. Bethuel, G. Huisken, S. Müller, K. Steffen, Calculus of Variations and Geometric Evolution Problems. Cetraro, 1996. Editors: S. Hildebrandt, M. Struwe. VII, 293 pages. 1999.

Vol. 1714: O. Diekmann, R. Durrett, K. P. Hadeler, P. K. Maini, H. L. Smith, Mathematics Inspired by Biology. Martina Franca, 1997. Editors: V. Capasso, O. Diekmann. VII, 268 pages. 1999.

Vol. 1715: N. V. Krylov, M. Röckner, J. Zabczyk. Stochastic PDE's and Kolmogorov Equations in Infinite Dimensions. Cetraro, 1998. Editor: G. Da Prato. VIII, 239 pages. 1999.

Vol. 1716: J. Coates, R. Greenberg, K. A. Ribet, K. Rubin, Arithmetic Theory of Elliptic Curves. Cetraro, 1997. Editor: C. Viola. VIII, 260 pages. 1999.

Vol. 1717: J. Bertoin, F. Martinelli, Y. Peres. Lectures on Probability Theory and Statistics. Saint-Flour, 1997. Editor: P. Bernard. IX, 291 pages. 1999.

Vol. 1718: A. Eberle,Uniqueness and Non-Uniqueness of Semigroups Generated by Singular Diffusion Operators. VIII, 262 pages. 1999.

Vol. 1719: K. R. Meyer, Periodic Solutions of the N-Body Problem. IX, 144 pages. 1999.

Vol. 1720: D. Elworthy, Y. Le Jan. X.-M. Li, On the Geometry of Diffusion Operators and Stochastic Flows. IV, 118 pages. 1999.